UNIT 1 – EXPRESSIONS AND FUNCTIONS

KU-610-890

CONTENTS

Leckie×Leckie
Scotland's leading educational publishers

CfE Higher
MATHS
STUDENT BOOK

Craig Lowther • Claire Anderson • Robin Christie
• Andrew Thompson • Stuart Welsh

© 2014 Leckie & Leckie Ltd

001/08052014

10 9 8 7 6 5 4 3 2 1

All rights reserved. No part of this publication may be reproduced, stored in a retrieval system, or transmitted in any form or by any means, electronic, mechanical, photocopying, recording or otherwise, without the prior written permission of the Publisher or a licence permitting restricted copying in the United Kingdom issued by the Copyright Licensing Agency Ltd., 90 Tottenham Court Road, London W1T 4LP.

ISBN 9780007549269

Published by
Leckie & Leckie Ltd
An imprint of HarperCollins*Publishers*
Westerhill Road, Bishopbriggs, Glasgow, G64 2QT
T: 0844 576 8126   F: 0844 576 8131
leckieandleckie@harpercollins.co.uk   www.leckieandleckie.co.uk

Special thanks to
Jouve (layout); Ink Tank (cover design); Peter Lindsay (answers); Laura Rigon (answers), Felicity Kendall (proofread); Project One Publishing Solutions (project management and editing)

A CIP Catalogue record for this book is available from the British Library.

Acknowledgements
Whilst every effort has been made to trace the copyright holders, in cases where this has been unsuccessful, or if any have inadvertently been overlooked, the Publishers would gladly receive any information enabling them to rectify any error or omission at the first opportunity.

Printed in Italy by Lego S.P.A

CONTENTS

www.leckieandleckie.co.uk/hmaths

Introduction

About this book

This book provides a resource to practise and assess your understanding of the mathematics covered for the CfE Higher qualification. There is a separate chapter for each of the skills specified in the Expressions and Functions, Relationships and Calculus, and Applications units, and most of the chapters use the same features to help you progress. You will find a range of worked examples to show you how to tackle problems, and an extensive set of exercises to help you develop the whole range of operational and reasoning skills needed for your Higher assessment.

You should not work through the book from page 1 to the end. Your teacher will choose a range of topics throughout the school year and teach them in the order they think works best for your class, so you will use different parts of the book at different times of the year.

Features

CHAPTER TITLE

The chapter title shows the operational skill covered in the chapter.

1 Logarithms and Exponentials

THIS CHAPTER WILL SHOW YOU HOW TO:

Each chapter opens with a list of topics covered in the chapter, and tells you what you should be able to do when you have worked your way through the chapter.

This chapter will show you how to:

- simplify a numerical expression using one or more of the laws of logarithms and exponents
- solve logarithmic and exponential equations

YOU SHOULD ALREADY KNOW:

After the list of topics covered in the chapter, there is a list of topics you should already know before you start the chapter. Some of these topics will have been covered before in Maths, and others will depend on having worked through different chapters in this book.

You should already know:

- how to simplify expressions using the laws of exponents/indices.

EXAMPLE

Each new topic is demonstrated with at least one worked Example, which shows how to go about tackling the questions in the following Exercise. Each Example breaks the question and solution down into steps so you can see what calculations are involved, what kind of rearrangements are needed and how to work out the best way of answering the question. Most Examples have comments, which help explain the processes. These comments are separated into three different types:

- (S) **Select a strategy**: what method are you adopting to tackle this problem?
- (P) **Process data**: the steps involved in working through the problem
- (C) **Interpret and communicate**: the clear expression of the answer, including the correct units and decimal places

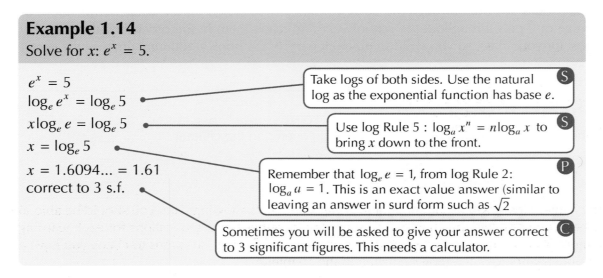

Example 1.14

Solve for x: $e^x = 5$.

$e^x = 5$

$\log_e e^x = \log_e 5$ •————— Take logs of both sides. Use the natural (S) log as the exponential function has base e.

$x\log_e e = \log_e 5$ •————— Use log Rule 5 : $\log_a x^n = n\log_a x$ to (S) bring x down to the front.

$x = \log_e 5$

$x = 1.6094... = 1.61$

correct to 3 s.f. •—————

Remember that $\log_e e = 1$, from log Rule 2: (P) $\log_a a = 1$. This is an exact value answer (similar to leaving an answer in surd form such as $\sqrt{2}$

Sometimes you will be asked to give your answer correct (C) to 3 significant figures. This needs a calculator.

EXERCISE

The most important parts of the book are the Exercises. The questions in the Exercises are carefully graded in difficulty, so you should be developing your skills as you work through an Exercise. If you find the questions difficult, look back at the Example for ideas on what to do. Use Key questions, marked with a star ★, to assess how confident you can feel about a topic. Questions which require reasoning skills are marked with a cog icon ⚙.

In a number of chapters, questions have been marked as either being suitable for solving with the assistance of a calculator ▦ or non-calculator based (▦).

ANSWERS

Answers to all Exercise questions are provided online at
www.leckieandleckie.co.uk/Hmaths.

CHALLENGE

Some chapters feature Challenge activities that are designed to help extend your understanding of the topic and get you thinking more deeply and independently about the skills and techniques involved.

 Challenge

   a Show that $(x + y)$ is a factor of $x^3 + y^3$ and that $(x - y)$ is a factor of $x^3 - y^3$

   b Hence, or otherwise, show that:

     i $\sin^3 x + \cos^3 x = (\sin x + \cos x)\left(1 - \frac{1}{2}\sin 2x\right)$.

HINTS

Where appropriate, Hints are given in the text and in Examples, to help give extra support.

To show that a triangle is right-angled using gradients, we must prove that two sides are perpendicular (meet at 90°).

END-OF-CHAPTER SUMMARY

Each chapter closes with a summary of learning statements showing what you should be able to do when you complete the chapter. The summary identifies the Key questions for each learning statement. You can use the End-of-chapter summary and the Key questions to check you have a good understanding of the topics covered in the chapter.

- I can evaluate expressions involving exponential functions. ★ Exercise 1A Q2
- I can convert between logarithmic and exponential form. ★ Exercise 1B Q3
- I can simplify expressions using the laws of logarithms. ★ Exercise 1C Q3

ASSESSMENTS

Preparation for assessment sections are provided for each of the three units. These cover the minimum competence for the unit content and are a good preparation for your unit assessment.

Integrated questions that draw from a number of different topics are provided online at www.leckieandleckie.co.uk/Hmaths

1 Logarithms and Exponentials

This chapter will show you how to:

- simplify a numerical expression using one or more of the laws of logarithms and exponents
- solve logarithmic and exponential equations
- solve logarithmic and exponential equations using more than one of the laws of logarithms and exponents.

You should already know:

- how to simplify expressions using the laws of exponents/indices.

Logarithms and exponentials

You should be familiar with **exponential functions** and their associated rules.

$y = a^x$ is an exponential function with base a and exponent (or index) x. The graph of an exponential function increases rapidly when $a > 1$ (exponential growth) and decreases rapidly when $0 < a < 1$ (exponential decay) which can make it difficult to interpret.

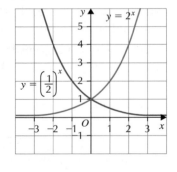

Consider the graphs of $y = 2^x$ and $y = \left(\frac{1}{2}\right)^x$

Note that $y = \left(\frac{1}{2}\right)^x = \left(2^{-1}\right)^x = 2^{-x}$ which is a reflection of $y = 2^x$ across the y-axis. You will come across this in more detail in Chapter 3.

Exponential functions have a range of real-life applications including analysis of population growth, radioactive decay, compound interest and bacterial growth or decay.

Exponential population growth

An experiment monitoring the population growth of mice starts with 6 mice. It is noted that the population doubles every six weeks.

The graph of the mice population is an example of **exponential growth**.

The function for this graph is $y = 6(2)^{\frac{x}{6}}$ where

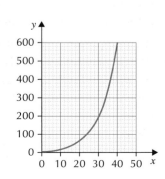

- 6 is the initial population
- the 2 doubles the population
- x is the number of weeks since the beginning of the experiment.

As the population doubles every six weeks, we have to divide x by 6.

We can use the function and the graph to predict the population after, for example, 36 weeks or to determine how long it will take for the population to reach a certain level.

An example of **exponential decay** is shown here. It might represent radioactive decay or bacterial decay. We can see from the graph that the substance is halving every hour.

You can use your graphic calculator to investigate different exponential functions of the form $y = ab^x$, varying the values of a and b.

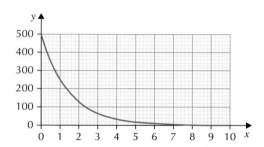

An introduction to e

The function $y = e^x$ is a specific exponential function which plays an enormous role in mathematics. The function can also be written as $y = \exp(x)$. This function, and related functions, can be plotted using a graphic calculator as shown.

Like π, e is a **transcendental** number, and is approximately equal to 2.71828. It is defined as:

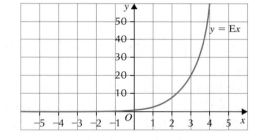

$$e = \lim_{n \to \infty} \left(1 + \frac{1}{n}\right)^n$$

The notation $lim_{n \to \infty}$ describes the **limit** value that we obtain as we substitute values for n which approach infinity. Try putting some very large values for n (such as 1000, 1 000 000, 1 000 000 000) into the expression $\left(1 + \frac{1}{n}\right)^n$. You should see that as n gets larger, $\left(1 + \frac{1}{n}\right)^n$ gets closer to $2.71828\ldots$

The letter e was first used by the mathematician Euler in his calculations. He is generally accepted to be the first to prove that e is irrational in the 18th century but he was not the first to work with this concept. e is also referred to as Napier's constant, after the Scottish mathematician John Napier, because of its crucial role in logarithms.

A major application of e is in compound interest calculations. You have probably performed compound interest calculations based on an APR (annual percentage rate) or an AER (annual equivalent rate). Using APR or AER means we can only calculate the interest earned after a full year. In reality, interest is often calculated daily, requiring the use of the function $y = e^x$. The exponential function features widely in real-life applications of mathematics.

Example 1.1

The atmospheric pressure P_h at h km is given by $P_h = P_0 e^{-kh}$ where P_0 is the pressure at sea level.

The pressure at sea level is 760 mmHg and $k \approx -0.126$. Calculate the pressure at 3 km above sea level.

$$P_h = P_0 e^{-kh}$$

$$P_h = 760 \times e^{0.126 \times 3}$$

Substitute the values for P_0, h and k into the formula. ⓟ

$$= 1109.115\ldots$$

Calculate using the e^x or e^0 key on your calculator. ⓟ

The pressure at 3 km is approximately 1109.1 mmHg.

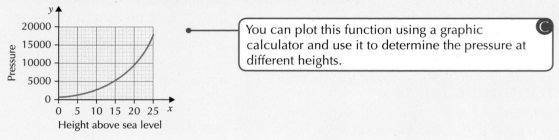

You can plot this function using a graphic calculator and use it to determine the pressure at different heights. Ⓒ

Exercise 1A

1 Calculate y, correct to 3 significant figures.

    a $y = e^3$ **b** $y = e^{-3}$ **c** $y = 2e^{0.5}$ **d** $y = e^{-0.2 \times 53}$

★ 2 A population of mice is increasing according to the law $P_t = P_0 e^{0.132t}$ where P_0 is the initial population and t is the time in weeks.

    a Given $P_0 = 100$, calculate the number of mice after:

        i 1 week   **ii** 5 weeks.

    b After how many weeks does the number of mice first exceed 250?

3 25 g of a radioactive substance decays according to the formula $m_t = 25e^{-0.018t}$ where m_t is the mass remaining after t years. Calculate the mass remaining after:

    a 10 years   **b** 20 years   **c** 50 years

4 An investment of £1000 grows according to the law $I(t) = 1000e^{rt}$ where r is the interest rate and t is the number of years.

    a Given a compound interest rate of 5% we obtain $I(t) = 1000e^{0.05t}$ calculate the value of the investment after 10 years.

    b Show that the value of the investment doubles between year 13 and year 14.

    c Calculate the value of a £1000 investment after 5 years at an interest rate of 3.2%.

5   The number of bacteria in a culture increases according to the formula
     $B(t) = B_0e^{0.14t}$ where B_0 is the initial number of bacteria and t is the number of
     hours since the experiment began.

   a   Given that $B_0 = 500$ calculate the number of bacteria present after 8 hours.

   b   After how many hours will the number of bacteria exceed 2000?

6   After 10 hours, an antibiotic is added to the bacterial culture which starts to kill
     bacteria according to the formula $B(t) = 2000e^{-0.21t}$

   a   Calculate the number of bacteria in the culture:

     i   4 hours after the antibiotic is added

     ii   10 hours after the antibiotic is added.

7   A patient has a drug injected which then begins to break down in the body
     according to the formula $D(t) = 25e^{-0.04t}$ where $D(t)$ is the amount of the drug in
     mg remaining in the body after t hours.

   a   What mass of the drug was injected?

   b   Calculate the amount of the drug remaining after 6 hours.

An introduction to the logarithmic function

The **logarithmic function** is the inverse of the exponential function:

If $y = a^x$ then $x = \log_a y$

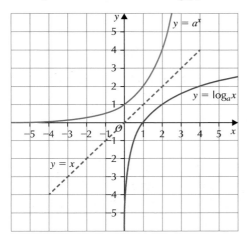

$y = \log_a x$ is a reflection of $y = a^x$ in the line $y = x$.

It will be useful to be able to convert between these two forms when we come to
solve equations involving exponential and logarithmic functions in. We have already
considered some of the applications of exponential functions in real life situations
and logarithms are key to solving problems. Consider Question 5b in Exercise 1A: we
will look at how we can use logarithms to solve this problem rather than relying on
trial and error or repetitive calculations.

We will also consider the use of logarithms in the context of earthquakes, measuring
sound or voltage or even particle size.

Example 1.2

Express $y = 3^x$ in logarithmic form.

$y = 3^x \Leftrightarrow x = \log_3 y$ •————— $\Leftrightarrow$ means 'if and only if', that is, if one is true, so is the other.

Note that $y = 3^x$ has base 3 and exponent x and in the logarithmic form we have base 3 and the exponent, x, is now the subject of the expression.

One of the uses of logarithms is to reduce very large or very small numbers to something more manageable (particularly useful before the invention of calculators). For example, the pH scale in chemistry is a log based scale, where

$$pH = -\log_{10} \text{(concentration of } H^+ \text{ ions)}$$

If the concentration of H^+ ions in a solution is 0.0001 mol $l^{-1}\left(1 \times 10^{-4}\right)$, then we have $y = 10^{-4}$. This can be written as $-4 = \log_{10} y$, so $pH = -(-4) = 4$. This is a much less cumbersome number to deal with.

Example 1.3

Express $4 = \log_2 16$ in exponential form.

$4 = \log_2 16$

$16 = 2^4$ •————— Both the exponential and log functions have base 2.

Exercise 1B

1 Express each exponential function as a logarithm.

    a $y = 5^3$ **b** $p = 4^t$ **c** $f = g^h$

    d $128 = 2^7$ **e** $y = e^x$

2 Express each log function as an exponential.

    a $3 = \log_2 8$ **b** $5 = \log_3 243$ **c** $y = \log_5 4$

    d $x = \log_m t$ **e** $3 = \log_4 y$

★ 3 Calculate the value of a in each equation.

    a $4 = \log_2 a$ **b** $3 = \log_9 a$ **c** $5 = \log_{10} a$ **d** $a = \log_4 16$

Rules of logarithms

Recall the rules of exponential functions:

$$a^m \times a^n = a^{m+n} \qquad a^m \div a^n = a^{m-n} \qquad (a^m)^n = a^{mn}$$

$$a^{-m} = \frac{1}{a^m} \qquad\qquad \sqrt[n]{a^m} = a^{\frac{m}{n}} \qquad\qquad a^0 = 1 \qquad\qquad a^1 = a$$

As with exponential functions, there are rules which we can use to simplify expressions involving logs.

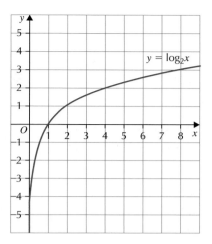

Rule 1 $\quad \log_a 1 = 0$ — A straightforward logarithmic graph will cross the x-axis at $x = 1$.

$a^0 = 1 \Rightarrow \log_a 1 = 0$

Rule 2 $\quad \log_a a = 1$ — From the graph, $\log_2 2 = 1$

$a^1 = a \Rightarrow \log_a a = 1$

Rule 3 $\quad \log_a x + \log_a y = \log_a xy$ — The base must be the same in each log term.

Rule 3 can be proved:

Let $\log_a x = m$ and $\log_a y = n$.

Writing these in exponential form, we obtain $x = a^m$ and $y = a^n$.

$\quad$ So, $xy = a^m a^n$

$\quad \therefore \ xy = a^{m+n}$ — Convert this to logarithmic form. **S**

$\quad \log_a xy = m + n$ — Replace m and n using the original statement. **S**

$\therefore \log_a xy = \log_a x + \log_a y$

Example 1.4

Simplify $\log_3 8 + \log_3 2$

$\log_3 8 + \log_3 2 = \log_3 8 \times 2 = \log_3 16$

Rule 4 $\quad \log_a \dfrac{x}{y} = \log_a x - \log_a y$

Rule 4 can be proved:

Let $\log_a x = m$ and $\log_a y = n$

Writing these in exponential form, we obtain $x = a^m$ and $y = a^n$

$\quad$ So, $\dfrac{x}{y} = \dfrac{a^m}{a^n}$

$$\therefore \frac{x}{y} = a^{m-n}$$

Convert this to logarithmic form. Ⓢ

$$\log_a \frac{x}{y} = m - n$$

$$\therefore \log_a \frac{x}{y} = \log_a x - \log_a y$$

Example 1.5
Simplify $\log_{10} 50 - \log_{10} 5$

$$\log_{10} 50 - \log_{10} 5 = \log_{10} \frac{50}{5}$$
$$= \log_{10} 10$$
$$= 1$$

We know the value of $\log_{10} 10$ from Rule 2. Ⓟ

Rule 5 $\log_a x^n = n\log_a x$

Rule 5 can be proved:

Let $\log_a x = m$ and so $x = a^m$

Then $x^n = (a^m)^n$

$$x^n = a^{mn}$$

Convert this to logarithmic form.

$$\log_a x^n = mn$$

$$\therefore \log_a x^n = \log_a x \times n$$
$$= n\log_a x$$

Example 1.6
Simplify $\log_4 4^{12}$

$$\log_4 4^{12} = 12\log_4 4$$
$$= 12 \times 1$$
$$= 12$$

Example 1.7
Simplify $\frac{1}{2}\log_5 625$

$$\frac{1}{2}\log_5 625 = \log_5 625^{\frac{1}{2}}$$
$$= \log_5 25$$
$$= \log_5 5^2$$
$$= 2\log 5^5$$
$$= 2$$

Example 1.8

Simplify $\log_3 \frac{1}{9}$

$$\log_3 \frac{1}{9} = \log_3 \frac{1}{3^2}$$

Express this using negative indices. **S**

$$= \log_3 3^{-2}$$

Bring the index to the front and simplify. **P**

$$= -2$$

or:

$$\log_3 1 - \log_3 9 = 0 - \log_3 9 = -\log_3 3^2 = -2$$

Example 1.9

Simplify $3\log_8 2 + \log_2 28 - \log_2 7$

$$3\log_8 2 + \log_2 28 - \log_2 7 = \log_8 2^3 + \log_2 \frac{28}{7}$$

There are two different bases involved here. Apply Rule 5 to the first term and Rule 3 to the second and third terms. **S**

$$= \log_8 8 + \log_2 4$$

$$= 1 + 2$$

$$\log_2 4 = \log_2 2^2 = 2\log_2 2$$ **P**

$$= 3$$

Exercise 1C

1 Simplify

a $\log_7 3 + \log_7 9$

b $\log_5 100 - \log_5 4$

c $\log_4 12 - \log_4 3$

d $\log_{12} 36 + \log_{12} 4$

e $\log_6 72 + \log_6 2 - \log_6 4$

f $\log_9 24 - \log_9 8 - \log_9 21 + \log_9 7$

g $\log_4 12 + \log_4 8 - \log_4 3 - \log_4 2$

h $\log_a 3 - \log_a 5 + \log_a 20 - \log_a 12$

2 Evaluate

a $\frac{1}{3}\log_3 27$

b $\log_2 \frac{1}{4}$

c $\log_p p^3$

d $\log_4 8^0$

e $4\log_{16} 2$

★ **3** Simplify

a $\log_2 8 + 3\log_2 4 - \log_2 16$

b $3\log_{10} 5 + \log_{10} 8$

c $\log_4 36 - 2\log_4 12$

d $2\log_3 9 + \log_3 5 - \log_3 15$

e $\log_5 \frac{1}{24} - \log_5 \frac{1}{8} + \log_5 15$

f $3\log_2 \frac{1}{4} + 2\log_2 16$

g $2\log_3 9 - 3\log_2 8$

h $\log_4 8 + \log_4 2 - 3\log_7 7$

i $\log_5 25 + 3\log_2 \sqrt[3]{2}$

j $\log_6 18 - \log_6 3 + \log_3 \sqrt{3}$

k $2\log_4 16 + 3\log_2 8 - \log_2 \sqrt{8}$

4 The absorbance of a particular wavelength of light passing through a substance is given by $A = -\log_{10} 8 + \log_{10} 80$. Show that $A = 1$.

5 In acoustics, decibels are used to measure sound pressure according to the law $L = 10\log_{10} P_1^2 - 20\log_{10} P_0$. Given that $P_o = 2 \times 10^{-5}$ and $P_1 = 2000$ for a rocket launcher, calculate the sound pressure L, created by the rocket launcher.

The natural logarithm

The inverse of the exponential function $y = e^x$ is called the **natural logarithm**, $y = \log_e x$. $y = \log_e x$ can also be written as $\ln x$. (ln is an abbreviation of natural logarithm.)

Scientific calculators all have an 'ln' key for calculating the natural log of a number and a log key which calculates '$\log_{10}$' of a number. The facility to calculate using alternative bases is becoming increasingly common.

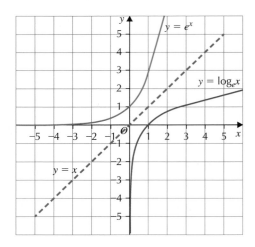

Example 1.10

Express $y = e^3$ in logarithmic form.

$y = e^3 \Leftrightarrow \log_e y = 3$

Example 1.11

Calculate $\log_e \sqrt{e}$

$\log_e \sqrt{e} = \log_e e^{\frac{1}{2}}$ ● ———— (Rearrange from root form to index form. ⓟ

     $= \frac{1}{2}\log_e e$ ● ———— (Use log rules. Bring the index to the front. ⓟ

     $= \frac{1}{2}$ ● ———— (Remember $\log_e e = 1$. ⓟ

Example 1.12

Calculate $\log_e 5$ correct to 3 significant figures.

$\log_e 5 = 1.6094... = 1.61$ correct to 3 s.f.   ● ——(Use the ln key on your calculator. ⓟ

Exercise 1D

1 Express in exponential form.

    a $y = \log_e 3$ **b** $4 = \log_e x$ **c** $p = \log_e q$

    d $y = \log_{10} 5$ **e** $3 = \log_{10} x$

2 Express in log form.

    a $y = e^5$ **b** $2 = e^x$ **c** $f = e^g$ **d** $y = 10^x$ **e** $x = 10^y$

3 Calculate, correct to 3 significant figures.

    a $\log_e 8$ **b** $\log_{10} 25$ **c** $e^{0.8}$ **d** $10^{-0.2}$

★ **4** Simplify:

    a $\log_e e^2$ **b** $\log_e \dfrac{1}{e}$ **c** $6\log_e \sqrt[3]{e}$

    d $\log_e e^3 + 2\log_e \sqrt[3]{e} - 4\log_e \sqrt{e}$

5 Simplify:

    a $\dfrac{\log_3 4p^2}{\log_3 2p}$, $p > 0$ **b** $\dfrac{\log_k 8t^3}{\log_k 2t}$, $t > 0,\ k > 0$ **c** $\dfrac{\log_a 27q^3}{\log_a 9q^2}$, $a > 0,\ q > 0$

Solving simple logarithmic and exponential equations

These laws of exponents and logarithms can be used to solve equations.

Example 1.13

Solve for x: $\log_{10} 2x = 3$

$\log_{10} 2x = 3 \Rightarrow 2x = 10^3$ ▢ Express in exponential form **S**

$2x = 1000$

$\therefore x = 500$

Example 1.14

Solve for x: $e^x = 5$.

$e^x = 5$

$\log_e e^x = \log_e 5$ ▢ Take logs of both sides. Use the natural **S**
                                      log as the exponential function has base e.

$x\log_e e = \log_e 5$ ▢ Use log Rule 5 : $\log_a x^n = n\log_a x$ to **S**
                                      bring x down to the front.

$x = \log_e 5$

$x = 1.6094... = 1.61$ ▢ Remember that $\log_e e = 1$, from log Rule 2: **P**

correct to 3 s.f. $\log_a a = 1$. This is an exact value answer (similar to
                                    leaving an answer in surd form such as $\sqrt{2}$

                                ▢ Sometimes you will be asked to give your answer correct **C**
                                to 3 significant figures. This needs a calculator.

Example 1.15
Solve for x : $8^x = 3$

$8^x = 3$ — Take logs of both sides Ⓢ

$\log_{10} 8^x = \log_{10} 3$ — Use log rules to bring x down to the front Ⓢ

$x\log_{10} 8 = \log_{10} 3$

$x = \dfrac{\log_{10} 3}{\log_{10} 8}$ — Here we used $\log_{10}$ because this is available on all scientific calculators. If your calculator can work in $\log_8$ then that would make the calculation slightly simpler. Ⓒ

$x = 0.5283\ldots$

$x = 0.528$ correct to 3 s.f.

Example 1.16
Solve for x : $\log_x 4913 = 3$

$\log_x 4913 = 3$ — Express in exponential form. Ⓢ

$\Rightarrow x^3 = 4913$

$x = \sqrt[3]{4913}$

$x = 17$

Exercise 1E

1 Solve for x.

    a $\log_3 x = 5$ **b** $\log_{10} x = 4$ **c** $\log_5 x = 3$

    d $\log_e x = 2$ **e** $3\log_e x = 12$ **f** $\frac{1}{2}\log_4 x = 3$

★ 2 Solve for x.

    a $\log_2 8x = 5$ **b** $\log_4 3x = 2$ **c** $\log_3 \frac{x}{4} = 5$

    d $\log_e \frac{x}{3} = 4$ **e** $3\log_6 2x = 12$ **f** $5\log_e 3x = 3$

    g $\log_4(x - 1) = 3$ **h** $\log_5(2x + 3) = 2$

★ 3 Solve for x.

    a $10^x = 3$ **b** $e^x = 8$ **c** $4^x = 262144$ **d** $12^x = 20736$

4 Solve for x.

    a $4e^x = 16$ **b** $3(2^x) = 48$ **c** $\dfrac{3^x}{2} = 40.5$ **d** $10^{4x} = 1000$

    e $2^{3x} = 512$ **f** $5^{\frac{x}{2}} = 15625$ **g** $3^{\frac{2x}{5}} = 81$ **h** $2e^{3x} = 8$

★ 5 Solve for $x, x > 0$

    a $\log_x 25 = 2$ **b** $\log_x 256 = 8$ **c** $\log_x 6561 = 4$

    d $\log_{2x} 1024 = 5$ **e** $\log_{x-1} 2401 = 4$

Solving logarithmic and exponential equations requiring the use of the laws of logarithms and exponents

---

Example 1.17

Solve for x: $\log_5 x + \log_5 3 = \log_5 21$

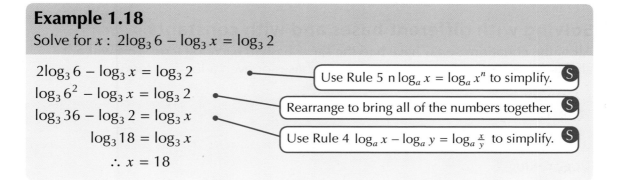

$\log_5 x + \log_5 3 = \log_5 21$ •——————— Use Rule 3: $\log_a x + \log_a y = \log_a xy$ to simplify. Ⓢ

$\log_5 3x = \log_5 21$ •———————

$3x = 21$ ——————— Take the anti-log of both sides since the bases are equal. Ⓢ

$x = 7$

Example 1.18

Solve for x: $2\log_3 6 - \log_3 x = \log_3 2$

$2\log_3 6 - \log_3 x = \log_3 2$ •——————— Use Rule 5 $n\log_a x = \log_a x^n$ to simplify. Ⓢ

$\log_3 6^2 - \log_3 x = \log_3 2$ •———————

$\log_3 36 - \log_3 2 = \log_3 x$ •——————— Rearrange to bring all of the numbers together. Ⓢ

$\log_3 18 = \log_3 x$ ——————— Use Rule 4 $\log_a x - \log_a y = \log_a \frac{x}{y}$ to simplify. Ⓢ

$\therefore x = 18$

Example 1.19

Solve for $x > 1$: $\log_8(x + 1) + \log_8(x - 1) = \log_8 15$

$\log_8(x + 1) + \log_8(x - 1) = \log_8 15$

$\log_8(x + 1)(x - 1) = \log_8 15$

$(x + 1)(x - 1) = 15$

$x^2 - 1 = 15$

$x^2 - 16 = 0$

$x = \pm 4$ but since $x > 1, x = 4$.

Exercise 1F

1 Solve for $x > 0$.

   a $\log_a x - \log_a 7 = \log_a 4$ **b** $\log_b x + \log_b 5 = \log_b 15$

   c $\log_t 3 + \log_t x = \log_t 7$ **d** $\log_a 12 - \log_a x = \log_a 4$

   e $\log_k x - \log_k 4 + \log_k 12 = \log_k 2$ **f** $\log_a x + 2\log_a 3 = \log_a 72$

   g $\log_b 18 - \log_b(x + 1) = \log_b 6$

★ **2** Solve for x (remember to ensure that your final answer takes careful account of the domain of each term).

a $\log_a 4x + \log_a(x + 1) = \log_a 8$ b $\log_2 x + \log_2(x - 5) = \log_2 6$

c $\log_{10}(x - 2) + \log_{10}(x - 3) = \log_{10} 2$ d $\log_p(x - 3) - \log_p 4 = \log_p 5$

e $\log_q(x^2 - 4) - \log_q(x + 2) = \log_q 7$ f $\log_r(2x + 1) + \log_r(x - 3) = \log_r(x + 5)$

g $\log_s x + \log_s(x + 1) + \log_s(x - 1) = \log_s 6$

h $\log_t(x^2 + 1) + \log_t(x + 2) = \log_t(2x + 4)$

3 The point $A(a, t)$ lies on the curve $y = \log_2 x$.

The point $B(b, t)$ lies on the curve $y = \frac{2}{3}\log_2 x$.

Determine a relationship between a and b and hence find a when $b = 8$.

Solving with different bases and with constants

All of the equations so far have had the same base in each log term. This will not always be the case. There may be different bases or even a log-free term.

Example 1.20

Solve $\log_3 x - \log_3 5 = 2$

$\log_3 x - \log_3 5 = 2$

$\log_3 \frac{x}{5} = 2$ Simplify using log rules. Ⓢ

$\frac{x}{5} = 3^2$ Express in exponential form. Ⓢ

$x = 5 \times 9 = 45$

Example 1.21

Solve for $x > 1$: $\log_2(x - 1) + \log_2(x + 2) = \log_4 16$

$\log_2(x - 1) + \log_2(x + 2) = \log_4 16$ Simplify the left- and right-hand side of Ⓢ

$\log_2(x - 1)(x + 2) = \log_4 4^2$ the equation using log Rules 3 and 5.

$\log_2(x - 1)(x + 2) = 2$ Express in exponential form. Ⓢ

$(x - 1)(x + 2) = 2^2$

$x^2 + x - 2 = 4$ Solve the resulting quadratic equation. Ⓟ

$x^2 + x - 6 = 0$ One side of the quadratic equation must equal zero before factorising.

$(x - 2)(x + 3) = 0$

$x = 2, x = -3$ but, as $x > 1, x = 2$ Remember to take into account Ⓒ the domain of the function.

Exercise 1G

Remember to consider a suitable domain for each equation.

1 Solve for x.

a $\log_4 x + \log_4 8 = 2$

b $\log_2 x - 2\log_2 6 = 3$

c $\log_3 x + \log_2 8 = 4$

d $\log_3(x - 1) + \log_7 49 = 2$

e $2\log_5 x - 3\log_8 4 = 1$

f $\log_4(2x + 1) - 3\log_9 3 = 1$

★ 2 Solve for x.

a $\log_4 x + \log_4(x + 12) = 3$

b $\log_7(x + 1) - \log_7(x - 1) = 1$

c $\log_2(2x - 1) + \log_2(2x + 1) = 3$

d $\log_{15} x + \log_{15}(x - 2) + \log_{15}(x + 3) = 1$

e $2\log_2(x + 1) - \log_2(2x) = 1$

Applications of exponential functions

Knowledge of exponential and logarithmic functions can be used to solve a wide range of real-life problems, from financial calculations involving interest rates, bacteria cell growth and death, spread of diseases and radioactive decay.

Example 1.22

£3000 is invested in an ISA with an interest rate of 2.25% per annum with the interest added at the end of each year.

a Calculate the value of the investment after 10 years.

b How long would it take the investment to increase by 50%?

a After 10 years: investment $= 3000 \times 1.0225^{10}$

$= 3747.610...$

To increase by 2.25% we multiply by 1.0225 ten times, so the index is 10. Ⓢ

The investment is worth £3747.61 after 10 years.

b $3000 \times 1.5 = 4500$

We want to know when the investment will be worth £4500. Ⓢ

$4500 = 3000 \times 1.0225^n$

$1.5 = 1.0225^n$

$\log 1.5 = \log 1.0225^n$

$= n\log 1.0225$

n represents the number of years. Simplify to isolate n. The first step is to divide by 3000. Ⓢ

$\therefore \dfrac{\log 1.5}{\log 1.0225} = n$

Bring n to the front using log Rule 5. Ⓢ

$n = 18.222....$

The investment will increase by 50% after 19 years. (18 would be insufficient).

Calculate the value of n using a calculator. Use any base, but it must be the same for both the numerator and the denominator. Base 10 is readily available on your calculator. Ⓟ

Example 1.23

In a biology experiment measuring the rate at which cells are dying, it is found that the cells are dying according to the formula $C_t = C_0 e^{-kt}$ where C_t is the number of cells after t days and C_0 is the initial number of cells.

a   Given that the experiment began with 250 000 cells and half of the cells had died after 8 days, determine the value of k.

b   Calculate the time taken for the cells to reduce to 20% of the initial population.

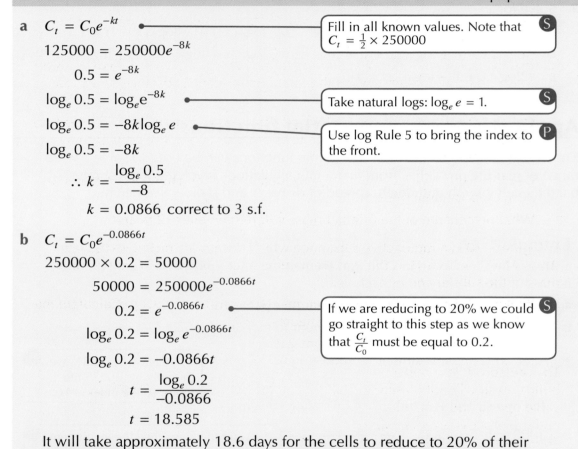

a $C_t = C_0 e^{-kt}$

> Fill in all known values. Note that $C_t = \frac{1}{2} \times 250000$ Ⓢ

$125000 = 250000 e^{-8k}$

$0.5 = e^{-8k}$

$\log_e 0.5 = \log_e e^{-8k}$

> Take natural logs: $\log_e e = 1$. Ⓢ

$\log_e 0.5 = -8k \log_e e$

> Use log Rule 5 to bring the index to the front. Ⓟ

$\log_e 0.5 = -8k$

$\therefore k = \dfrac{\log_e 0.5}{-8}$

$k = 0.0866$ correct to 3 s.f.

b $C_t = C_0 e^{-0.0866t}$

$250000 \times 0.2 = 50000$

$50000 = 250000 e^{-0.0866t}$

$0.2 = e^{-0.0866t}$

> If we are reducing to 20% we could go straight to this step as we know that $\frac{C_t}{C_0}$ must be equal to 0.2. Ⓢ

$\log_e 0.2 = \log_e e^{-0.0866t}$

$\log_e 0.2 = -0.0866t$

$t = \dfrac{\log_e 0.2}{-0.0866}$

$t = 18.585$

It will take approximately 18.6 days for the cells to reduce to 20% of their original number.

Exercise 1H

1   The number of people infected by a virus is increasing according to the formula $I_t = I_0 e^{0.43t}$ where I_t is the number of people infected after t days and I_0 is the initial number of people infected.

a   Initially, 7 people are known to be infected. Calculate the number of people infected after 1 week.

b   How long will it take for 770 people to become infected?

2   £20000 is invested in an ISA paying an APR of 2.4%.

    a   Calculate the value of the investment after 5 years.

    b   How long will it take for the investment to be worth £25000?

3   The number of bacteria in a sample is being monitored and is determined to be increasing according to the formula $B_t = 200e^{0.8t}$ where B_t is the number of bacteria after t hours.

    a   Calculate the number of bacteria present at the start of the monitoring process.

    b   How long, in hours and minutes, will it take for the number of bacteria to triple?

★ 4   Radioactive substances decay according to the law $N_t = N_0e^{-kt}$ where N_t is the mass remaining after t years, N_0 is the initial mass of the substance and k is a constant.

    a   Cobalt-60 has a half-life of 5.27 years. That means that 100 kg will reduce to 50 kg in 5.27 years. Determine the value of k for cobalt-60, correct to 3 significant figures.

    b   Use this value of k to determine how long cobalt-60 will take to reduce to 30% of its original mass.

    c   What percentage of the original mass will remain after 12 years?

5   Polonium-210 is a radioactive substance which decays according to the law $M_t = M_0e^{-kt}$ where M_t is the mass remaining after t days and M_0 is the initial mass of the substance.

    a   It takes 10 days for 100 g of Polonium-210 to reduce to 95.1 g. Calculate the value of k correct to 3 significant figures.

    b   Determine the half-life of Polonium-210.

6   The pressure P in a boiler builds up slowly until a safety valve opens. The valve closes and the process is repeated. The pressure P is given by $P(t) = P_0e^{kt}$

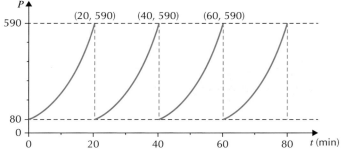

    a   Find the values of P_0 and k.

    b   Find the equations for $P(t)$ for $20 \le t \le 40$ and $40 \le t \le 60$.

7   A small meteor passes through a dust shower. It picks up particles and gains weight (kg) in time t (hours), according to the expression $W(t) = 1.2e^{0.05t}$

    a   What did the meteor weigh initially?

    b   How long would it take to double in weight?

    c   Sketch the graph of $W(t)$ against t showing the main points and features.

Applications of logarithms

Logarithms were first developed by John Napier, a Scottish mathematician, in the late 16th and early 17th century. Since then they have been developed by numerous mathematicians and applied to a wide variety of real life problems, including financial calculations such as compound interest, population growth, radioactive decay, pH, the decibel scale and the Richter scale.

One measure of the magnitude of an earthquake is the Richter scale which uses a logarithmic scale of base 10. That means that an earthquake measuring 4 on the Richter scale is ten times stronger than one measuring 3.

Example 1.24

The formula used to determine the magnitude of an earthquake on the Richter Scale is $M = \log_{10} \frac{I}{I_0}$, where M is the magnitude, I is the intensity and I_0 is a constant (the intensity of an earthquake measuring 0 on the Richter Scale).

In May 2013 there was an earthquake in Acharacle in the west of Scotland which measured 2.9 on the Richter scale. In the same month, there was also an earthquake on the Llyn Peninsula in Wales which was 8 times stronger (more intense). What did the Welsh earthquake measure on the Richter Scale?

The intensity of the Welsh earthquake is 8 times that of the Scottish one: $I_w = 8I_S$

Also, $M_s = \log_{10} \frac{I_S}{I} = 2.9$ (where M_s is the magnitude of the Scottish earthquake)

$$M_W = \log_{10} \frac{I_w}{I}$$

Use the formula and the value given to produce a relationship between the intensity and the magnitude. **S**

$$= \log_{10} \frac{8I_s}{I}$$

$I_w = 8I_s$. Information is given about the magnitude in terms of I_s so replace I_w with $8I_s$ **S**

$$= \log_{10} 8 + \log_{10} \frac{I_s}{I}$$

$$= \log_{10} 8 + 2.9$$

Use log Rule 3 to separate the 8 from the known value. **S**

$$= 3.8$$

The Welsh earthquake measured 3.8 on the Richter Scale.

3.8 on the Richter Scale is classed as a minor earthquake. You would probably feel it and might notice objects shaking in your home but it is unlikely to cause much damage. Noticeable damage to well-built buildings generally occurs when the earthquake measures above 6 on the Richter Scale.

Exercise 1I

1 To calculate the pH of ethanoic acid, the expression

$$pH = -\frac{1}{2}\log_{10} Ka + \frac{1}{2}\log_{10} c$$

can be used where Ka is a constant and c is the original concentration of the acid.

  a Calculate the pH of ethanoic acid given that $Ka = 1.8 \times 10^{-5}$ and $c = 2$.

  b Given that $pH = -\log_{10} H$, show that $\log_{10} c = \log_{10} \frac{Ka}{H^2}$

⚙ **2** Decibels are used to measure the difference in intensity between two sounds according to the rule

$$D = 10\log_{10}\frac{I_1}{I_2}$$

Where D is the number of decibels, I_1 is the intensity of the first sound and I_2 is the intensity of the second sound.

Determine the difference, in decibels, between a jet engine taking off which has an intensity of $10\,000\,W$ and a motorcycle accelerating which has an intensity of $0.1\,W$.

⚙ **3** In electrical circuits, the power gain is also measured in decibels, this time according to the rule

$$D = 10\log_{10}\frac{V_1^2}{V_0^2}$$

where V_1 is the voltage being measured and V_0 is a reference voltage.

a Show that $D = 20\log_{10}\frac{V_1}{V_0}$

b Determine the voltage being measured in a circuit where the power gain is 6 decibels and the reference voltage, V_0, is 10.

⚙ **4** The Krumbein Phi (ϕ) scale is used to define the size of particles. For example, a pebble would have a ϕ value of between −2 and −6 whereas silt would have a ϕ value between 4 and 8. The ϕ value is calculated according to the rule

$$\phi = -\log_2\frac{D}{D_0}$$

where D is the diameter of the particle and D_0 is a reference diameter (a constant).

Particle 1 with diameter $= D_1$ has $\phi = 2$. Calculate ϕ for a second particle with a diameter 4 times greater than particle 1.

⚙ **5** The absorbance, A, of a particular wavelength of light by a material can be calculated using the formula

$$A = -\log_{10}\frac{I_1}{I_0}$$

where I_0 is the intensity of the light before it passes through the material and I_1 is the intensity afterwards.

The absorbance of material P is 2. For the same wavelength of light, calculate the absorbance of material Q for which the intensity of the light (after it had passed through) was 3 times that of the intensity for material P.

Interpreting experimental data

Often, in scientific experiments – for example, measuring the rate of reaction between a metal and an acid, or determining the rate at which a body metabolises a drug – plotting the data obtained gives what looks like an exponential growth or decay function. The rapid changes in these functions can make them difficult to interpret and so logarithms are applied to make it easier to determine the equation of the function.

These functions can be of the form:

- $y = kx^n$

- $y = ab^x$

Example 1.25

Results from an experiment are shown in the graph.

a Show that this graph represents a relationship of the form $y = kx^n$

b Determine the values of k and n.

Note that there are logs on both axes. Ⓢ

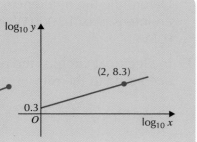

a $y = kx^n$

$\log_{10} y = \log_{10} kx^n$

$\log_{10} y = \log_{10} k + \log_{10} x^n$

$\log_{10} y = \log_{10} k + n\log_{10} x$

$\log_{10} y = n\log_{10} x + \log_{10} k$

This is a straight line with gradient n and y-intercept $\log_{10} k$ (as shown in the graph) and so this graph represents a relationship of the form $y = kx^n$

> The graph uses base 10, so apply $\log_{10}$ to the equation. Ⓢ

> Use log rules to split the right-hand side. Ⓢ

> Rearrange so that it resembles the equation of a straight line. Ⓢ

> This can be written this as $Y = mX + c$ where $Y = \log_{10} y$ and $X = \log_{10} x$ Ⓒ

b $n = $ gradient: $\quad m = \dfrac{8.3 - 0.3}{2 - 0} = 4$

$\therefore n = 4$

$\log_{10} k = 0.3 \quad \therefore k = 10^{0.3} \approx 2$

So the relationship is $y = 2x^4$

Example 1.26

a Show that this graph represents a relationship of the form $y = ab^x$

b Determine the values of a and b.

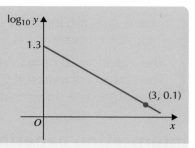

a $y = ab^x$

$\log_{10} y = \log_{10} ab^x$

> The graph uses base 10, so apply $\log_{10}$ to the equations. Ⓢ

> Use log Rules 3 and 5 to split the right-hand side. Ⓢ

$\log_{10} y = \log_{10} a + \log_{10} b^x$

$\log_{10} y = \log_{10} a + x\log_{10} b$

> Note that this time, only the y-axis is expressed in terms of $\log_{10}$. This corresponds with the graph.

$\log_{10} y = x\log_{10} b + \log_{10} a$ which is a straight line as shown.

b $\log_{10} b$ is the gradient of the straight line.

$$m = \frac{1.3 - 0.1}{0 - 3} = \frac{1.2}{-3} = -0.4$$

$\therefore \log_{10} b = -0.4$

$\qquad b = 10^{-0.4}$

$\qquad b = 0.398...$

$\qquad b \approx 0.4$

$\log_{10} a$ is the y-intercept.

$\log_{10} a = 1.3$

$\qquad a = 10^{1.3}$

$\qquad a = 19.952...$

$\qquad a \approx 20$

The relationship is $y = 20(0.4^x)$.

Exercise 1J

1 Show that each graph represents a function of the form $y = kx^n$ and determine the values of k and n.

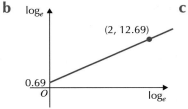

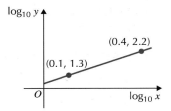

2 Show that each graph represents a function of the form $y = ab^x$ and determine the values of a and b.

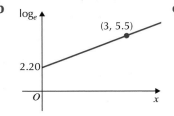

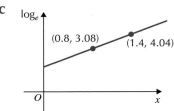

★ **3** Determine the equation of each function in the form $y = kx^n$ or $y = ab^x$.

a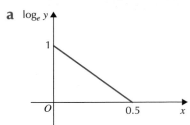

b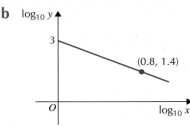

⚙ **4** An experiment resulted in the following data set:

x	0.2	0.4	0.6	0.8
y	625	78.125	23.148	9.766

When the points were plotted, it looked as though the curve was of the form $y = kx^n$.

Using logarithms, show that this data does plot a curve of the form $y = kx^n$ and determine the values of k and n

---

- I can evaluate expressions involving exponential functions. ★ Exercise 1A Q2
- I can convert between logarithmic and exponential form. ★ Exercise 1B Q3
- I can simplify expressions using the laws of logarithms. ★ Exercise 1C Q3
- I can simplify expressions involving the natural logarithm and e. ★ Exercise 1D Q4
- I can solve simple equations involving logarithmic and exponential terms. ★ Exercise 1E Q2, 3, 5
- I can solve more complex equations requiring the use of the laws of logarithms. ★ Exercise 1F Q2, Exercise 1G Q2, Exercise 1H Q4
- I can use logarithms to determine the equations of functions of the form $y = kx^n$ and $y = ab^x$. ★ Exercise 1J Q3

For further assessment opportunities, see the Preparation for Assessment for Unit 1 on pages 132–135.

2 Manipulating trigonometric expressions

This chapter will show you how to:

- find **exact value ratios** for trigonometric ratios
- convert between **radians** and degrees
- use the **addition formula** for the sum and difference of two angles
- use the **double angle formula**
- find exact value ratios using the addition and double angle formulae
- prove **trigonometric identities**
- work with the **wave function** $a\cos x + b\sin x$ where a and b are constants.

You should already know:

- how to identify related angles
- exact value ratios for the sine, cosine and tangent of 30°, 60°, 45° and multiples of 90°
- how to prove trigonometric identities using: $\sin^2 A + \cos^2 A = 1$ and $\tan A = \dfrac{\sin A}{\cos A}$
- how to sketch and identify graphs of the form $y = k\cos(x \pm a)$ or the sine equivalent
- what is meant by the term **phase angle**.

Working with the sine, cosine and tangent ratios of an angle x

As well as being fundamental to the further study of mathematics, the trigonometric functions, sine and cosine in particular, are used in many areas of science and engineering. The periodicity of the sine and cosine functions, collectively known as sinusoidal or wave functions, makes them ideal for the mathematical modelling of many phenomena of our natural world, including the motion of the planets around the sun, the seasons and the tides. Trigonometric functions are also used extensively to model all kinds of waves from light and sound, to electricity and the distribution of electrons within an atom.

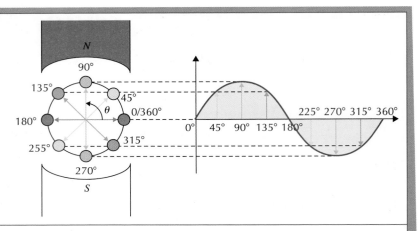

When a coil is rotated in a magnetic field an alternating current is generated in the coil. The graph shows the voltage of the current.

The blue waves show wind-blown waves with high frequency and short wavelength. The red wave shows a tsunami wave brought on by an earthquake when it strikes a coast, and the green wave shows the deep ocean tsunami wave, with a very long wavelength and a low frequency. All the waves can be modelled as sine waves.

Wind-blown waves. Wavelength about 100 m   10 feet

Coastal tsunami, Wavelength about 3 km   100 feet

Wavelength about 100 km   3 feet

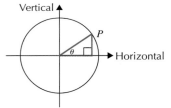

The unit circle (with $r = 1$) is used to define the three trigonometric functions, as shown in the diagram.

$\sin \theta°$ = vertical displacement of P from the x-axis

$\cos \theta°$ = horizontal displacement of P from the y-axis

$\tan \theta°$ = the gradient of OP (M_{OP})

The quadrant diagram is used to identify a group of related angles in the four quadrants.

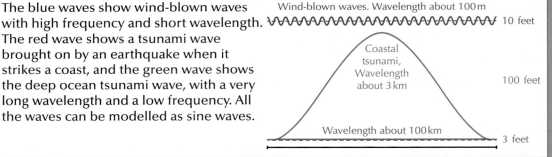

1st quadrant	2nd quadrant	3rd quadrant	4th quadrant
$a°$	$(180 - a)°$	$(180 + a)°$	and $(360 - a)°$

This group of angles can be extended to include angles bigger than 360° and less than 0°, for example $(360 + a)°$; $(540 - a)°$; $-a°$; and $-(180 - a)°$ which, as can be seen by going round the quadratic diagram, are also related to $a°$.

The unit circle is also used to develop important trigonometric identities linking:

- sin, cos and tan
- $a°$ and $(-a)°$
- $a°$ and $(90 - a)°$

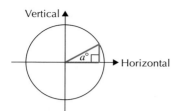

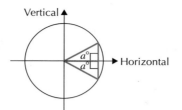

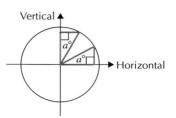

$$\sin^2 a° + \cos^2 a° = 1$$

$$\tan a° = \frac{\sin a°}{\cos a°}$$

$$\sin(-a)° = -\sin a°$$

$$\cos(-a)° = \cos a°$$

$$\tan(-a)° = -\tan a°$$

$$\sin(90-a)° = \cos a°$$

$$\cos(90-a)° = \sin a°$$

Remember, anti-clockwise angles are positive and clockwise angles are negative.

There is a group of angles for which there are exact value ratios for sin, cos and tan. Exact value ratios are expressed as fractions or surds, not the decimal approximations which you can get from a calculator. For example, the **exact value** of $\sin 45°$ is $\frac{1}{\sqrt{2}}$, while 0.866 is simply a decimal approximation.

The triangles shown in the table give the exact values for the angles 30°, 60° and 45°.

	angle	sin	cos	tan
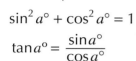	30°	$\frac{1}{2}$	$\frac{\sqrt{3}}{2}$	$\frac{1}{\sqrt{3}}$
	60°	$\frac{\sqrt{3}}{2}$	$\frac{1}{2}$	$\sqrt{3}$
	45°	$\frac{1}{\sqrt{2}}$	$\frac{1}{\sqrt{2}}$	1

For 0°, 90° and all other multiples of 90°, use the appropriate graphs to read off the values of sin, cos and tan.

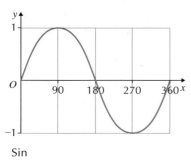

Sin

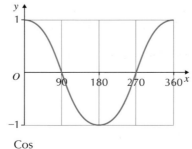

Cos

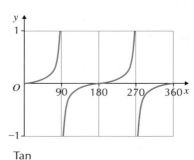

Tan

Finding exact value ratios for angles related to 30°, 60°or 45° and for angles which are multiples of 90°

Exact values for the sine, cosine and tangent of any angle related to 30°, 60° or 45° are found using the triangles above, along with the quadrant diagram. Exact values for the sine, cosine or tangent of multiples of 90° are taken from appropriate graphs.

Example 2.1

Find the exact value.

a $\sin 240°$ **b** $\cos 330°$ **c** $\tan (-45)°$

d $\sin 270°$ **e** $\tan 570°$ **f** $\cos (-225)°$

a $\sin 240° = \sin 60°$

$= -\dfrac{\sqrt{3}}{2}$

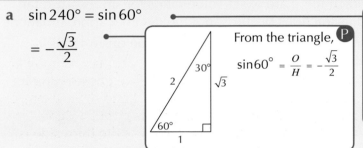

From the triangle, **P**

$\sin 60° = \dfrac{O}{H} = -\dfrac{\sqrt{3}}{2}$

240° is in the 3rd quadrant and so $\sin 240°$ is negative.

240 is 180 + 60 so the related acute angle is 60°.

b $\cos 330° = \cos 30°$

$= \dfrac{\sqrt{3}}{2}$

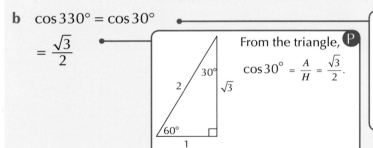

From the triangle, **P**

$\cos 30° = \dfrac{A}{H} = \dfrac{\sqrt{3}}{2}.$

330° is in the 4th quadrant and so $\cos 330°$ is positive.

330 is 360 − 30 so the related acute angle is 30°.

c $\tan (-45)° = -\tan 45°$

$= -1$

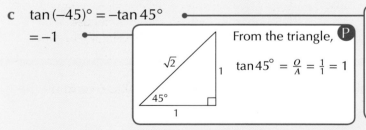

From the triangle, **P**

$\tan 45° = \dfrac{O}{A} = \dfrac{1}{1} = 1$

The angle is negative, so it represents a clockwise turn. −45° is in the 4th quadrant and is related to 45°.

d $\sin 270° = -1$

270 is a multiple of 90. Read the value of $\sin 270°$ from the sine curve.

e $\tan 570° = \tan 30°$

   $= \dfrac{1}{\sqrt{3}}$

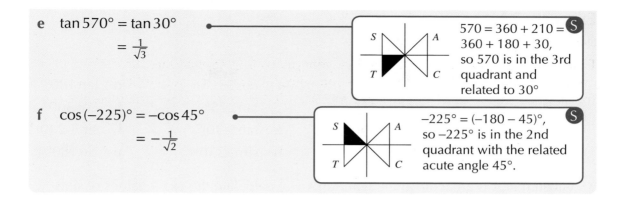

$570 = 360 + 210 =$ **S**
$360 + 180 + 30,$
so 570 is in the 3rd quadrant and related to 30°

f $\cos(-225)° = -\cos 45°$

   $= -\dfrac{1}{\sqrt{2}}$

$-225° = (-180 - 45)°,$ **S**
so −225° is in the 2nd quadrant with the related acute angle 45°.

Finding exact value ratios for acute angles when one of the sine, cosine or tangent value is given as an exact ratio.

In addition to angles related to 30°, 60° or 45° and multiples of 90°, there is another group of angles for which exact value ratios can be found. These are angles where one of sine, cosine or tangent is given as an exact ratio. You may be told, for example, that $\sin A = \frac{3}{5}$ where A is an acute angle. When one exact ratio is given, a right-angled triangle can be formed using that information. The length of the third side is then calculated using Pythagoras and the exact values of the other two ratios read from the triangle.

Example 2.2

Given that P is an acute angle with $\sin P = \frac{3}{5}$ calculate the exact values of $\cos P$ and $\tan P$.

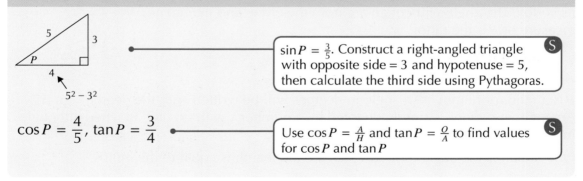

$\sin P = \frac{3}{5}$. Construct a right-angled triangle **S** with opposite side = 3 and hypotenuse = 5, then calculate the third side using Pythagoras.

$\cos P = \dfrac{4}{5},\ \tan P = \dfrac{3}{4}$

Use $\cos P = \frac{A}{H}$ and $\tan P = \frac{O}{A}$ to find values **S** for $\cos P$ and $\tan P$

In Example 2.2, if P had been an **obtuse angle** with $\sin P = \frac{3}{5}$, you would have worked with the same triangle for its related acute angle, then noted that for an obtuse angle both the cosine and tangent are negative, to give $\cos P = -\frac{4}{5}$ and $\tan P = -\frac{3}{4}$.

Exercise 2A

Do not use a calculator for this Exercise.

★ 1 Find exact values.

a	$\cos 30°$	**b**	$\sin 150°$	**c**	$\cos 330°$	**d**	$\tan 180°$
e	$\sin 240°$	**f**	$\cos 135°$	**g**	$\tan 225°$	**h**	$\cos 300°$
i	$\sin 540°$	**j**	$\cos 510°$	**k**	$\tan(-30)°$	**l**	$\sin(-240)°$
m	$\cos 270°$	**n**	$\tan 750°$	**o**	$\sin(-330)°$	**p**	$\sin 180°$

2 Given that A is an acute angle with $\tan A = \frac{3}{2}$ calculate the exact values of $\sin A$ and $\cos A$.

3 If P is an acute angle with $\sin P = \frac{5}{6}$ calculate the exact values of $\cos P$ and $\tan P$.

★ 4 X is an acute angle with $\sin X = \frac{5}{13}$.

    a Calculate the exact value of $\cos X$ and $\tan X$.

    b Suppose X is obtuse, rather than acute. What would the values of $\cos X$ and $\tan X$ be then?

Working with angles measured in radians

Until now angles have been measured in degrees but they can also be measured in **radians**. Many important formulae involving trigonometric functions are simpler when the angles are measured in radians rather than degrees. This is particularly the case when dealing with the differentiation and integration of the sine and cosine functions (see Chapters 9 and 11).

To find the radian measure of an angle θ, draw any circle round its vertex. The angle will cut an arc from the circle and the radian measure of θ is the ratio:

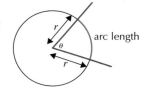

$$\boldsymbol{\theta} = \frac{\textbf{arc length}}{\textbf{radius}}$$

θ is a ratio, or number. An angle in degrees will be written with the degree sign ($x°$). An angle in radians is simply denoted by a number x with no unit attached although in certain contexts you would write 'x radians'.

One radian is the angle formed when the arc length is equal to the radius.

Making the link between radians and degrees

If the radius of the circle is r then:

- the circumference $= 2\pi r$

- half of the circumference is πr

- the radian measure corresponding to $180°$ is $\frac{\pi r}{r} = \pi$

  You must learn the link $180° = \pi$ radians

With this link established you should be able to convert between degrees and radians starting with the exact ratio angles shown in the table.

Angle in degrees	Angle in radians	
30°	$\dfrac{\pi}{6}$	$30° = \dfrac{1}{6}$ of $180° = \dfrac{1}{6}$ of π
45°	$\dfrac{\pi}{4}$	$45° = \dfrac{1}{4}$ of $180° = \dfrac{1}{4}$ of π
60°	$\dfrac{\pi}{3}$	$60° = \dfrac{1}{3}$ of $180° = \dfrac{1}{3}$ of π
90°	$\dfrac{\pi}{2}$	$90° = \dfrac{1}{2}$ of $180° = \dfrac{1}{2}$ of π
180°	π	
270°	$\dfrac{3\pi}{2}$	$270° = 3 \times 90° = 3 \times \dfrac{\pi}{2}$
360°	2π	$360° = 2 \times 180° = 2 \times \pi$

Example 2.3

Change these angles from degrees to radians.

a 120°   **b** 225°   **c** 330°

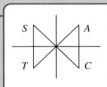

When converting from degrees to radians, use the quadrant diagram along with the fractional conversions above.

a $120° = \dfrac{2\pi}{3}$

$120° = (180 - 60)° = \pi - \dfrac{\pi}{3} = \dfrac{3\pi}{3} - \dfrac{\pi}{3}$ ⓟ

b $225° = \dfrac{5\pi}{4}$

$225° = (180 + 45)° = \pi + \dfrac{\pi}{4} = \dfrac{4\pi}{4} + \dfrac{\pi}{4}$ ⓟ

c $330° = \dfrac{11\pi}{6}$

$330° = (360 - 30)° = 2\pi - \dfrac{\pi}{6} = \dfrac{12\pi}{6} - \dfrac{\pi}{6}$ ⓟ

Example 2.4

Change these angles from radians to degrees.

a $\dfrac{5\pi}{6}$ b $\dfrac{3\pi}{4}$ c $\dfrac{7\pi}{3}$ d $-\dfrac{3\pi}{2}$

a $\dfrac{5\pi}{6} = 150°$ $\dfrac{5\pi}{6} = 5 \times \frac{\pi}{6} = 5 \times 30°$ **P**

b $\dfrac{3\pi}{4} = 135°$ $\dfrac{3\pi}{4} = 3 \times \frac{\pi}{4} = 3 \times 45°$ **P**

c $\dfrac{7\pi}{3} = 420°$ $\dfrac{7\pi}{3} = 7 \times \frac{\pi}{3} = 7 \times 60°$ **P**

d $-\dfrac{3\pi}{2} = -270°$ $-\dfrac{3\pi}{2} = -3 \times -3\frac{\pi}{2} = -3 \times 90°$ **P**

Other angles which are fractions or multiples of 180°

Any angle which is a fraction or multiple of 180° can be converted to the same fraction or multiple of π.

Example 2.5

a Convert 20° to radians.   b Convert $\frac{\pi}{5}$ radians to degrees.

a $20 = \dfrac{1}{9}$ of 180 so $20° = \dfrac{\pi}{9}$ b $\dfrac{\pi}{5} = \dfrac{180°}{5} = 36°$

Converting angles using direct proportion

For those angles which are not easily expressed in terms of fractions or multiples of 180, the conversions are made using direct proportion. In these cases, a calculator can be used.

Example 2.6

a Convert 256° to radians.   b Convert 3.68 radians to degrees.

a $180° = \pi$

$\Rightarrow 1° = \dfrac{\pi}{180}$

$\Rightarrow 256° = \dfrac{\pi}{180} \times 256 \approx 4.47$ radians

b $\pi = 180°$

$\Rightarrow 1$ radian $= \dfrac{180°}{\pi}$

$\Rightarrow 3.68$ radians $= \dfrac{180°}{\pi} \times 3.68 \approx 210.8°$

Exact value ratios for angles measured in radians

Example 2.7

Find the exact value of:

a $\cos\dfrac{2\pi}{3}$

b $\sin\dfrac{3\pi}{2}$

a $\cos\dfrac{2\pi}{3} = -\cos\dfrac{\pi}{3}$

  $= -\dfrac{1}{2}$

$\dfrac{2\pi}{3}$ is in the 2nd quadrant so $\cos\dfrac{2\pi}{3}$ is negative.

$\dfrac{2\pi}{3} = \pi - \dfrac{\pi}{3}$ so related acute angle is $\dfrac{\pi}{3}$ (or 60°).

b $\sin\dfrac{3\pi}{2} = -1$

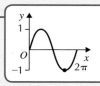

$\dfrac{3\pi}{2}$ is a multiple of $\dfrac{\pi}{2}$ (or 90°) so the value of $\sin\dfrac{3\pi}{2}$ is read from the sine curve.

Exercise 2B

Do not use a calculator for this exercise unless stated in the question

★ 1   Convert these angles to radians giving your answer as a fraction or multiple of π.

Sketch the quadrant diagram before starting to answer this question.

a   30°   b   150°   c   300°   d   135°   e   315°

f   120°   g   240°   h   450°   i   −60°   j   720°

★ 2   Convert these angles to degrees.

a $\dfrac{\pi}{4}$ b $\dfrac{\pi}{12}$ c $\dfrac{\pi}{5}$ d $\dfrac{3\pi}{2}$ e $\dfrac{5\pi}{4}$

f $\dfrac{3\pi}{4}$ g $\dfrac{7\pi}{6}$ h $\dfrac{7\pi}{3}$ i $-\dfrac{\pi}{6}$ j 3π

3   Use a calculator to complete these conversions, rounding to 1 decimal place for degrees and 2 for radians.

a   Convert to radians: i   47°   ii   324°

b   Convert to degrees: i   4.36 radians   ii   1.57 radians

★ 4   Give exact value ratios for each of these ratios.

a $\cos\dfrac{\pi}{4}$ b $\cos\dfrac{3\pi}{4}$ c $\sin\dfrac{5\pi}{3}$ d $\tan\dfrac{11\pi}{6}$

e $\cos(-\pi)$ f $\tan\dfrac{7\pi}{3}$ g $\sin\dfrac{5\pi}{6}$ h $\sin\dfrac{3\pi}{2}$

Working with the sine and cosine and tangent ratios of compound angles

Compound angles are angles expressed as functions of x, such as $(x + 30)°$, $\left(x - \frac{\pi}{3}\right)$, $2x$ and $3x°$. Angles such as $(x + 30)°$, $\left(x - \frac{\pi}{3}\right)$ involve the **sum** and **difference** of two angles. You will learn to work with such angles using the **addition formulae**.

The addition formulae for sine and cosine

The four addition formulae for sine and cosine are given here without proof. Proofs are given or derived in Chapter 6.

The addition formulae for any angles A and B are:

- $\sin(A + B) = \sin A \cos B + \cos A \sin B$
- $\sin(A - B) = \sin A \cos B - \cos A \sin B$
- $\cos(A + B) = \cos A \cos B - \sin A \sin B$
- $\cos(A - B) = \cos A \cos B + \sin A \sin B$

These four formulae are very important and you must learn them. Every other formula in this chapter is derived from them. It may help to note that, when comparing the left-hand side with the right-hand side of each:

'**S**igns agree for **S**in, **C**ontradict for **C**os'

The addition formulae, or identities, work both ways. For example, you must be able to expand $\sin(A + B)$, but it is equally important that you are able to recognise its expansion and be able to convert back.

To keep the addition formulae independent of particular letters and so avoid any confusion, they will often be referred to as $\sin(\text{sum})$ instead of $\sin(A + B)$, or $\cos(\text{difference})$ instead of $\cos(A - B)$. You must remember that 'difference' always refers to the first angle minus the second.

> ## Example 2.8
> Rewrite these expressions in terms of a single angle.
>
> **a** $\cos 25° \cos 30° - \sin 25° \sin 30°$
>
> **b** $\sin \frac{\pi}{3} \cos \frac{\pi}{5} - \cos \frac{\pi}{3} \sin \frac{\pi}{5}$

a $\cos 25° \cos 30° - \sin 25° \sin 30°$ —————(Recognise as $\cos(\text{sum})$.)

$= \cos(25 + 30)°$

$= \cos 55°$

b $\sin \frac{\pi}{3} \cos \frac{\pi}{5} - \cos \frac{\pi}{3} \sin \frac{\pi}{5}$ —————(Recognise as $\sin(\text{difference})$.)

$= \sin\left(\frac{\pi}{3} - \frac{\pi}{5}\right) = \sin \frac{2\pi}{15}$

Example 2.9

Expand and simplify $\sin(P + 30°)$ using the appropriate formula and the exact value ratios for 30°.

$$\sin(P + 30°) = \sin P \cos 30° + \cos P \sin 30°$$

Use the formula for sin(sum).

$$= \sin P \times \frac{\sqrt{3}}{2} + \cos P \times \frac{1}{2}$$

Any time you get an angle for which an exact value ratio can be given, you will be expected to use it.

$$= \frac{\sqrt{3}\sin P + \cos P}{2}$$

Exercise 2C

Do not use a calculator for this Exercise

Remember that the difference is always (first angle – second angle).

★ 1 Rewrite these expressions in terms of a single angle.

a $\sin P \cos Q + \cos P \sin Q$

b $\sin M \cos N - \cos M \sin N$

c $\cos 75° \cos 30° - \sin 75° \sin 30°$

d $\sin 25° \cos 40° - \cos 25° \sin 40°$

e $\sin \frac{\pi}{3} \cos \frac{\pi}{5} - \cos \frac{\pi}{3} \sin \frac{\pi}{5}$

f $\cos 140° \cos 65° - \sin 140° \sin 35°$

g $\cos 70° \cos 85° + \sin 70° \sin 85°$

h $\cos \frac{2\pi}{3} \cos \frac{\pi}{4} - \sin \frac{2\pi}{3} \sin \frac{\pi}{4}$

★ 2 Expand these compound angles. Use exact value ratios to simplify in parts **e** to **h**.

a $\sin(P + Q)$ b $\cos(R - S)$ c $\cos(A + 48°)$ d $\sin(B - 15°)$

e $\cos(x - 60)°$ f $\sin\left(x + \frac{\pi}{4}\right)$ g $\sin(t - 120)°$ h $\cos\left(x + \frac{5\pi}{6}\right)$

3 Noting that $\tan A = \frac{\sin A}{\cos A}$ show that:

a if $\sin(30 + t)° = 2\sin(30 - t)°$ then $\tan t° = \frac{1}{3\sqrt{3}}$

b if $2\cos\left(x + \frac{\pi}{4}\right) = \cos\left(x - \frac{\pi}{4}\right)$ then $\tan x = \frac{1}{3}$

Finding exact value ratios using the addition formulae for sine and cosine

Example 2.10

Find the exact value of:

a $\cos 160° \cos 20° - \sin 160° \sin 20°$

b $\sin 40° \cos 70° - \cos 40° \sin 70°$

(continued)

a $\cos 160° \cos 20° - \sin 160° \sin 20° = \cos(160 + 20)°$

> Recognise as $\cos$ (sum). Ⓢ

$$= \cos 180°$$
$$= -1$$

b $\sin 40° \cos 70° - \cos 40° \sin 70° = \sin(40 - 70)°$

> Recognise as $\sin$ (difference). Ⓢ Remember it is always (first angle – second angle).

$$= \sin(-30)°$$
$$= -\frac{1}{2}$$

Example 2.11

Find the exact value of $\cos 15°$.

$\cos 15° = \cos(45 - 30)°$

> To find the exact value of $\cos 15°$, Ⓢ change it to $\cos(45 - 30)°$ and expand using the $\cos$ (difference) formula.

$$= \cos 45° \cos 30° + \sin 45° \sin 30°$$
$$= \frac{1}{\sqrt{2}} \times \frac{\sqrt{3}}{2} + \frac{1}{\sqrt{2}} \times \frac{1}{2}$$
$$= \frac{1 + \sqrt{3}}{2\sqrt{2}}$$

Example 2.11 could also have been answered by changing 15 to (60 – 45). This method works when the angle can be expressed as the sum or difference of angles for which you already know the exact ratios.

Example 2.12

Given that $\sin x = \frac{4}{5}$ and $\cos y = \frac{5}{13}$, where x and y are both acute angles, find the exact value of $\cos(x - y)$.

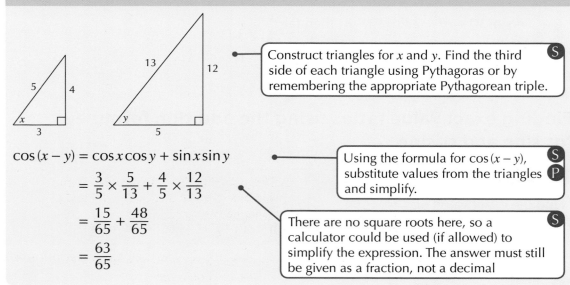

> Construct triangles for x and y. Find the third Ⓢ side of each triangle using Pythagoras or by remembering the appropriate Pythagorean triple.

$\cos(x - y) = \cos x \cos y + \sin x \sin y$

> Using the formula for $\cos(x - y)$, Ⓢ substitute values from the triangles Ⓟ and simplify.

$$= \frac{3}{5} \times \frac{5}{13} + \frac{4}{5} \times \frac{12}{13}$$
$$= \frac{15}{65} + \frac{48}{65}$$
$$= \frac{63}{65}$$

> There are no square roots here, so a Ⓢ calculator could be used (if allowed) to simplify the expression. The answer must still be given as a fraction, not a decimal

Example 2.13

If P is an acute angle with $\tan P = \frac{2}{3}$, evaluate $\sin\left(P - \frac{\pi}{6}\right)$

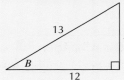

> Construct triangles for P and $\frac{\pi}{6} (= 30°)$ **S**

$$\sin\left(P - \frac{\pi}{6}\right) = \sin P \cos\frac{\pi}{6} - \cos P \sin\frac{\pi}{6}$$

$$= \frac{2}{\sqrt{13}} \times \frac{\sqrt{3}}{2} - \frac{3}{\sqrt{13}} \times \frac{1}{2}$$

> Even if a calculator had been allowed here, you would still be required to give the fraction in terms of square roots. **P**

$$= \frac{2\sqrt{3} - 3}{2\sqrt{13}}$$

Example 2.14

$\cos B = \frac{12}{13}$, where B is an acute angle. Find the exact value of $\cos(B + 90°)$.

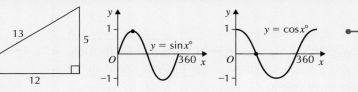

> Construct triangle for B and sketch (or visualise) the graphs of $y = \sin x°$ and $y = \cos x°$ for 90°. **S**

$$\cos(B + 90°) = \cos B \cos 90° - \sin B \sin 90°$$

$$= \frac{12}{13} \times 0 - \frac{5}{13} \times 1$$

$$= -\frac{5}{13}$$

> The answer is negative so $(B + 90°)$ must be in the 2nd quadrant (obtuse angle). **C**

Example 2.15

Show that $\sin x° = \frac{6(\sqrt{10} - 2)}{35}$.

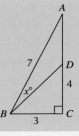

$$x° = A\hat{B}C - D\hat{B}C = P - Q$$

> Recognise $x°$ as the difference between two angles, which are renamed P and Q to simplify working. **S**

 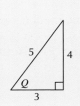

> Redraw each triangle separately to avoid confusion. **S**

(continued)

$$\sin x° = \sin(P - Q)$$
$$= \sin P \cos Q - \cos P \sin Q$$
$$= \frac{\sqrt{40}}{7} \times \frac{3}{5} - \frac{3}{7} \times \frac{4}{5}$$
$$= \frac{3\sqrt{40} - 12}{35}$$
$$= \frac{6\sqrt{10} - 12}{35}$$
$$= \frac{6(\sqrt{10} - 2)}{35}$$

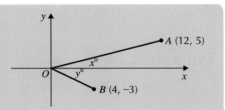

$3\sqrt{40} = 3 \times \sqrt{4} \times \sqrt{10} = 3 \times 2 \times \sqrt{10}.$ ⓟ

Example 2.16

A is the point $(12, 5)$ and B is the point $(4, -3)$.
Calculate the exact value of $\sin(x + y)°$.

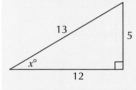

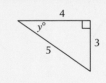

Construct triangles for each angle using the coordinates given. Ⓢ

$$\sin(x + y)° = \sin x° \cos y° + \cos x° \sin y°$$
$$= \frac{5}{13} \times \frac{4}{5} + \frac{12}{13} \times \frac{3}{5}$$
$$= \frac{20}{65} + \frac{36}{65}$$
$$= \frac{56}{65}$$

Exercise 2D

Do not use a calculator for this Exercise

★ **1** Use addition formulae to calculate exact values for each of the following.

   a $\cos 140° \cos 50° + \sin 140° \sin 50°$ **b** $\sin\frac{\pi}{5} \cos\frac{2\pi}{15} + \cos\frac{\pi}{5} \sin\frac{2\pi}{15}$

   c $\cos 80° \cos 20° + \sin 80° \sin 20°$ **d** $\sin 25° \cos 85° - \cos 25° \sin 85°$

   e $\sin 75°$ **f** $\cos 105°$

2 Find the exact value of $\sin(x + 60)° + \sin(x + 240)°$.

3 Given that $\sin A = \frac{3}{5}$ and $\tan B = \frac{5}{12}$, where A and B are both acute angles, find the exact value of $\sin(A - B)$.

★ **4** If P and Q are acute angles with $\tan P = \frac{3}{4}$ and $\tan Q = \frac{1}{7}$, show that $\cos(P + Q) = \frac{1}{\sqrt{2}}$

5 If x is an acute angle with $\cos x = \frac{3}{5}$, calculate the exact value of $\sin(x + 30°)$.

6 If A is an acute angle with $\sin A = \frac{2}{3}$, calculate the exact value of $\cos\left(A + \frac{3\pi}{2}\right)$.

7 P and Q are acute angles with $\sin P = \frac{2}{3}$ and $\cos Q = \frac{3}{4}$. Calculate the exact value of $\tan(P + Q)$.

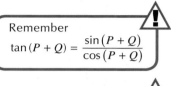

Remember
$\tan(P + Q) = \dfrac{\sin(P + Q)}{\cos(P + Q)}$

★ **8** Calculate the exact value of $\sin PQR$.

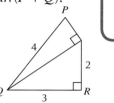

PQR is the sum of two angles.

9 Calculate the exact value of $\cos(\angle ABC)$.

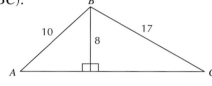

10 Show that the exact value of $\cos x° = \dfrac{6 + \sqrt{13}}{10}$

11 A is the point (a, b) and B is the point (b, a). Calculate the exact value of $\cos AOB$ in terms of a and b.

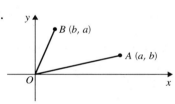

The double angle formulae

Double angle formulae are formulae for $\sin 2A$ and $\cos 2A$. The formulae can be derived from the addition formulae.

For any angle A:

$$\begin{aligned}\sin 2A &= \sin(A + A) \\ &= \sin A \cos A + \cos A \sin A \\ &= 2\sin A \cos A\end{aligned}$$

$$\begin{aligned}\cos 2A &= \cos(A + A) \\ &= \cos A \cos A - \sin A \sin A \\ &= \cos^2 A - \sin^2 A\end{aligned}$$

In the case of $\cos 2A$, the identity $\cos^2 A + \sin^2 A = 1$ leads to two alternative formulae:

$$\cos^2 A - \sin^2 A = \cos^2 A - (1 - \cos^2 A) = 2\cos^2 A - 1$$

$$\cos^2 A - \sin^2 A = (1 - \sin^2 A) - \sin^2 A = 1 - 2\sin^2 A$$

- $\sin 2A = 2\sin A \cos A$

- $\cos 2A = \cos^2 A - \sin^2 A$
  $\cos 2A = 2\cos^2 A - 1$
  $\cos 2A = 1 - 2\sin^2 A$

These formulae, including all three variations for cosine should be learned. The formulae work for any angle which can be expressed as a double. For example:

$$\sin 6A = 2 \sin 3A \cos 3A \quad \sin A = 2\sin\left(\tfrac{1}{2}A\right)\cos\left(\tfrac{1}{2}A\right)$$

Finding exact value ratios using the double angle formulae for sine and cosine

Example 2.17

Find the exact value of each of the following:

a $2 \sin 15° \cos 15°$ **b** $1 - 2\sin^2\left(\tfrac{\pi}{8}\right)$

a $2 \sin 15° \cos 15° = \sin(2 \times 15)°$

               $= \sin 30° = \dfrac{1}{2}$

> Recognise as $\sin$(double angle). **S**

b $1 - 2\sin^2\left(\tfrac{\pi}{8}\right) = \cos\left(2 \times \tfrac{\pi}{8}\right)$

               $= \cos\dfrac{\pi}{4} = \dfrac{1}{\sqrt{2}}$

> Recognise as $\cos$(double angle). **S**

Example 2.18

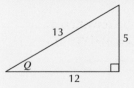

Find the exact value of $\sin 2\theta$ where θ is an acute angle with $\tan\theta = \tfrac{5}{12}$.

> Construct the triangle using Pythagoras. **S**

$\sin 2\theta = 2 \sin\theta \cos\theta$

     $= 2 \times \dfrac{5}{13} \times \dfrac{12}{13}$

     $= \dfrac{120}{169}$

Example 2.19

Find the exact value of $\cos 2x$ where x is an acute angle with $\cos x = \tfrac{2}{5}$

$\cos 2x = \cos^2 x - \sin^2 x$

     $= \left(\dfrac{2}{5}\right)^2 - \left(\dfrac{\sqrt{21}}{5}\right)^2$

     $= \dfrac{4}{25} - \dfrac{21}{25}$

     $= -\dfrac{17}{25}$

> Any of the three formulae for $\cos 2x$ could have been used here, but $\cos^2 x - \sin^2 x$ has been chosen as it is easier to simplify if both terms have the same denominator. **S**

Example 2.20

If A and B are acute angles with $\cos A = \frac{4}{5}$ and $\sin B = \frac{5}{13}$, find the exact value of:

a $\sin 2A$ **b** $\cos 2A$ **c** $\sin(2A + B)$

 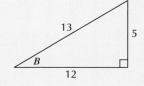

a $\sin 2A = 2\sin A \cos A$

$$= 2 \times \frac{3}{5} \times \frac{4}{5} = \frac{24}{25}$$

b $\cos 2A = \cos^2 A - \sin^2 A$

$$= \left(\frac{4}{5}\right)^2 - \left(\frac{3}{5}\right)^2$$

$$= \frac{16}{25} - \frac{9}{25} = \frac{7}{25}$$

c $\sin(2A + B) = \sin 2A \cos B + \cos 2A \sin B$

$$= \frac{24}{25} \times \frac{12}{13} + \frac{7}{25} \times \frac{5}{13}$$

$$= \frac{288}{325} + \frac{35}{325} = \frac{323}{325}$$

Example 2.21

If P is an acute angle with $\cos P = \frac{2}{3}$, show that $\sin 4P = \frac{8\sqrt{5}}{81}$

Using the double angle formula, $\sin 4P$ can be rewritten as $2\sin 2P \cos 2P$, so the first step is to find $\sin 2P$ and $\cos 2P$.

$\sin 2P = 2\sin P \cos P$ $\cos 2P = \cos^2 P - \sin^2 P$

$$= 2 \times \frac{\sqrt{5}}{3} \times \frac{2}{3} = \frac{4\sqrt{5}}{9} \qquad\qquad = \left(\frac{2}{3}\right)^2 - \left(\frac{\sqrt{5}}{3}\right)^2 = \frac{4}{9} - \frac{5}{9}$$

$$= -\frac{1}{9}$$

$\sin 4P = 2\sin 2P \cos 2P$

$$= 2 \times \frac{4\sqrt{5}}{9} \times \left(-\frac{1}{9}\right)$$

$$= -\frac{8\sqrt{5}}{81}$$

Example 2.22

If A is an acute angle with $\cos A = \frac{3}{5}$, calculate $\sin 3A$.

There is no direct formula for $\sin 3A$ so rewrite it as $\sin (2A + A)$. This is equal to $\sin 2A \cos A + \cos 2A \sin A$, so the first step is to calculate $\sin 2A$ and $\cos 2A$.

$$\sin 2A = 2 \sin A \cos A$$
$$= 2 \times \frac{4}{5} \times \frac{3}{5}$$
$$= \frac{24}{25}$$

$$\cos 2A = \cos^2 A - \sin^2 A$$
$$= \left(\frac{3}{5}\right)^2 - \left(\frac{4}{5}\right)^2$$
$$= \frac{9}{25} - \frac{16}{25}$$
$$= -\frac{7}{25}$$

$$\sin 3A = \sin (2A + A)$$
$$= \sin 2A \cos A + \cos 2A \sin A$$
$$= \frac{24}{25} \times \frac{3}{5} + \left(-\frac{7}{25}\right) \times \frac{4}{5}$$
$$= \frac{72}{125} - \frac{28}{125}$$
$$= \frac{44}{125}$$

Exercise 2E

Do not use a calculator for this Exercise.

★ **1** Find exact values for the following.

    a $\cos^2 15° - \sin^2 15°$ **b** $2\cos^2 \frac{\pi}{8} - 1$

    c $2 \sin 75° \cos 75°$ **d** $2 \sin 105° \cos 105°$

2 If $\cos x = \frac{2}{5}$, where x is an acute angle find the exact value of these ratios.

    a $\sin 2x$ **b** $\cos 2x$ **c** $\tan 2x$

3 If $\tan Y = \frac{3}{2}$, where Y is an acute angle, find the exact value of $\sin 4Y$.

★ **4** P is an acute angle with $\sin P = \frac{2}{\sqrt{5}}$. Calculate the value of $\cos 3P$.

5 If A is an acute angle with $\cos A = \frac{3}{5}$, calculate the exact value of $\sin (2A + 60°)$.

⚙ **6** If $\cos 2A = \frac{3}{5}$, where A is an acute angle, show that $\cos A = \frac{2\sqrt{5}}{5}$

7 If $\cos 2A = \frac{7}{25}$, where A is an acute angle, calculate the exact value of $\sin A$.

8 If A is the point $(4, 3)$. Find the exact gradient of OB.

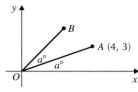

9 Find the exact value of $\cos TPQ$.

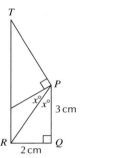

10 Triangle PQR is isosceles, with angle $PQR = x°$.
Calculate the area of triangle PQR.

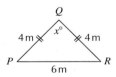

11 a Show that for the isosceles triangle shown in the
diagram that $\frac{b}{\sin x°} = \frac{a}{\cos\left(\frac{1}{2}x\right)°}$

 b Hence show that $b = 2a\sin\left(\frac{1}{2}x\right)°$

Finding the addition and double angle formulae for the tangent function

The wave nature of the sine and cosine functions makes them central to the study of trigonometry. Although they have fewer direct practical applications, there are also addition and double angle formulae for the tangent function. These formulae can be derived from the addition formulae for sin and cos as shown in Example 2.23.

Example 2.23

Using the relationship $\tan(A + B) = \frac{\sin(A+B)}{\cos(A+B)}$, show that:

a $\tan(A + B) = \frac{\tan A + \tan B}{1 - \tan A \tan B}$ **b** $\tan(A - B) = \frac{\tan A - \tan B}{1 + \tan A \tan B}$

c $\tan 2A = \frac{2\tan A}{1 - \tan^2 A}$

(continued)

a $\tan(A + B) = \dfrac{\sin(A + B)}{\cos(A + B)}$

$= \dfrac{\sin A \cos B\ +\ \sin B \cos A}{\cos A \cos B\ -\ \sin A \sin B}$

> Expand $\sin(A + B)$ and $\cos(A + B)$. Ⓢ

$= \dfrac{\dfrac{\sin A}{\cos A} + \dfrac{\sin B}{\cos B}}{1 - \dfrac{\sin A \sin B}{\cos A \cos B}}$

> Divide each term of the numerator and denominator by $\cos A \cos B$. Ⓢ

$= \dfrac{\tan A + \tan B}{1 - \tan A \tan B}$

b $\tan(A - B) = \tan(A + (-B))$

> Rewrite $\tan(A - B)$ as $\tan(A + (-B))$. Ⓢ

$= \dfrac{\tan(A) + \tan(-B)}{1 - \tan(A)\tan(-B)}$

> $\tan(-B) = -\tan B$ Ⓟ

$= \dfrac{\tan(A) - \tan(B)}{1 + \tan(A)\tan(B)}$

c $\tan 2A = \tan(A + A)$

> Rewrite $\tan 2A$ as $\tan(A + A)$. Ⓢ

$= \dfrac{\tan A + \tan A}{1 - \tan A \tan A}$

$= \dfrac{2\tan A}{1 - \tan^2 A}$

Exercise 2F

1 If X and Y are acute angles with $\tan X = 3$ and $\tan Y = 2$, use the formulae derived in Example 2.23 to calculate the exact values of:

   a $\tan(X + Y)$ **b** $\tan(X - Y)$ **c** $\tan 2X$

2 If X and Y are acute angles with $\tan X = \frac{1}{2}$ and $\tan Y = \frac{1}{3}$ use the formulae derived in Example 2.23 to calculate the exact values of:

   a $\tan(X + Y)$ **b** $\tan(X - Y)$ **c** $\tan 2X$

Proving trigonometric identities

Trigonometric identities are statements of equality involving trigonometric expressions which are true for every possible value of the angles involved. You will be expected to prove the truth of trigonometric identities using the relationships shown in the table on the next page. You will prove trigonometric identities by working with the left-hand side (LHS) and/or the right-hand side (RHS) of each expression separately until the sides are identical.

For any angle A		
$\sin(-A) = -\sin(A)$ $\cos(-A) = \cos A$ $\tan(-A) = -\tan A$	$\sin(90° - A) = \cos A$ $\cos(90° - A) = \sin A$	$\sin^2 A + \cos^2 A = 1$ $\tan A = \dfrac{\sin A}{\cos A}$

For any angles A and B		
$\sin(A + B) = \sin A \cos B + \cos A \sin B$ $\sin(A - B) = \sin A \cos B - \cos A \sin B$ $\cos(A + B) = \cos A \cos B - \sin A \sin B$ $\cos(A - B) = \cos A \cos B + \sin A \sin B$	$\sin 2A = 2\sin A \cos A$	$\cos 2A = \cos^2 A - \sin^2 A$ $\qquad = 2\cos^2 A - 1$ $\qquad = 1 - 2\sin^2 A$

Example 2.24

Show that $\cos(30 - x)° + \sin(120 + x)° = \sqrt{3}\cos x°$

$\text{LHS} = \cos(30 - x)° + \sin(120 + x)°$ ——— Separate the LHS and RHS. **S**

$= (\cos 30° \cos x° + \sin 30° \sin x°)$
$\quad + (\sin 120° \cos x° + \cos 120° \sin x°)$ ——— Expand $\cos(30 - x)°$ and $\sin(120 + x)°$ using the addition formula. **P** **S**

$= \dfrac{\sqrt{3}}{2} \times \cos x° + \dfrac{1}{2} \times \sin x° + \dfrac{\sqrt{3}}{2} \times \cos x° + \left(-\dfrac{1}{2}\right) \times \sin x°$ ——— Substitute exact values, noting that 120° is in the second quadrant and related to 60°. **P**

$= \sqrt{3}\cos x°$ ——— Simplify. **P**

$\text{RHS} = \sqrt{3}\cos x°$ ——— Make a final check and state the equality. **C**

$\text{LHS} = \text{RHS}$

Example 2.25

Prove that $\sin\left(x + \frac{\pi}{6}\right) = \frac{1}{2}\left(\cos x + \sqrt{3}\sin x\right)$

$\text{LHS} = \sin\left(x + \dfrac{\pi}{6}\right)$ ——— Expand $\sin\left(x + \frac{\pi}{6}\right)$. **S**

$= \sin x \cos\dfrac{\pi}{6} + \cos x \sin\dfrac{\pi}{6}$ ——— $\frac{\pi}{6}$ does not appear on the RHS so substitute values for the sin and cos of $\frac{\pi}{6}$. **P**

$= \sin x \times \dfrac{\sqrt{3}}{2} + \cos x \times \dfrac{1}{2}$

$= \dfrac{1}{2}\left(\cos x + \sqrt{3}\sin x\right)$ ——— Simplify. You must show the two sides to be identical, so write the LHS expression in the same order as that on the RHS. **P** **S**

$\text{RHS} = \dfrac{1}{2}\left(\cos x + \sqrt{3}\sin x\right)$

$\text{LHS} = \text{RHS}$ ——— Make a final check and state the equality. **C**

Example 2.26

Prove that $= \dfrac{\sin(\alpha + \beta)}{\cos\alpha\cos\beta} = \tan\alpha + \tan\beta$

$\text{LHS} = \dfrac{\sin(\alpha + \beta)}{\cos\alpha\cos\beta}$ $\qquad$ $\text{RHS} = \tan\alpha + \tan\beta$

$= \dfrac{\sin\alpha\cos\beta + \cos\alpha\sin\beta}{\cos\alpha\cos\beta}$ $\qquad$ ⟶ | Use the sin (sum) formula on the LHS. Ⓢ

$= \dfrac{\sin\alpha\cos\beta}{\cos\alpha\cos\beta} + \dfrac{\cos\alpha\sin\beta}{\cos\alpha\cos\beta}$ $\qquad$ ⟶ | The RHS contains two separate terms, so write LHS as two separate terms. Ⓢ

$= \dfrac{\sin\alpha}{\cos\alpha} + \dfrac{\sin\beta}{\cos\beta}$

$= \tan\alpha + \tan\beta$

$\text{LHS} = \text{RHS}$

Example 2.27

Show that $\sin^2 A = \frac{1}{2}(1 - \cos 2A)$

$\text{LHS} = \sin^2 A$ $\qquad$ $\text{RHS} = \dfrac{1}{2}(1 - \cos 2A)$

$= \dfrac{1}{2}(1 - (1 - 2\sin^2 A))$ $\qquad$ ⟶ | Substitute for $\cos 2A$ on the RHS. As $\sin^2 A$ appears on LHS use $\cos 2A = 1 - 2\sin^2 A$. Ⓢ

$= \dfrac{1}{2}(2\sin^2 A)$

$= \sin^2 A$

$\text{LHS} = \text{RHS}$

Example 2.28

Prove that $\cos 3B = 4\cos^3 B - 3\cos B$

$\text{LHS} = \cos 3B$ $\qquad$ $\text{RHS} = 4\cos^3 B - 3\cos B$

$= \cos(2B + B)$ $\qquad$ ⟶ | Change $3B$ to $(2B + B)$ on the LHS. Ⓢ

$= \cos 2B\cos B - \sin 2B\sin B$ $\qquad$ ⟶ | RHS contains B but not $2B$, so use the double angle formula to substitute $\cos 2B$ and $\sin 2B$. The RHS contains $\cos B$ but not $\sin B$, so use $\cos 2B = 2\cos^2 B - 1$. Ⓢ

$= (2\cos^2 B - 1)\cos B - 2\sin B\cos B\sin B$

$= 2\cos^3 B - \cos B - 2\sin^2 B\cos B$

$= 2\cos^3 B - \cos B - 2(1 - \cos^2 B)\cos B$ $\qquad$ ⟶ | Sin B does not appear on the RHS, so use the substitution $\sin^2 B = 1 - \cos^2 B$. Ⓢ

$= 4\cos^3 B - 3\cos B$

$\text{LHS} = \text{RHS}$

Exercise 2G

Do not use a calculator for this Exercise

★ **1** Show that:

  a $\sin(x + 45)° + \cos(x + 45)° = \sqrt{2}\cos x°$

  b $2\cos(x + 30)° - \sin x° = \sqrt{3}\cos x° - 2\sin x°$

  c $\sin(x - 60)° + \cos(x + 30)° = 0$

  d $\sin(x + 225)° - \cos(x + 135)° = 0$

  e $\sin\theta - \sin\left(\theta + \dfrac{\pi}{3}\right) + \cos\left(\theta + \dfrac{\pi}{6}\right) = 0$

2 Prove that:

  a $\cos(\alpha + \beta) + \cos(\alpha - \beta) = 2\cos\alpha\cos\beta$

  b $\sin(\alpha + \beta) - \sin(\alpha - \beta) = 2\cos\alpha\sin\beta$

3 Show that $\cos^2 x° = \frac{1}{2}(\cos 2x° + 1)$

4 Show that $\dfrac{\cos(x+y)°}{\cos x°\cos y°} = 1 - \tan x°\tan y°$

⚙ **5** Show that:

  a $(\cos x - \sin x)(\cos x + \sin x) = \cos 2x$

  b $\cos 2x + \cos x = (2\cos x - 1)(\cos x + 1)$

⚙ **6** Show that $2\cos 2x - \cos^2 x = 1 - 3\sin^2 x$

⚙ **7** Show that $\sin 3A = 3\sin A - 4\sin^3 A$

⚙ **8** Show that $\cos 4\theta = 8\cos^4\theta - 8\cos^2\theta + 1$

⚙ **9** Show that $\cos^4 A - \sin^4 A = \cos 2A$

⚙ **10** Show that:

  a $\sin(x + y) + \sin(x - y) = 2\sin x\cos y$

  b Hence, by letting $x + y = A$ and $x - y = B$ derive a formula for $\sin A + \sin B$ in terms of x and y.

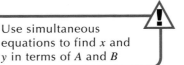
Use simultaneous equations to find x and y in terms of A and B

  c Derive a similar formula for $\cos A + \cos B$

  d Hence show that $\dfrac{(\sin 5x + \sin 3x)}{(\cos 3x + \cos x)} = 2\sin 2x$

Introducing and working with the wave function
a cos *x* + *b* sin *x* where *a* and *b* are constant

There are many examples in real life of waves, represented by trigonometric functions, being added together. The music of an orchestra is produced by adding together multiple sound waves; scientists investigating wave power study the effects of adding water waves and an earthquake, as illustrated on a seismograph, is also the result of adding different wave functions. The photograph relates to the 2011 Japanese tsunami.

A seismologist poses for the media as he points to a seismographic graph showing the magnitude of the earthquake in Japan, on a monitor at the British Geological Survey office in Edinburgh, Scotland March 11, 2011. The biggest earthquake on record to hit Japan struck the northeast coast on Friday, triggering a 10-metre tsunami that swept away everything in its path, including houses, ships, cars and farm buildings.

While these are more complex situations, study here will be restricted to adding two functions of the forms $f(x) = a\cos x$ and $f(x) = b\cos x$ respectively, which are in phase and have the same period. The result of adding such functions can be investigated fully using a graphic calculator or other graphing package.

The diagram on the left shows the shows the graphs of $y = 3\cos x°$ and $y = 4\sin x°$. The diagram on the right shows the graph of $y = 3\cos x° + 4\sin x°$.

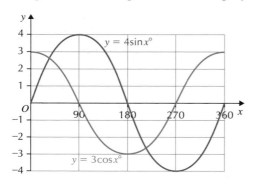

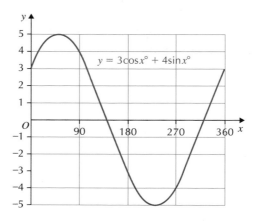

It can be seen from the graph that the resultant of adding the two functions is a cosine function with the same period, 360°, as the original two functions but with an increased amplitude, 5, and a phase angle of approximately 53°. Adding the two functions gives

$$y = 3\cos x° + 4\sin x° \approx 5\cos(x - 53)°$$

In general it can be shown that, for any constants a and b, the sum of $a\cos x°$ and $b\sin x°$ can be expressed in a similar way with:

$a\cos x° + 4\sin x° = k\cos(x - \alpha)°$ for some constants k and α where $k > 0$ and $0 \le \alpha < 360$

Functions of the form $f(x) = a \cos x + b \sin x$

The sum of the two waves, $a\cos x + b\sin x$, can also be expressed in the form $k\cos(x + \alpha)$ or $k\sin(x \pm \alpha)$.

$a\cos x + b\sin x$ can be rewritten in the forms $k\cos(x \pm \alpha)$ or $k\sin(x \pm \alpha)$ by calculating the values of k and α algebraically rather than graphically, but before doing this, it is necessary to learn how to solve a particular form of simultaneous equations for k and α where $k > 0$ and $0 \leq \alpha < 360$. These are equations of the form:

- $k\cos\alpha° = a$ \qquad (1)

- $k\sin\alpha° = b$ \qquad (2)

To find k use the identity $\sin^2\alpha + \cos^2\alpha = 1$ as follows:

$(k\cos\alpha°)^2 + (k\sin\alpha°)^2 = a^2 + b^2$

$\Rightarrow k^2(\cos^2\alpha° + \sin^2\alpha°) = a^2 + b^2$

$\Rightarrow k^2 = a^2 + b^2$

$\Rightarrow k = \sqrt{(a^2 + b^2)}$

To find α use the identity $\tan a° = \frac{\sin a°}{\cos a°}$ as follows:

$\tan\alpha° = \frac{k\sin\alpha°}{k\cos\alpha°}$

$\Rightarrow \tan\alpha° = \frac{b}{a}$

The value of $\alpha°$ depends on which quadrant it is in and that is established by looking at the signs of all three of sin, cos and tan (as will be illustrated in the examples that follow).

Example 2.29

Solve each of the following pairs of simultaneous equations for $k > 0$ and $0 \leq \alpha < 360$.

a $k\cos\alpha° = 3$ and $k\sin\alpha° = 4$

b $k\cos\alpha° = -2$ and $k\sin\alpha° = 5$

c $k\cos\alpha° = -3$ and $k\sin\alpha° = -2$

a $k = \sqrt{(3^2 + 4^2)} = 5$

$\tan\alpha° = \frac{4}{3}$ where $\alpha°$ is in the first quadrant.

$\Rightarrow \alpha° = \tan^{-1}\left(\frac{4}{3}\right)$

$\Rightarrow \alpha = 53.1$

> P
> $\tan\alpha° = \frac{k\sin a°}{k\cos a°}$. All of $\sin\alpha°$, $\cos\alpha°$ and $\tan\alpha°$ are positive, so α is in the first quadrant.

(continued)

b $k = \sqrt{\left((-2)^2 + 5^2\right)} = \sqrt{29}$

$\tan\alpha° = \frac{5}{-2} = -\frac{5}{2}$ where $\alpha°$ is in the second quadrant

$\Rightarrow \alpha° = 180° - \tan^{-1}\left(\frac{5}{2}\right)$

$\Rightarrow \alpha = 111.8$

> **P** Only $\sin\alpha°$ is positive; $\cos\alpha°$ and $\tan\alpha°$ are negative, so α is in the second quadrant.

c $k = \sqrt{\left((-3)^2 + (-2)^2\right)} = \sqrt{13}$

$\tan\alpha° = \frac{-2}{-3} = \frac{2}{3}$ where $\alpha°$ is in the third quadrant

$\Rightarrow \alpha° = 180° + \tan^{-1}\left(\frac{2}{3}\right)$

$\Rightarrow \alpha = 213.7$

> **P** Only $\tan\alpha°$ is positive; $\cos\alpha°$ and $\sin\alpha°$ are negative, so α is in the third quadrant.

Exercise 2H

If necessary, calculators may be used for finding α but k should be left in exact form as a surd.

1 Solve for $k > 0$ and $0 \le \alpha < 360$

   a $k\cos\alpha = 4$ and $k\sin\alpha = 3$

   b $k\cos\alpha = 5$ and $k\sin\alpha = -3$

   c $k\cos\alpha = -1$ and $k\sin\alpha = 2$

2 Solve for $k > 0$ and $0 \le \alpha < 2\pi$

   a $k\cos\alpha = 1$ and $k\sin\alpha = 1$

   b $k\cos\alpha = -\sqrt{3}$ and $k\sin\alpha = -1$

   c $k\cos\alpha = -4$ and $k\sin\alpha = 5$

Expressing $a\cos x + b\sin x$, where a and b are constants, in the form $k\cos(x \pm \alpha)$ or $k\sin(x \pm \alpha)$

Now that you can solve the above simultaneous equations you can find k and α algebraically in order to express $a\cos x + b\sin x$ in the form $k\cos(x - \alpha)$.

You already know how to expand $k\cos(x - \alpha)$ where k and α are constants using the addition formula.

$k\cos(x - \alpha) = k\cos x\cos\alpha + k\sin x\sin\alpha$

$= (k\cos\alpha)\cos x + (k\sin\alpha)\sin x$

$= a\cos x + b\sin x$, where a and b are constants

> ⚠ As k and α are both constants, so are $k\cos\alpha$ and $k\sin\alpha$

Working in reverse, to find the particular values of k and α which allow you to state that $a\cos x + b\sin x = k\cos(x - \alpha)$, you will start by assuming the equality and working backwards:

If $a\cos x + b\sin x = k\cos(x - \alpha)$ it follows that:

$a\cos x + b\sin x = k\cos x\cos\alpha + k\sin x\sin\alpha$ and so:

$a\underline{\cos x} + b\underline{\underline{\sin x}} = (k\cos\alpha)\underline{\cos x} + (k\sin\alpha)\underline{\underline{\sin x}}$

Comparing sides gives:

$k\cos\alpha = a$ and $k\sin\alpha = b$

This gives a pair of simultaneous equations of the type solved in Example 2.29.

Expand $k\cos(x - \alpha)°$

Rearrange the RHS and single and double underline $\cos x$ and $\sin x$ to make the comparison of the two sides easier.

Example 2.30

Write $3\cos x° + 4\sin x°$ in the form $k\cos(x - \alpha)°$ where $k > 0$ and $0 \leq \alpha < 360$.

If $3\cos x° + 4\sin x° = k\cos(x - \alpha)°$ it follows that:

$3\cos x° + 4\sin x° = k\cos x°\cos\alpha° + k\sin x°\sin\alpha°$

$3\underline{\cos x°} + 4\underline{\underline{\sin x°}} = (k\cos\alpha°)\underline{\cos x°} + (k\sin\alpha°)\underline{\underline{\sin x°}}$

Comparing sides gives:

$k\cos\alpha° = 3$ and $k\sin\alpha° = 4$

leading to:

$k = \sqrt{3^2 + 4^2} = 5$

$\tan\alpha° = \frac{4}{3}$ and α is in the first quadrant.

$\alpha° = \tan^{-1}\frac{4}{3}$

$\alpha = 53.1$

And so: $3\cos x° + 4\sin x° = 5\cos(x - 53.1)°$

> Expand $k\cos(x - \alpha)°$. **S**

> Rearrange RHS and use single and double underlining to help compare sides. **S**

> $\tan\alpha = \frac{k\sin\alpha}{k\cos\alpha}$. All of $\sin\alpha$, $\cos\alpha$ and $\tan\alpha$ are positive, so α is in the first quadrant. **P**

It might sometimes be tempting to take shortcuts in completing this type of question, but you should avoid doing so. These questions are typically worth several marks in an exam, and need evidence that you have a complete understanding of what you are doing and that you are able to communicate it clearly. You must show evidence of all **four stages** of the solution:

1   compare $a\cos x + b\sin x$ with the expanded form of $k\cos(x - \alpha)$

2   list the values of $k\cos\alpha$ and $k\sin\alpha$

3   calculate k using these values as shown

4   calculate α using these values as shown.

If you do not show working for all stages, you will score only one mark out of all the possible marks for calculating k even if your answer is correct.

Example 2.31

Write $\cos x - \sin x$ in the form $R\cos(x - \theta)$ where $k > 0$ and $0 \leq \theta < 2\pi$

If $\cos x - \sin x = R\cos(x - \theta)$ it follows that:

$\cos x - \sin x = R\cos x \cos\theta + R\sin x \sin\theta$ •————— Expand $R\cos(\theta - \alpha)$. Ⓢ

$\underline{\cos x} - \underline{\underline{\sin x}} = (R\cos\theta)\underline{\cos x} + (R\sin\theta)\underline{\underline{\sin x}}$

Comparing sides gives:

$R\cos\theta = 1$ and $R\sin\theta = -1$

leading to:

$$R = \sqrt{1^2 + (-1)^2} = \sqrt{2}$$

$\tan\theta = -1$ and θ is in the 4th quadrant. •————

Only $\cos\theta$ is positive, while $\sin\theta$ and $\tan\theta$ are negative, so θ is in the 4th quadrant. Ⓟ

$$\theta = 2\pi - \tan^{-1}(1) = 2\pi - \frac{\pi}{4} = \frac{8\pi}{4} - \pi = \frac{7\pi}{4}$$

And so: $\cos x - \sin x = \sqrt{2}\cos\left(x - \frac{7\pi}{4}\right)$

Example 2.32

Write $3\sin x° + 11\cos x°$ in the form $k\cos(x - \alpha)°$ where $k > 0$ and $0 \leq \alpha < 360$.

If $3\sin x° + 11\cos x° = k\cos(x - \alpha)°$ it follows that:

$3\sin x° + 11\cos x° = k\cos x° \cos\alpha° + k\sin x° \sin\alpha°$

$3\underline{\sin x°} + 11\underline{\underline{\cos x°}} = (k\cos\alpha°)\underline{\underline{\cos x°}} + (k\sin\alpha°)\underline{\sin x°}$ •———

When comparing sides, note that $\sin x°$ and $\cos x°$ do not come in the same order on both sides. You must always be careful. This is where the single and double underlining helps. Ⓢ

Comparing sides gives:

$k\cos\alpha° = 11$ and $k\sin\alpha° = 3$

leading to:

$$k = \sqrt{11^2 + 3^2} = \sqrt{130}$$

$\tan\alpha° = \frac{3}{11}$ and α is in the first quadrant

$$\alpha° \tan^{-1} = \left(\frac{3}{11}\right) = 15.3°$$

$\alpha = 15.3$

And so $3\sin x° + 11\cos x° = \sqrt{130}\cos(x - 15.3)°$

Examples 2.33 and 2.34 show variations but it is important to note that the overall method is no different.

Example 2.33

Write in the form $R\sin(x+\theta)$ where $k>0$ and $0 \le \theta < 2\pi$

If $\sin x - \sqrt{3}\cos x = R\sin(x+\theta)$ it follows that:

$\sin x - \sqrt{3}\cos x = R\sin x\cos\theta + R\cos x\sin\theta$

$\underline{\sin x} - \sqrt{3}\,\underline{\cos x} = (R\cos\theta)\underline{\sin x} + (R\sin\theta)\underline{\cos x}$

> Start with the expansion for $\sin(x+\theta)$. **S**

Comparing sides gives:

$R\cos\theta = 1$ and $R\sin\theta = -\sqrt{3}$

leading to:

$R = \sqrt{(1^2 + (-\sqrt{3})^2)} = 2$

$\tan\theta = -\dfrac{\sqrt{3}}{1} = -\sqrt{3}$ and θ is in the 4th quadrant

$\theta = 2\pi - \tan^{-1}\left(\sqrt{3}\right) = 2\pi - \dfrac{\pi}{3} = \dfrac{5\pi}{3}$

And so: $\sin x - \sqrt{3}\cos x = 2\sin\left(x - \dfrac{5\pi}{3}\right)$

Example 2.34

Write $5\cos 2x° - 12\sin 2x°$ in the form $k\sin(2x-\alpha)°$ where $k>0$ and $0 \le \alpha < 360$

If $5\cos 2x° - 12\sin 2x° = k\sin(2x-\alpha)°$ it follows that:

$5\cos 2x° - 12\sin 2x° = k\sin 2x°\cos\alpha° - k\cos 2x°\sin\alpha°$

$5\,\underline{\cos 2x°} - 12\,\underline{\sin 2x°} = (k\cos\alpha°)\underline{\sin 2x°} - (k\sin\alpha°)\underline{\cos 2x°}$

Comparing sides gives:

$k\cos\alpha° = -12$ and $k\sin\alpha° = -5$

leading to:

$k = \sqrt{\left((-12)^2 + (-5)^2\right)} = 13$

$\tan\alpha° = \dfrac{5}{12}$ and α is in the 3rd quadrant

$\alpha° = 180° + \tan^{-1}\left(\dfrac{5}{12}\right)$

$\alpha = 202.6$

And so $5\cos 2x° - 12\sin 2x° = 13\sin(2x - 202.6)°$

Exercise 2I

Calculators can be used in this exercise unless stated otherwise

★ **1** Express $\cos x° + 2\sin x°$ in each of the following forms for $k > 0$ and $0 \le \alpha < 360$:

   a $k\cos(x - \alpha)°$ **b** $k\cos(x + \alpha)°$

   c $k\sin(x + \alpha)°$ **d** $k\sin(x - \alpha)°$

2 Express $5\cos x° - 4\sin x°$ in the form $R\cos(x + \beta)°$.

 3 Express $\sqrt{2}\cos\theta + \sqrt{2}\sin\theta$ in the form $k\sin(\theta - \alpha)$ for $0 \le \theta < 2\pi$.

 4 Express $\sin x - \cos x$ in the form $R\sin(x - \beta)$ for $0 \le x < 2\pi$.

 5 Express $\sqrt{3}\cos x - \sin x$ in the form $k\cos(x + \alpha)$ for $0 \le x < 2\pi$.

 6 Express $3\cos x + 3\sin x$ in the form $R\sin(x + \alpha)$ for $0 \le x < 2\pi$.

7 Express $4\sin 2x° - 3\cos 2x°$ in the form $k\sin(2x - \alpha)°$ for $0 \le x < 360$.

8 Express $2\sin(3x)° + 3\cos(3x)°$ in the form $k\cos(3x - \alpha)°$ for $0 \le x < 360$.

9 Express $2\sin\left(x + \frac{\pi}{3}\right)$ in the form $R\cos(x - \beta)$ for $0 \le x < 2\pi$.

- I can find exact value trigonometric ratios for all angles related to 30°, 60° and 45° and for multiples of 90°. ★ Exercise 2A Q1

- I can calculate exact values for the remaining trigonometric ratios when one of sine, cosine or tangent is given in as fraction or exact ratio. ★ Exercise 2A Q4

- I know that 180° = π radians and can convert between radians and degrees. ★ Exercise 2B Q1, 2

- I can find exact value ratios for angles measured in radians. ★ Exercise 2B Q4

- I can work with the addition formulae for the sum and difference of two angles. ★ Exercise 2C Q1, 2

- I can find exact value ratios using the addition formulae. ★ Exercise 2D Q1, 4, 8

- I can work with the double angle formulae and use them to find exact value ratios. ★ Exercise 2E Q1, 4, 10

- I can prove trigonometric identities. ★ Exercise 2G Q1

- I can work with the wave function. ★ Exercise 2I Q1

For further assessment opportunities, see the Preparation for Assessment for Unit 1 on pages 132–135.

3 Identifying and sketching related functions

This chapter will show you how to:

- identify or sketch a function after a transformation of the form $af(x)$, $f(bx)$, $f(x) + c$, $f(x + d)$ or any combination of the four
- complete the square of a quadratic function and use the completed square form to sketch or determine equations and to find maximum or minimum value of a function
- sketch the graph of an exponential or logarithmic function and find the equation when given the graph
- sketch the graph of, or determine the equation of, a trigonometric function of the form $y = a\sin(bx + d)° + c$ or the cosine equivalent including examples derived from the wave function
- find the maximum and minimum values of a trigonometric function of the form $f(x) = a\sin(bx + d)° + c$ or the cosine equivalent including examples derived from the wave function
- sketch the graph of the derived function $f'(x)$ when given the graph of $f(x)$.

You should already know:

- how to complete the square of a quadratic function with unitary x^2 coefficient and use the completed square form to sketch the graph
- how to sketch or determine the equation of the graphs of trigonometric functions of the form $y = a\sin(bx)° + c$ or $y = \sin(x + d)°$ and their cosine equivalent.

You must be able to answer every type of questions in this chapter without the use of a calculator. Graphic calculators or other graphing packages are advised, however, both to investigate transformations and to back up and check your solutions.

Sketching the graphs of related functions

Suppose you start with any graph, $y = f(x)$, with certain key points noted.

From this you will be expected to be able to sketch the graphs of the related functions:

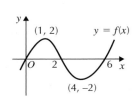

$y = af(x)$ $\qquad$ $y = f(bx)$

$y = f(x) + c$ $\qquad$ $y = f(x + d)$

along with any combinations of the four.

While this can be explored using a graphic calculator or other graphing package, it is sometimes easier to remember new facts if they are based on something you are already familiar with. Start by looking at the graph of $y = \sin x°$ where $0 \leq x \leq 360$ and investigate the effect of these transformations on it.

$y = a\sin x°$ where $0 \leq x \leq 360$ (amplitude = a)

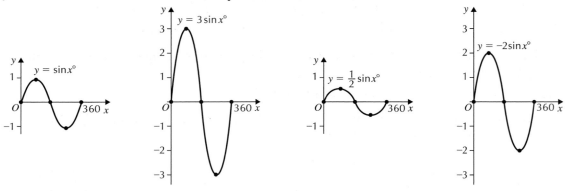

- There has been no horizontal transformation on the original graph of $y = \sin x°$.
- The graph of $y = \sin x°$ has been stretched or compressed vertically depending on whether $|a| > 1$ or $|a| < 1$.
- In the case of $a < 0$ the graph has been reflected in the x–axis.

$y = \sin(bx)°$ where $0 \leq x \leq 360$ (b cycles in 360°; period $= \frac{1}{b}$)

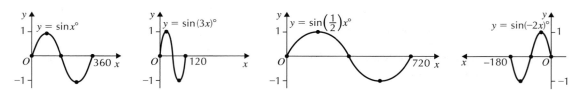

- The original graph has been stretched or compressed horizontally depending on whether $|b| < 1$ or $|b| > 1$.
- In the case of $b < 0$ the graph has been reflected in the y–axis.
- There has been no vertical transformation.

$y = \sin x° + c$ where $0 \leq x \leq 360$

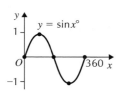 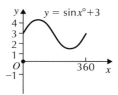 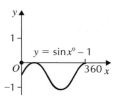

- There has been no horizontal transformation.
- The original graph has been slid up or down by $|c|$ depending on whether c is positive or negative.

$y = \sin(x + d)°$ where $0 \le x \le 360$ (phase angle $-d°$)

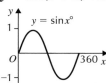 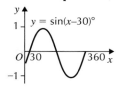 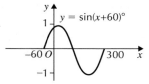

- The original graph has been slid left or right by $|d|$ depending on whether or not d is positive or negative.
- There has been no vertical transformation.

These transformations can be checked out on different functions using a graphic calculator or other graphing package.

Related function	Transformation in words	Transformation in coordinates
$y = af(x)$	no horizontal transformation graph stretched or compressed vertically according to $|a| > 1$ or $|a| < 1$ graph reflected in the x-axis if $a < 0$	$(x, y) \to (x, ay)$
$y = f(bx)$	graph compressed or stretched horizontally according to $|b| > 1$ or $|b| < 1$ graph reflected in the y axis if $b < 0$ no vertical transformation	$(x, y) \to \left(\frac{x}{b}, y\right)$
$y = f(x) + c$	no horizontal transformation when $c > 0$: $f(x) + c$ means the graph slides up c units $f(x) - c$ means the graph slides down c units	$(x, y) \to (x, y + c)$
$y = f(x + d)$	when $d > 0$ $f(x + d)$ means the graph slides back d units $f(x - d)$ means graph slides forward by d units no vertical transformation	$(x, y) \to (x - d, y)$

The order in which transformations are applied follows similar rules to the BODMAS rules for the order of operations. In the same way that you **multiply** or **divide before** you **add** or **subtract**, you **reflect**, **stretch** or **compress before** you **slide horizontally** or **vertically**.

Example 3.1

Part of the graph of $y = f(x)$ is shown here.

Use this to sketch:

a $y = 2f(x)$ **b** $f(2x)$ **c** $-3f(2x)$

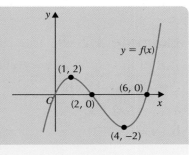

a $y = 2f(x)$ so the transformation is:

horizontal: $(x) \Rightarrow$ no horizontal change

(continued)

vertical: $2 \times f(\ ) \Rightarrow$ multiply y-coordinate by 2 (stretch by a factor of 2)

so $(x, y) \rightarrow (x, 2y)$

$(0, 0)\ \rightarrow (0, 0)$

$(1, 2)\ \rightarrow (1, 4)$

$(2, 0)\ \rightarrow (2, 0)$

$(4, -2) \rightarrow (4, -4)$

$(6, 0)\ \rightarrow (6, 0)$

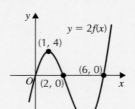

> **P** Apply this to the key points of the original graph.

> **P** Check that the graph has been stretched vertically by a factor of 2.

b $y = f(2x)$ so the transformation is:

horizontal: $(2x) \Rightarrow$ divide x-coordinate by 2 (compress by a factor of $\frac{1}{2}$)

vertical: $f(\ ) \Rightarrow$ no vertical change

so $(x, y) \rightarrow \left(\frac{x}{2}, y\right)$

$(0, 0)\ \rightarrow (0, 0)$

$(1, 2)\ \rightarrow \left(\frac{1}{2}, 2\right)$

$(2, 0)\ \rightarrow (1, 0)$

$(4, -2) \rightarrow (2, -2)$

$(6, 0)\ \rightarrow (3, 0)$

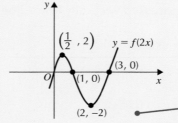

> **P** Check that the graph has been compressed horizontally by a factor of $\frac{1}{2}$.

c $y = -3f(2x)$

horizontal: $(2x) \Rightarrow$ divide x-coordinate by 2 (compress by a factor of $\frac{1}{2}$)

vertical: $-3 \times f(\ ) \Rightarrow$ multiply y-coordinate by -3 (reflect in the x-axis and stretch by a factor of 3)

so $(x, y) \rightarrow \left(\frac{x}{2}, -3y\right)$

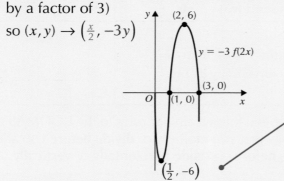

> **P** Check that the graph has been compressed horizontally by a factor of $\frac{1}{2}$, stretched vertically by a factor of 3 and reflected in the x axis.

Example 3.2

Part of the graph of $y = f(x)$ is shown here.

Use this to sketch:

a $y = f(x) + 3$

b $y = f(x + 3)$

c $y = 2 - 3f(x + 1)$

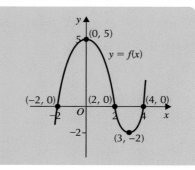

a $y = f(x) + 3$

horizontal: $(x) \Rightarrow$ no change

vertical: $f(\ ) + 3 \Rightarrow$ add 3 to y-coordinate (slide up by 3 units)

so $(x, y) \rightarrow (x, y + 3)$

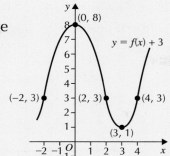

> Check that the graph has been slid up 3. ⓟ

b $y = f(x + 3)$

horizontal: $(x + 3) \Rightarrow$ subtract 3 from x-coordinate (slide back by 3 units)

vertical: $f(\ )$ no change

so $(x, y) \rightarrow (x - 3, y)$

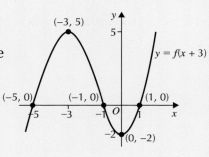

> Check that the graph has been slid 3 back. ⓟ

c Rearrange $y = 2 - 3f(x + 1)$ as $y = -3f(x + 1) + 2$

horizontal: $(x + 1) \Rightarrow$ subtract 1 from x-coordinate (slide back by 1 unit)

vertical: $-3 \times f(\ ) + 2 \Rightarrow$ multiply y-coordinate by -3 then add 2 (reflect in the x-axis and stretch by a factor of 3 before sliding up by 2 units)

so $(x, y) \rightarrow (x - 1, -3y + 2)$

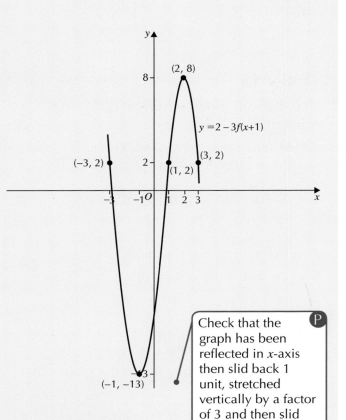

> Check that the graph has been reflected in x-axis then slid back 1 unit, stretched vertically by a factor of 3 and then slid down 2. ⓟ

Example 3.3

Part of the graph of $y = f(x)$ is shown here.
Use it to sketch the graph of $y = f(2x - 6) + 1$

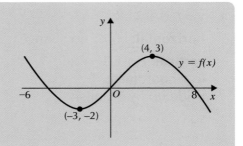

$y = f(2x - 6) + 1 \Rightarrow f(2(x - 3)) + 1$ ●——————⟨ Rewrite $2x - 6$ as $2(x - 3)$. ⟩ **S**

horizontal: $(2 \times (x - 3)) \Rightarrow$ divide the x-coordinate
by 2 then add 3 (compress by a factor of $\frac{1}{2}$ then slide forward 3 units)

vertical: $f(\ ) + 1 \Rightarrow$ add 1 to the y-coordinate (slide up 1 unit)

so $(x, y) \to \left(\frac{x}{2} + 3,\ y + 1\right)$

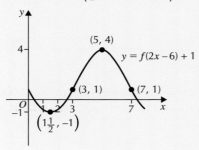

Check that the graph has been compressed by a factor of $\frac{1}{2}$ horizontally **then** slid forward 3 and up 1. **P**

Exercise 3A

1 The diagram shows the graph of $y = f(x)$.

On separate diagrams, sketch the graphs of these functions.

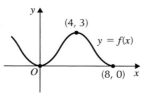

⚠ Make sure you clearly mark the new position of your key points in each diagram

a $y = f(x - 2)$	**b** $y = f(2x) + 3$	**c** $y = 2f(x) + 1$	**d** $y = f(x + 4) - 5$
e $y = -f(x)$	**f** $y = f(-x)$	**g** $2 - f(x)$	**h** $y = 2f(-x) - 3$

2 The diagram shows the graph of $y = f(x)$.
Hence sketch the graphs of:

a $y = 2 + f(x)$ **b** $y = 3f(x - 2)$ **c** $y = f(2x) + 3$

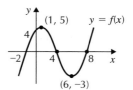

★ **3** The diagram shows the graph of $y = g(x)$.

On separate diagrams, sketch the graphs of these functions.

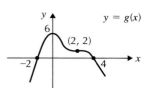

a $y = 2g(x)$ **b** $y = g(x + 1) - 1$ **c** $y = 2g(3x)$

d $y = g(x - 5)$ **e** $y = -4g(2x) + 3$ **f** $y = g(2x - 4) + 1$

4 The diagram shows the graph of $y = f(x)$.

On separate diagrams, sketch the graphs of these functions.

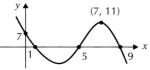

a $y = 2f(x - 3) + 1$ **b** $y = 2 + 3f(x)$ **c** $y = 3f(x + 2) - 4$

d $y = 3f(-x) + 2$ **e** $y = 8 - 2f(x + 3)$ **f** $y = -f(2x - 1)$

5 The diagram shows the graph of $y = x^2 + 2x - 5$.

Hence sketch the graph of:

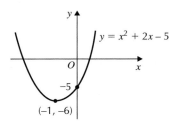

a $y = x^2 + 2x - 2$

b $y = 2x^2 + 4x - 10$

c $y = (x - 1)^2 + 2(x - 1) - 5$

d $y = 5 - 2x - x^2$

Expressing a quadratic function of the form $ax^2 + bx + c$ in completed square form

Sometimes when asked to sketch the graph of a quadratic function or asked to find its maximum or minimum value you must first express it in the form $a(x + p)^2 + r$. This is known as the completed square form.

Example 3.4

Complete the square to express each of the following quadratic expressions in completed square form, that is in the form $p(x + q)^2 + r$.

a $x^2 + 6x + 2$ **b** $4x^2 + 8x + 3$ **c** $7 + 8x - 2x^2$

a $x^2 + 6x + 2 = (x^2 + 6x) + 2$ •———— Isolate ($x^2 \pm x$ term) in brackets. **S**

$= (x^2 + 6x + 9) - 9 + 2$ •——

$= (x + 3)^2 - 7$

$x^2 + 6x$ is part of the square $(x + 3)(x + 3) = x^2 + 6x + 9$. To complete the square add 9 inside the bracket and then compensate by subtracting 9. **P**

b $4x^2 + 8x + 3 = 4(x^2 + 2x) + 3$ •

$= 4(x^2 + 2x + 1) - 4 + 3$

$= 4(x + 1)^2 - 1$

Isolate ($x^2 \pm x$ term) in bracket: $4x^2 + 8x = 4(x^2 + 2x)$ **S**

$x^2 + 2x$ is part of the square $(x + 1)(x + 1) = x^2 + 2x + 1$. By adding 1 inside the bracket you have actually added $4 \times (+1) = +4$ and so must compensate by subtracting 4. **P**

c $7 + 8x - 2x^2 = -2(x^2 - 4x) + 7$ •————

$= -2(x^2 - 4x + 4) + 8 + 7$ •——

$= -2(x - 2)^2 + 15$

$8x - 2x^2 = -2(x^2 - 4x)$ **S**

$-2 \times (+4) = -8$ so to compensate add 8. **P**

Example 3.5

Express $(2x + 1)(2x - 3)$ in the form $a(x + b)^2 + c$

The expression cannot be converted to completed square form as it stands so the first step must be to multiply out the brackets

$(2x + 1)(2x - 3)$

$= 4x^2 - 4x - 3$

$= 4(x^2 - x) - 3$

$= 4\left(x^2 - x + \dfrac{1}{4}\right) - 1 - 3$

$= 4\left(x - \dfrac{1}{2}\right)^2 - 4$

Exercise 3B

1 Express in completed square form.

     a $x^2 + 4x$ **b** $x^2 - 6x + 10$ **c** $x^2 + 8x - 1$ **d** $5 + 2x + x^2$

     e $10 + 4x - x^2$ **f** $7 - 6x - x^2$ **g** $6x - 3 - x^2$ **h** $10 - 3x - x^2$

★ 2 Express in the form $a(x - b)^2 + c$

     a $2x^2 + 8x + 3$ **b** $2x^2 - 4x + 15$ **c** $3x^2 + 12x - 5$ **d** $5x^2 - 30x + 36$

     e $4x^2 - 12x + 1$ **f** $2x^2 + 6x - 9$ **g** $3x^2 + 15x + 7$ **h** $4x^2 + 28x + 7$

★ 3 Write in the form $p - q(x + r)^2$

     a $15 - 4x - 2x^2$ **b** $12x - x^2$ **c** $3 + 8x - 4x^2$ **d** $-3x^2 + 12x - 8$

     e $1 + 6x - 2x^2$ **f** $19 - 20x - 4x^2$ **g** $4 + 7x - 7x^2$ **h** $-2x^2 - 6x - 3$

⚙ 4 Write in the form $a(x + b)^2 + c$

     a $(2 - x)(3 + 2x)$ **b** $(3x + 2)(2x + 1)$ **c** $3x(x + 2) - 5$ **d** $(x + 3)^2 - 2x + 5$

⚙ 5 By expressing the quadratic function $ax^2 + bx + c$ in completed square form, derive the quadratic formula for the solution of the quadratic equation $ax^2 + bx + c = 0$.

Working with the graphs of quadratic functions expressed in completed square form

The graph of $y = x^2$ is a parabola with a **key point** at $(0, 0)$, the minimum turning point.

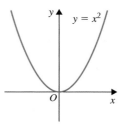

To sketch the graph of $y = 3(x - 1)^2 + 2$, consider the transformation on the turning point $(0, 0)$ of the graph $y = f(x)$ where $f(x) = x^2$

$y = 3 \times (x - 1)^2 + 2 = 3 \times f(x - 1) + 2$ so $(x, y) \rightarrow (x + 1, 3y + 2)$

$(0, 0) \rightarrow (0 + 1, 0 + 2) = (1, 2)$

The point $(0, 0)$ is slid forward 1 and up 2, so the turning point of $y = 3(x - 1)^2 + 2$ is $(1, 2)$.

x^2 terms and turning points are related:

- positive x^2 term $\Rightarrow$ minimum turning point
- negative x^2 term $\Rightarrow$ maximum turning point

and so $(1, 2)$ will be a minimum turning point.

The final step will be to find where the graph cuts the y-axis.

$x = 0 \Rightarrow y = 3 \times (0 - 1)^2 + 2 = 5$ and so the graph cuts the y-axis at $(0, 5)$.

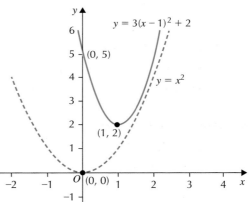

Example 3.6

Find the turning points of these quadratic functions stating their nature.

a $y = 2(x - 5)^2 - 3$ **b** $y = 7 - 5(x + 2)^2$

a $y = 2(x - 5)^2 - 3$ so $(x, y) \rightarrow (x + 5, y - 3)$

$(0, 0) \rightarrow (0 + 5, 0 - 3) = (5, -3)$

$(0, 0)$ on the graph of $y = x^2$ is slid forward 5 and down 3. Ⓢ

The minimum turning point is $(5, -3)$

Positive x^2 term $\Rightarrow$ minimum turning point. Ⓢ

b $y = -5(x + 2)^2 + 7$

so $(0, 0) \rightarrow (-2, 7)$

$(0, 0)$ on the graph of $y = x^2$ is slid back 2 and up 7. Ⓢ

The maximum turning point is $(-2, 7)$

Negative x^2 term $\Rightarrow$ maximum turning point. Ⓢ

Example 3.7

Sketch the graph of $y = 2(x - 1)^2 + 3$

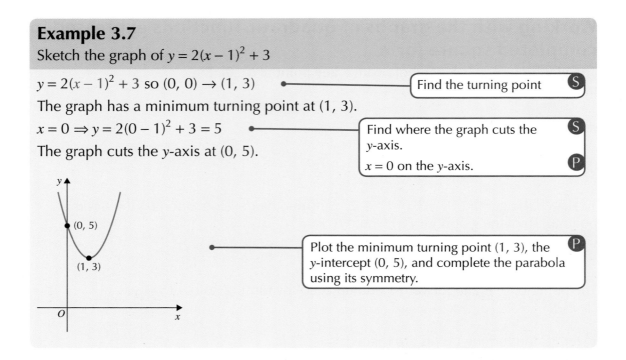

$y = 2(x - 1)^2 + 3$ so $(0, 0) \rightarrow (1, 3)$ — Find the turning point **S**

The graph has a minimum turning point at $(1, 3)$.

$x = 0 \Rightarrow y = 2(0 - 1)^2 + 3 = 5$ — Find where the graph cuts the y-axis. **S**

The graph cuts the y-axis at $(0, 5)$.

$x = 0$ on the y-axis. **P**

Plot the minimum turning point $(1, 3)$, the y-intercept $(0, 5)$, and complete the parabola using its symmetry. **P**

Using turning points and points of intersection to state equations of the form $y = a(x + b)^2 + c$

Starting with a graph of a parabola which identifies the turning point and the point of intersection with the y-axis, it is possible to state the equation in the form $y = a(x + b)^2 + c$.

The graph of the parabola on the right has turning point $(2, -3)$ and cuts the y-axis at 5. This gives enough information to find the equation in the form $y = a(x + b)^2 + c$ by once again comparing it with the graph of $y = x^2$ and the transformation on the turning point $(0, 0)$.

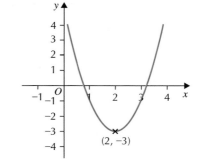

$(0, 0)$ has been slid forward 2 and down 3 so:

$y = a(x - 2)^2 - 3$

The next step is to find a.

$(0, 5)$ lies on the parabola and so $y = 5$ when $x = 0$. This gives:

$5 = a(0 - 2)^2 - 3 \Rightarrow 5 = 4a - 3 \Rightarrow 4a = 8 \Rightarrow a = 2$ and so:

The equation of the parabola is $y = 2(x - 2)^2 - 3$

Example 3.8

Find the equation of these graphs.

a

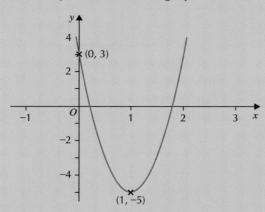

b

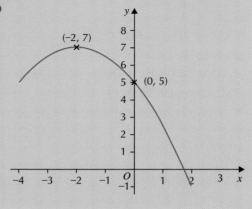

a Turning point at $(1, -5) \Rightarrow$ equation is of the form $y = a(x - 1)^2 - 5$

> $(0, 0)$ has slid forward S 1 and down 5.

Now find a

$(0, 3)$ lies on the parabola $\Rightarrow 3 = a \times (0 - 1)^2 - 5$

$\Rightarrow 3 = a - 5$

$\Rightarrow a = 8$

The equation is $y = 8(x - 1)^2 - 5$

b Turning point at $(-2, 7) \Rightarrow$ equation is of the form $a(x + 2)^2 + 7$

> $(0, 0)$ has slid back 2 and up 7. S

Now find a

$(0, 5)$ lies on the parabola $\Rightarrow 5 = a \times (0 + 2)^2 + 7$

$\Rightarrow 5 = 4k + 7$

$\Rightarrow 4k = -2$

$\Rightarrow k = -\frac{1}{2}$

Equation is $y = -\frac{1}{2}(x + 2)^2 + 7$ or $y = 7 - \frac{1}{2}(x + 2)^2$

Finding maximum and minimum values of a quadratic function by expressing it in completed square form

Example 3.9

a Express $6x - x^2 - 8$ in the form $a - b(x - c)^2$

b Hence sketch the graph of $y = 6x - x^2 - 8$

c What is the maximum value of $6x - x^2 - 8$ and for what value of x does it occur?

(continued)

a $6x - x^2 - 8 = -(x^2 - 6x) - 8$

 $-(+9) = -9$ so compensate by adding 9.

$= -(x^2 - 6x + 9) + 9 - 8$

$= 1 - (x - 3)^2$

b Find the turning point:

$y = -(x - 3)^2 + 1$

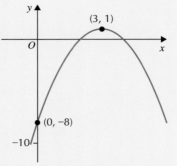

$(0, 0) \rightarrow (3, 1)$ and so the graph has a maximum turning point at $(3, 1)$.

Find where the graph cuts the y-axis:

$x = 0 \Rightarrow y = 1 - (0 - 3)^2 = -8$, and so graph cuts the y-axis at the point $(0, -8)$.

c The graph of $y = 6x - x^2 - 8$ has a maximum turning point $(3, 1)$ so the maximum value of $6x - x^2 - 8$ is 1 and it occurs when $x = 3$.

Example 3.10

a Write $f(x) = 5 + 8x - 2x^2$ in the form $p - q(x + r)^2$

b Hence state the minimum value of $\frac{1}{5 + 8x - 2x^2}$

a $5 + 8x - 2x^2 = -2(x^2 - 4x) + 5$

$= -2(x^2 - 4x + 4) + 8 + 5$

 $-2 \times (+4) = -8$ so compensate by adding 8.

$= -2(x - 2)^2 + 13$

$= 13 - 2(x - 2)^2$

b $f(x) = 5 + 8x - 2x^2 = -2(x - 2)^2 + 13$

The graph of $y = f(x)$ has a maximum turning point at $(2, 13)$ and so the maximum value of $5 + 8x - 2x^2 = 13$.

It follows that the minimum value of $\frac{1}{5 + 8x - 2x^2}$ is $\frac{1}{13}$

Example 3.11

Show that $2x^2 - 6x + 5$ is always greater than 0.

$2x^2 - 6x + 5$ will always be greater than 0 if its minimum value is greater than 0.

$2x^2 - 6x + 5 = 2(x^2 - 3x) + 5$

 Find the minimum value of $2x^2 - 6x + 3$ by first completing the square.

$$= 2\left(x^2 - 3x + \frac{9}{4}\right) - \frac{9}{2} + 5$$

$$= \left(x - \frac{3}{2}\right)^2 + \frac{1}{2}$$

The minimum turning point is $\left(\frac{3}{2}, \frac{1}{2}\right)$ and so the minimum value of $2x^2 - 6x + 5$ is $\frac{1}{2}$

This means that $2x^2 - 6x + 5$ will always be greater than or equal to $\frac{1}{2}$ and so will always be greater than 0.

Exercise 3C

★ **1** Write down the coordinates of the turning point of each of these graphs, stating whether each is a maximum or minimum.

  a $y = (x - 3)^2 + 2$ **b** $y = 2(x + 3)^2 - 7$ **c** $y = 12 - 2(x + 4)^2$

  d $y = 3 + 2(x - 5)^2$ **e** $y = 5 - 3(x - 5)^2$ **f** $y = -4(x + 5)^2 + 7$

  g $y = 9 + (x - 2)^2$ **h** $y = 13 - (x + 7)^2$

2 Sketch the graphs, showing the turning point and the intersection with the y-axis.

  a $y = (x - 2)^2 + 3$ **b** $y = 5 - (x + 2)^2$ **c** $y = 2(x - 1)^2 + 5$

  d $y = -3(x - 2)^2 - 9$ **e** $y = 5 + 3(x - 4)^2$ **f** $y = 3(x + 1)^2 + 7$

  g $y = 4 - 2(x + 3)^2$ **h** $y = 4(x + 1)^2 - 5$

★ **3** **a** Write the expression $3x^2 + 6x + 10$ in the form $a(x + b)^2 + c$.

  b Hence sketch the graph of $y = 3x^2 + 6x + 10$

★ **4** **a** Write the expression $15 - 4x - 2x^2$ in the form $p - q(x + r)^2$

  b Hence sketch the graph of $y = 15 - 4x - 2x^2$

★ **5** Find the equation of each of these graphs.

a 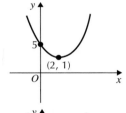   **b** 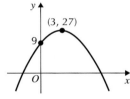   **c**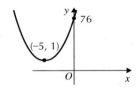

d   **e** 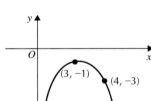   **f**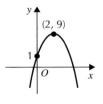

★ **6** **a** Write the function $f(x) = 3x^2 + 6x - 2$ in the form $a(x + b)^2 + c$

  b Hence state the minimum value of $f(x)$.

7 **a** Write the function $f(x) = 5 + 3x - x^2$ in the form $p - q(x - r)^2$

  b Hence state the maximum value of $5 + 3x - x^2$

🐾 **8** **a** Express $3x^2 + 12x + 54$ in the form $a(x + b)^2 + c$

  b Hence state the maximum value of $\dfrac{1}{3x^2 + 12x + 54}$

🐾 **9** **a** Express $(2x - 5)(2x + 3)$ in the form $a(x + b)^2 + c$

  b Hence find the maximum value of $\dfrac{1}{(2x - 5)(2x + 3)}$

🐾 **10** Show that $x^2 + 4x + 9$ is greater than 4 for all values of x.

🐾 **11** Show that the function $f(x) = 15 - 2x - x^2$ is always less than 20.

🐾 **12** Show that the function $f(x) = 2x^2 + 6x + 11$ can never be negative.

🐾 **13** By expressing the quadratic function $f(x) = px^2 + qx + r$ in completed square form, find the coordinates of the turning point of the graph $y = px^2 + qx + r$ in terms of p, q and r.

Working with the graphs of exponential and logarithmic functions

You should have completed Chapter 1 before starting this section.

Exponential functions are used to model and study many real life phenomena such as population growth; the growth of bacteria; the spread of a virus and radioactive decay. The uses of logarithmic functions include measuring the strength of an earthquake.

Working with the graphs of exponential functions

An exponential function is a function of the form $f(x) = a^x$ where $a > 0$ and $a \neq 1$.

More specifically it is called the exponential function to base a.

- 2^x is the exponential function to base 2.
- $\left(\frac{1}{2}\right)^x$ is the exponential function to base $\frac{1}{2}$

The properties of these functions for $a > 0$ and $0 < a < 1$ can be studied using a graphic calculator or other graphing package to verify the following:

For any a such that $a > 0$, $a \neq 1$ the graph of $y = a^x$ is defined for all real numbers and will pass through the key points:

- **(0, 1)**   **because $a^0 = 1$ for all a**
- **(1, a)**   **because $a^1 = a$ for all a**

It will also approach but never cut the x-axis making the x-axis a boundary line or asymptote for the graph.

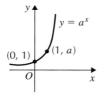

With $a > 1$ the graph illustrates
exponential growth

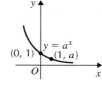

With $0 < a < 1$ the graph illustrates
exponential decay

Example 3.12

Sketch these graphs.

a $y = 5^x$ **b** $y = \left(\frac{1}{3}\right)^x$

a   The key points are (0, 1) and (1, 5); the boundary line is the x-axis.

b   The key points are (0, 1) and $\left(1, \frac{1}{3}\right)$; the boundary line is the x-axis.

Example 3.13

State the equation of each of these exponential graphs.

a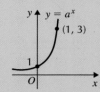

b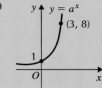

a The graph passes through (0, 1) and (1, 3) and so the equation is $y = 3^x$.

b The graph passes through (0, 1) and (3, 8).

so: $8 = a^3 \Rightarrow a = \sqrt[3]{8} = 2$ •————

> (3, 8) is not a key point but it tells you that $y = 8$ when $x = 3$ **P**

The equation is $y = 2^x$

Drawing the graph of any function related to $f(x) = a^x$

The graph of any function related to $f(x) = a^x$ can be drawn by considering the transformation on the key points (0, 1) and (1, a) and the boundary line or x-axis.

For example, if asked to sketch the graph of $y = 2^{(x-1)} + 3$ consider the transformation on the key points and boundary line of the graph of the function $f(x) = 2^x$

$y = 2^{(x-1)} + 3 = f(x-1) + 3$ where $f(x) = 2^x$ and so $(x, y) \rightarrow (x + 1, y + 3)$
(forward 1 and up 3)

The key points and the boundary line of the graph of $y = 2^x$ are transformed as follows:

$(0, 1) \rightarrow (1, 4)$

$(1, 2) \rightarrow (2, 5)$

The x-axis is slid up 3 to the line with equation $y = 3$.

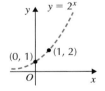 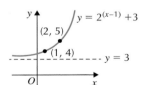

Example 3.14

Sketch the graph $y = y = 2^{(x-2)} + 1$ showing where it cuts the y-axis.

$y = 2^{(x-2)} + 1$ so $(x, y) \rightarrow (x + 2, y + 1)$

$(0, 1) \rightarrow (2, 2)$ •————

> The key points and boundary line of $y = 2^x$ are transformed. **P**

$(1, 2) \rightarrow (-1, 3)$

(continued)

The x-axis slides up 1 to the line with equation $y = 1$

$x = 0 \Rightarrow y = 2^{(0-2)} + 1$ ← $x = 0$ on the y-axis Ⓢ

$\qquad = \frac{1}{4} + 1 = \frac{5}{4}$

and so the curve cuts the y-axis at $\left(0, \frac{5}{4}\right)$

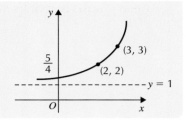

Example 3.15

Sketch the graph of $y = 2^{(2x)} - 8$ stating where it cuts the x-axis.

$y = 2^{(2x)} - 8$ so $(x, y) \to \left(\frac{x}{2}, y - 8\right)$

$(0, 1) \to (0, -7)$ ← The key points and boundary line of $y = 2^x$ are transformed. Ⓟ

$(1, 2) \to \left(\frac{1}{2}, -6\right)$

The x-axis slides down to the line with equation $y = -8$

$y = 0 \Rightarrow 0 = 2^{2x} - 8$ ← $y = 0$ on the x-axis Ⓢ

$\Rightarrow 2^{2x} = 8 \Rightarrow 2x = 3$

$\Rightarrow x = \frac{3}{2}$

and so the curve cuts the x-axis at $\left(\frac{3}{2}, 0\right)$.

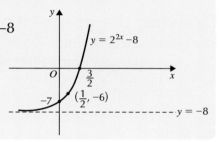

Example 3.16

The equation of the exponential function shown on the graph is of the form $y = a^x + b$. Calculate a and b and hence write down the equation of the curve.

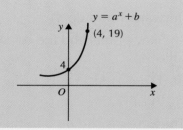

$y = a^x + b$ which is simply the graph of $y = a^x$ slid up b

$(0, 1)$ has been slid up to $(0, 4)$ and so $b = 3$

$(4, 19)$ lies on the curve so $19 = a^4 + 3 \Rightarrow a^4 = 16 \Rightarrow a = 2$

The equation is $y = 2^x + 3$.

Exercise 3D

★ 1   Sketch the graphs of these exponential functions by first plotting the key points.

a $y = 3^x$ b $y = 10^x$ c $y = 4^x$ d $y = 7^x$

e $y = \left(\frac{1}{2}\right)^x$ f $y = \left(\frac{1}{4}\right)^x$ g $y = \left(\frac{3}{5}\right)^x$ h $y = \left(\frac{2}{3}\right)^x$

★ **2** These graphs are of the form $y = a^x$. Write down their equations.

a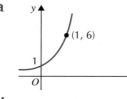

b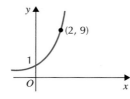

c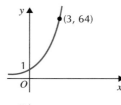

d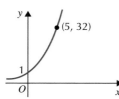

e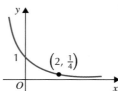

f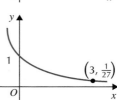

3 Sketch the graphs of these functions.

    a $y = 2^{(x+1)}$ **b** $y = 3^{(x-4)} + 5$ **c** $y = 2^{3x}$ **d** $y = 3^x + 2$

    e $y = 4^{(2x-3)}$ **f** $y = 3 - 6^{(x-1)}$ **g** $y = 4^{2x+1}$ **h** $y = 5^x - 2$

★ **4** Sketch the graph of $y = 2^{(x-2)} - 8$ and state where it cuts the x-axis.

5 Sketch the graph of $y = 3^{(x+1)} - 9$ and state where it cuts the x-axis.

6 The diagram shows the graph of $y = a^x + b$.

Find the values of a and b and hence state the equation of the curve.

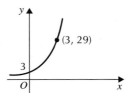

Sketching and finding the equation of the graph of the logarithmic function: $y = \log_a x$ where, $a > 0$, $a \neq 1$

The exponential function $f(x) = a^x$ has an inverse function $f^{-1}(x) = \log_a x$ such that

- $\log_a(a^x) = x$
- $a^{(\log_a x)} = x$

Consider the exponential function $f(x) = 2^x$, its inverse $f^{-1}(x) = \log_2 x$ and their graphs.

x	$f(x) = 2^x$
-2	$\frac{1}{4}$
-1	$\frac{1}{2}$
0	1
1	2
2	4
3	8

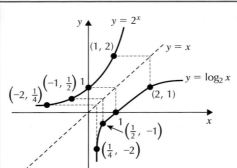

x	$f^{-1}(x) = \log_2 x$
$\frac{1}{4}$	-2
$\frac{1}{2}$	-1
1	0
2	1
4	2
8	3

With these two graphs shown together you should notice that:

1 The graphs are symmetrical about the line $y = x$. This property is true of all pairs of inverse functions.

2 The points corresponding to (0, 1) and (1, 2) are (1, 0) and (2, 1) respectively.

3 The boundary line or asymptote for the logarithmic function is the y-axis and so the logarithmic function is only defined for $x > 0$.

Using a graphic calculator or other graphing package, you can explore this symmetry for different values of a, leading to the following general statement:

For any a such that $a > 0$, $a \neq 1$ the graph of $y = \log_a x$ is defined for positive real numbers only and will pass through the key points:

- **(1, 0)   because $f(x) = \log_a x$ is the inverse of $f(x) = a^x$ with its key point (0, 1)**
- **(a, 1)   because $f(x) = \log_a x$ is the inverse of $f(x) = a^x$ with its key point (1, a)**

It will also approach but never cut the y-axis, making the y-axis a boundary line or asymptote of the graph.

$a > 1$ $0 < a < 1$

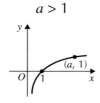   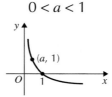

Example 3.17

Sketch the graphs of:

a $y = \log_3 x$ b $y = \log_{\frac{1}{2}} x$

a Key points: (1, 0) and (3, 1); boundary line: the y-axis

b Key points: (1, 0) and $\left(\frac{1}{2}, 1\right)$; boundary line: the y-axis

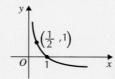

Example 3.18

Give the equation of this logarithmic function.

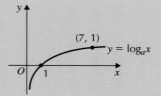

Key points: (1, 0) and (7, 1) can be read from the graph and so the equation is $y = \log_7 x$

Drawing the graph of any function related to $f(x) = \log_a x$

The graph of any function related to $f(x) = \log_a x$ can be drawn by considering the transformation on the key points $(1, 0)$ and $(a, 1)$ and the boundary line or y-axis.

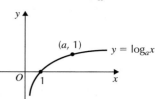

For example, if asked to sketch the graph of $y = 3\log_2(x + 1) - 5$ consider the transformation on key points and boundary line of the graph of the function $f(x) = \log_2 x$

$y = 3\log_2(x + 1) - 5 = 3 \times f(x + 1) - 5$ where $f(x) = \log_2 x$ and so $(x, y) \rightarrow (x - 1, 3y - 5)$

The key points and the boundary line of the graph of $y = \log_2 x$ are transformed as follows:

$(1, 0) \rightarrow (1 - 1, 3 \times 0 - 5) = (0, -5)$

$(2, 1) \rightarrow (2 - 1, 3 \times 1 - 5) = (1, -2)$

The y-axis slides back 1 to the line with equation $x = -1$.

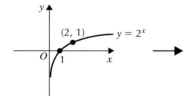 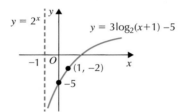

Example 3.19

Sketch the graph of $y = 2\log_3(x - 1)$

$y = 2\log_3(x - 1)$ and so $(x, y) \rightarrow (x + 1, 2y)$

so:

$(1, 0) \rightarrow (2, 0)$

$(3, 1) \rightarrow (4, 2)$

The y-axis slides forward 1 to the line with equation $x = 1$.

> The key points and the boundary line of the graph of $y = \log_3 x$ are transformed. **P**

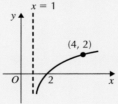

Example 3.20

The diagram shows the graph $y = \log_3 x$. Use this to:

a sketch the graph of $y = \log_3 x^2$

b sketch the graph of $y = \log_3\left(\frac{1}{x}\right)$

c sketch the graph of $y = \log_3 9x$

> ⚠ This example depends on knowledge of the logarithmic laws from Chapter 1.

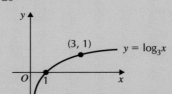

(*continued*)

a $y = \log_3 x^2 \Rightarrow y = 2\log_3 x$ and $(x, y) \to (x, 2y)$

so:

$(1, 0) \to (1, 0)$

$(3, 1) \to (3, 2)$

The boundary line remains the y-axis.

The key points and the boundary line of the graph of $y = \log_3 x$ are transformed. **P**

b $y = \log_3\left(\frac{1}{x}\right)$

$\Rightarrow y = -\log_3 x$ so $(x, y) \to (x, -y)$

so:

$(1, 0) \to (1, -0) = (1, 0)$

$(3, 1) \to (3, -1) = (3, -1)$

The boundary line remains the y-axis.

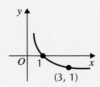

The key points and the boundary line of the graph of $y = \log_3 x$ are transformed. **P**

c $y = \log_5 25x \Rightarrow y = \log_5 25 + \log_5 x$

$= \log_5 5^2 + \log_5 x$

$= 2\log_5 5 + \log_5 x$

$= 2 + \log_5 x$

$y = \log_5 x + 2$ so $(x, y) \to (x, y + 2)$

so:

$(1, 0) \to (1, 2)$

$(5, 1) \to (5, 3)$

The boundary line remains the y-axis.

Remember that $\log_5 5 = 1$ **S**

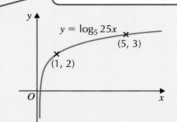

The key points and the boundary line of the graph of $y = \log_5 x$ are transformed. **P**

Example 3.21

The diagram shows a graph of the form $y = \log_a(x - b)$. Find the values of a and b and hence write the equation of the curve.

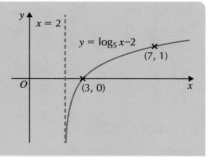

The graph of $\log_a x$ has been moved forward $2 \Rightarrow b = 2$

$(7, 1)$ lies on the curve so $1 = \log_a(7 - 2) \Rightarrow 1 = \log_a 5 \Rightarrow a = 5$

The equation is $y = \log_5(x - 2)$.

Exercise 3E

★ **1** Sketch the graph of these functions.

 a $y = \log_3 x$ **b** $y = \log_4 x$ **c** $y = \log_{\frac{1}{2}} x$ **d** $y = \log_{\frac{1}{3}} x$

★ **2** These graphs are of the form $y = \log_a x$. In each case state the value of a and write down the equation.

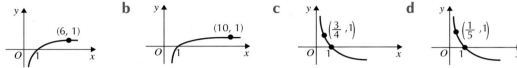

 a (6, 1) **b** (10, 1) **c** $\left(\frac{3}{4}, 1\right)$ **d** $\left(\frac{1}{5}, 1\right)$

★ **3** Sketch the graph of these logarithmic functions.

 a $y = \log_4(x - 2)$ **b** $y = \log_3(x + 4)$ **c** $y = 2\log_4 x$

 d $y = \log_2 x + 3$ **e** $y = 3\log_3(x - 1)$ **f** $y = 3 - 2\log_5 x$

 g $y = 2\log_3(x + 2) - 1$ **h** $y = 3 - \log_2 x$

 4 Use the logarithmic laws to help you sketch the graphs of these functions.

 a $y = \log_2 x^3$ **b** $y = \log_4(x^2)$ **c** $y = \log_4\left(\frac{1}{x}\right)$ **d** $y = \log_3 x^2$

 e $y = \log_2(8x)$ **f** $y = \log_2(32x)$ **g** $y = \log_5\left(\frac{5}{x}\right)$ **h** $y = \log_3\left(\frac{27}{x}\right)$

 5 The graph shows the function of $y = \log_a(x - b)$

 Write down the values of a and b.

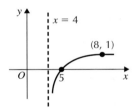

Working with trigonometric graphs

You should have completed Chapter 2 before starting this section. Previous work has covered graphs of the form $y = a\sin(bx)° + c$ and $y = \sin(x + d)°$ or the cosine equivalents. Now you must extend to more complex graphs of the form $y = a\sin(bx + d) + c$ or $y = a\cos(bx + d) + c$

Working with graphs of the form $y = a\sin(bx + d) + c$ or $y = a\cos(bx + d) + c$

The maths previously used to draw trigonometric graphs can be extended, but it is possibly easier to draw the graph of, for example, $y = a\sin(bx + d)° + c$, by considering the transformation on the key points of the graph of $y = \sin x°$. However, it is still advisable to note that:

• the amplitude $= a$
• there are b cycles in 360° so there is one cycle in $\frac{360°}{b}$ (the period is $\frac{360°}{b}$).

The graphs of $y = \sin x°$ and $y = \cos x°$ have key points as shown in the diagrams below.

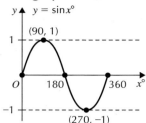

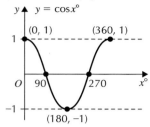

Suppose you are asked to sketch the graph of $y = 3\cos(2x - 30)° + 2$ where $0 \leq x \leq 180$

It is useful to start by noting the amplitude and the number of cycles in the interval for the graph:

- amplitude = 3
- 2 cycles in 360° $\Rightarrow$ one cycle in 180°, which is the interval for this graph.

Next consider the transformation on the key points of the graph $y = f(x)$ where $f(x) = \cos x°$

$y = 3\cos(2(x - 15))° + 2 = 3 \times f(2(x - 15))° + 2$ where $f(x) = \cos x°$

so $(x, y) \rightarrow (\frac{x}{2} + 15, 3 \times y + 2)$

The key points of the graph of $y = \cos x°$ are transformed as follows:

$(0, 1) \rightarrow (\frac{0}{2} + 15, 3 \times 1 + 2) = (15, 5)$

$(90, 0) \rightarrow (\frac{90}{2} + 15, 3 \times 0 + 2) = (60, 2)$

$(180, -1) \rightarrow \left(\frac{180}{2} + 15, 3 \times (-1) + 2\right) = (105, -1)$

$(270, 0) \rightarrow \left(\frac{270}{2} + 15, 3 \times 0 + 2\right) = (150, 2)$

The final step is to find where the graph cuts the y-axis

$x = 0 \Rightarrow y = 2\cos(0 - 30)° + 2 = 4.6$ (to one decimal place)

So the graph cuts the y-axis at $(0, 4.6)$

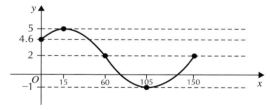

Horizontally the graph of $y = \cos x°$ has been compressed by a factor of $\frac{1}{2}$ then slid forward 15; vertically it has been stretched by a factor of 3 and slid up 2.

Example 3.22

Sketch the graph $y = 40\sin(10x - 60)°$ for $0 \leq x \leq 72$, showing clearly the points where the graph cuts the x-axis and any stationary points.

- amplitude = 40
- 10 cycles in 360° $\Rightarrow$ 1 cycle in 36° $\Rightarrow$ 2 cycles in 72°, the interval for this graph.
  $y = 40\sin(10(x - 6))°$ so $(x, y) \rightarrow \left(\frac{x}{10} + 6, 40 \times y\right)$

so:

$(0, 0) \rightarrow (6, 0)$

$(90, 1) \rightarrow (15, 40)$

The key points of the graph of $y = \sin x°$ are transformed. **P**

$(180, 0) \rightarrow (24, 0)$

$(270, -1) \rightarrow (33, -40)$

$x = 0 \Rightarrow y = 40 \sin (0 - 60)° = -20\sqrt{3}$

Find where the graph cuts the y-axis. **P**

The graph cuts the y-axis at $(0, -20\sqrt{3})$.

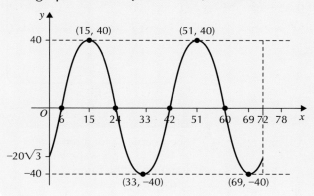

Example 3.23

The diagram shows a graph of the form $y = p \sin(x + q)° + r$. Find the values of p, q and r.

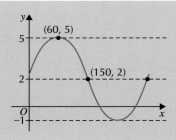

The graph $y = \sin x°$ has been stretched vertically by a factor of $3 \Rightarrow p = 3$

The graph has been slid up by $2 \Rightarrow r = 2$

The maximum turning point has been slid back by $30° \Rightarrow q = 30$

So $y = 3\sin(x + 30)° + 2$

Example 3.24

Given $f(x) = 2\sin(3x - 60)° + 5$ where $0 \le x \le 120$, find the greatest and least values of $f(x)$ and the values of x for which they occur.

$y = \sin x°$ has maximum turning point $(90, 1)$ and minimum turning point $(270, -1)$

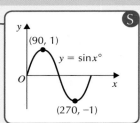

$y = 2 \sin(3(x - 20))° + 5$ so $(x, y) \rightarrow \left(\frac{x}{3} + 20, 2y + 5\right)$

$(90, 1) \rightarrow (50, 7)$ and $(270, -1) \rightarrow (110, 3)$

The maximum value of $f(x)$ is 7 and it occurs when $x = 50$.

The minimum value of $f(x)$ is 3 and it occurs when $x = 110$.

Example 3.25

Given $f(x) = 4 - 3\cos\left(x + \frac{\pi}{3}\right)$ where $0 \le x \le 2\pi$ find the greatest value of $f(x)$ and the value of $f(x)$ for which it occurs.

$y = -3\cos\left(x + \frac{\pi}{3}\right) + 4$ so $(x, y) \to \left(x - \frac{\pi}{3}, -3y + 4\right)$

$y = \cos x°$ has minimum turning point $(\pi, -1)$

$(\pi, -1) \to \left(\pi - \frac{\pi}{3}, -3 \times (-1) + 4\right) = \left(\frac{2\pi}{3}, 7\right)$

So the maximum value of $f(x)$ is 7 when $x = \frac{2\pi}{3}$

> **S** The function is $-3\cos(\ldots)$, so the maximum value of $f(x)$ will correspond to the minimum value of $\cos x$.
>
>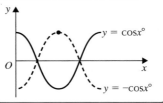

Example 3.26

a Express $\cos 30x° + \sqrt{3}\sin 30°$ in the form $k\cos(30x° - \alpha)°$, $k > 0$ and $0 \le \alpha < 360$

b Hence sketch the graph of $y = \cos 30x° + \sqrt{3}\sin 30° + 1$, $0 \le x \le 12$

> ⚠️ See Chapter 2 for a reminder of how to express the wave function in the forms $k\cos(x \pm \alpha)°$ or $k\sin(x \pm \alpha)$

a If $\cos 30x° + \sqrt{3}\sin 30x° = k\cos(30x° - \alpha)°$ it follows that:

$\cos 30x° + \sqrt{3}\sin 30x° = k\cos 30x° \cos\alpha° + k\sin 30x° \sin\alpha°$

$\Rightarrow \underline{\cos 30x°} + \sqrt{3}\underline{\underline{\sin 30x°}} = (k\cos\alpha°)\underline{\cos 30x°} + (k\sin\alpha°)\underline{\underline{\sin 30x°}}$

> **S** Expand $k\cos(30x - \alpha)°$

Comparing sides gives:

> **S** Rearrange RHS and use single and double underlining to help compare sides.

$k\cos\alpha° = 1 \qquad k\sin\alpha° = \sqrt{3}$

leading to:

$k = \sqrt{(1^2 + (\sqrt{3})^2)} = 2$

$\tan\alpha° = \frac{\sqrt{3}}{1}$ and α is in the first quadrant

$\Rightarrow \alpha° = \tan^{-1}(\sqrt{3})$

$\Rightarrow \alpha = 60$

> **S** $\tan\alpha = \frac{k\sin\alpha}{k\cos\alpha}$ All of $\sin\alpha$, $\cos\alpha$ and $\tan\alpha$ are all positive so α is in the first quadrant.

So $\cos 30x° + \sqrt{3}\sin 30° = 2\cos(30x - 60)°$

b $y = \cos 30x° + \sqrt{3}\sin 30° + 1$

$\Rightarrow y = 2\cos(30x - 60)° + 1$, $0 \le x \le 12$

amplitude = 2

30 cycles in 360° and so 1 cycle in 12°, the interval of this graph.

$y = 2\cos(30(x - 2))° + 1$ and so $(x, y) \to \left(\frac{x}{30} + 2, 2y + 1\right)$

$(0, 1) \to (2, 3)$

$(90, 0) \to (5, 1)$

$(180, -1) \to (8, -1)$

> **P** You could go on to find more points, but the cyclical nature of the cosine graph means that this may be enough.

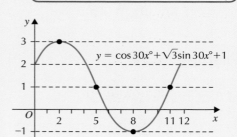

$x = 0 \Rightarrow y = 2 \times \cos(30 \times 0 - 60)° + 1 = 2$ so the graph cuts the y-axis at 2

Exercise 3F

★ **1** Sketch the graphs of these functions.

    a $y = 3\sin(x - 30)° + 1, \ 0 \le x \le 360$ **b** $y = 2\cos(x + \frac{\pi}{6}) - 1, \ 0 \le x \le 2\pi$

    c $y = 5 - 3\cos(2x + 60)°, \ 0 \le x \le 180$ **d** $y = 2 + 4\sin(3x - \pi), \ 0 \le x \le \frac{2\pi}{3}$

    e $y = 2\sin(5x - 6)° + 1, \ 0 \le x \le 72$ **f** $y = 1 - 3\sin\left(2x + \frac{\pi}{2}\right), \ 0 \le x \le \pi$

⚙ **2** The displacement, d, of a wave after t seconds is approximated by the formula $d = 3\cos(200t - 60)°$.

    Sketch the graph $d = 3\cos(200t - 60)°. \ 0 \le t \le 3.6$

⚙ **3** The average high temperature, $T°C$, in a holiday resort can be approximated by the formula; $T = 25 - 9.5\sin(30m - 120)°$ where m is the month of the year starting with January = 1.

    By first sketching the graph of $T = 25 - 9.5\sin(30m - 120)°, \ 0 \le m \le 12$, calculate the highest and lowest average temperatures and the months in which they occur.

★ ⚙ **4** The graph is of the form $y = a\cos(x - b)° + c$.
Calculate the values of a, b and c.

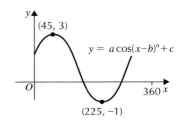

⚙ **5** The graph is of the form $y = p\sin(x + q)° + r$.
Find the values of p, q and r.

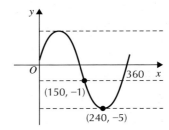

⚙ **6** The depth, $d\,$m, of water in a west coast inlet t hours after midnight can be approximated by $d = a\sin(30t - b)° + c$.

    The water reaches a depth of 5.8 m at high tide falling to 1.4 m at low tide.

    If high tide occurs at 2 am and low tide at 8 am, calculate the values of a, b and c.

★ **7** Find the maximum and minimum values of the following functions and the values of x at which they occur.

    a $3\cos(x - 48)° + 1, \ 0 \le x \le 360$ **b** $20 + 7\sin(x - 30)°, \ 0 \le x \le 360$

    c $5 + 20\sin(x + \frac{\pi}{4}), \ 0 \le x \le 2\pi$ **d** $30\sin(x - \frac{\pi}{3}) + 5, \ 0 \le x \le 2\pi$

    e $4 - 7\cos(2x - 72)°, \ 0 \le x \le 180$ **f** $2 - 4\sin(10x - 30)°, \ 0 \le x \le 36$

    g $10 - 50\sin(30x - 60)°, \ 0 \le x \le 12$ **h** $5\cos\left(12x - \frac{\pi}{2}\right) - 1, \ 0 \le x \le \frac{\pi}{6}$

8 The depth, d cm, in a harbour is approximated by the formula $d = 200 + 160\cos(30t - 60)°$ where t is the time in hours after midnight.

What is the level of water at low tide and at what time does this occur?

9 A person's blood pressure at time t minutes varies with the beat of the heart and can be approximated by the function $p(t) = 20\sin(160\pi t) + 100$

   a Calculate the maximum and minimum blood pressure

   b The heart beats once for every cycle of the graph. Use this to calculate the persons heart beat in beats per minute.

10 a Express $4\cos x° + 3\sin x°$ in the form $y = k\cos(x - \alpha)°$ where $k > 0$, and $0 \le \alpha < 360$

   b Hence sketch the graph of $y = 3\cos x° + 2\sin x° + 2$ where $0 \le x \le 360$

11 a Express $5\sin x° + 12\cos x°$ in the form $k\sin(x + \alpha)°$ where $k > 0$, and $0 \le \alpha < 360$

   b Sketch the graph of $f(x) = 3 - 2(5\sin x° + 12\cos x°)$ where $0 \le x \le 360$

   c Hence state the maximum and minimum values of $f(x)$ and when they occur.

★ **12 a** Express $\cos(2x)° + \sqrt{3}\sin(2x)°$ in the form $R\sin(2x + \alpha)°$.

   b Hence state the maximum and minimum values of $f(x) = 3(\cos(2x)° + \sqrt{3}\sin(2x)°) - 5$ and the value of x between 0 and π for which they occur.

13 a Express $3\sin 12t° - 4\cos 12t°$ in the form $k\sin(12t - \alpha)°$ where $k > 0$ and $0 \le \alpha < 360$

   b Hence sketch the graph of $10 - 2(3\sin 12t° - 4\cos 12t°)$ where $0 \le t < 12$

Sketching the graph of $y = f'(x)$, given the graph of $y = f(x)$

You should have completed Chapter 8 before starting this section

To sketch the graph of $y = f'(x)$ you must know:

- $f'(x) = 0$ when $f(x)$ is stationary
- $f'(x) > 0$ when $f(x)$ is increasing
- $f'(x) < 0$ when $f(x)$ is decreasing

and

- $f(x) = ax^2 + bx + c$ (quadratic) $\Rightarrow f'(x) = 2ax + b$ (linear)
- $f(x) = ax^3 + bx^2 + cx + d$ (cubic) $\Rightarrow f'(x) = 3ax^2 + 2bx + c$ (quadratic)

This tells you that:

- **if $f(x)$ is quadratic then the graph of $y = f'(x)$ will be a straight line**
- **if $f(x)$ is cubic then the graph of $y = f'(x)$ will be a parabola.**

In the same way, if $f(x)$ is a quartic function, then $f'(x)$ will be cubic and so on.

Example 3.27

The diagram shows the graph of $y = f(x)$ where $f(x)$ is a cubic function.

Sketch the graph of $y = f'(x)$

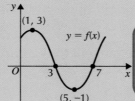

> $f(x)$ is cubic, so $f'(x)$ will be quadratic and the graph will be a parabola

$f'(x) = 0$ when $f(x)$ is stationary, so the graph of $y = f'(x)$ cuts the x–axis when $x = 1$ and $x = 5$.

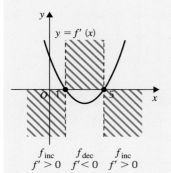

f_{inc} f_{dec} f_{inc}
$f' > 0$ $f' < 0$ $f' > 0$

> The graph will be a parabola, so it will be either
>
>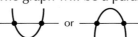
>
> Split the graph at $x = 1$ and $x = 5$ and check if $f(x)$ is increasing or decreasing in each of the three regions. Shade in the no-go areas then sketch the parabola.

Example 3.28

The diagram shows the graph of $y = f(x)$ where $f(x)$ is a quartic function.

Sketch the graph of $y = f'(x)$.

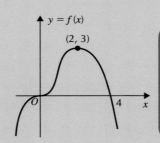

> $f(x)$ is quartic, so $f'(x)$ will be cubic. This time you cannot predict the shape in advance – only that it cannot have more than 2 stationary points.

$f'(x) = 0$ when $f(x)$ is stationary and so the graph of $y = f'(x)$ cuts the x–axis at 0 and 2.

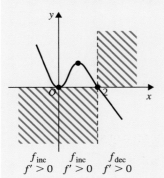

f_{inc} f_{inc} f_{dec}
$f' > 0$ $f' > 0$ $f' > 0$

> The graph must:
> • pass through (0, 0) and (2, 0)
> • remain in the non shaded regions
> • have no more than 2 turning points

Sketching the graph of $y = f'(x)$ when $f(x)$ is quadratic

When $f(x)$ is quadratic, the graph of $y = f'(x)$ will be a straight line. It must also be noted that:

- $f'(a)$ is the gradient of the tangent to the curve at the point where $x = a$.

Example 3.29

The diagram shows the graph of $y = f(x)$ where $f(x)$ is a quadratic function.

The maximum turning point is (1, 3) and the gradient of the tangent at the origin is 3.

Sketch the graph of $y = f'(x)$.

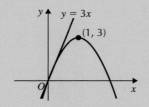

⚠ As $f(x)$ is quadratic, $f'(x)$ will be linear and so the graph of $y = f(x)$ will be a straight line.

$f'(x) = 0$ when $f(x)$ is stationary, so the graph cuts the x–axis when $x = 1$.

At the point where $x = 0$, the gradient of the tangent is 3.

This means that $f'(0) = 3$ and so (0, 3) lies on the line.

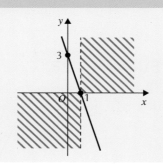

Exercise 3G

★ 1   For each of the following cubic functions, sketch the graph of $y = f'(x)$.

a
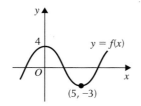

b
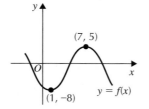

c
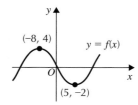

d
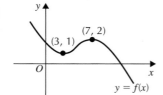

e
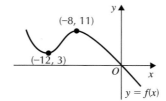

f
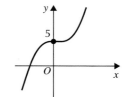

★ **2** Sketch the graph of $y = f'(x)$ for each of the following.

a
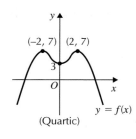
(Quartic)

b
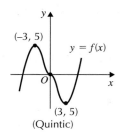
(Quintic)

c
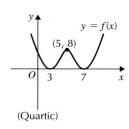
(Quartic)

d
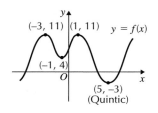
(Quintic)

e
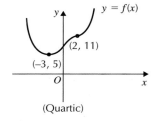
(Quartic)

f
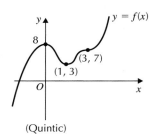
(Quintic)

★ **3** For each of the following **quadratic** functions, sketch the graph of $y = f'(x)$.

a
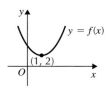

The gradient of the curve at the point (0, 3) is −1.

b
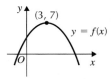

The gradient of the curve at the point (1, 5) is $\frac{1}{2}$

c
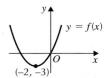

At the origin the curve has gradient 2.

d
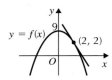

The gradient of the tangent to the curve at (2, 2) is −3.

e
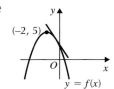

The gradient of the tangent to the curve at the point where it cuts the y–axis is $-\frac{1}{2}$

f
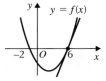

The gradient of the tangent to the curve at (6, 0) is 4.

- I can identify or sketch a function after a transformation of the form $af(x)$, $f(bx)$, $f(x) + c$, $f(x + d)$ or any combination of the four. ★ Exercise 3A Q3

- I can complete the square of a quadratic function and use the completed square form to sketch or determine equations and to find the maximum or minimum value of a quadratic function. ★ Exercise 3B Q2, 3 Exercise 3C Q1, 2, 5, 6

- I can sketch the graph of an exponential or logarithmic function and in limited cases find the equation when given the graph. ★ Exercise 3D Q1–3 Exercise 3E Q1–3

- I can sketch the graph of, or determine the equation of, a trigonometric function of the form $y = a\sin(bx + d)° + c$ or the cosine equivalent including examples involving the wave function. ★ Exercise 3FQ 1, 4

- I can find the maximum and minimum values of a trigonometric function of the form $f(x) = a\sin(bx + d)° + c$ or the cosine equivalent including examples derived from the wave function. ★ Exercise 3F Q7

- I can sketch the graph of the derived function $f'(x)$ when given the graph of $f(x)$. ★ Exercise 3G Q 1–3

For further assessment opportunities, see the Preparation for Assessment for Unit 1 on page 132–135.

4 Determining composite and inverse functions

This chapter will show you how to:

- identify the range of a function from its graph
- find restrictions on the domain of a function
- find a formula for a composite function $f(g(x))$
- find a formula for the inverse $f^{-1}(x)$ of a linear function.

You should already know:

- the notation for sets of numbers
- the definition of a function and the meaning of the terms domain and range
- how to sketch the graphs of quadratic functions with unitary x^2 term
- how to sketch the graphs of $y = a \sin bx°$ or $y = a \sin (x - d)° + c$

Determining the domain and range of a function

A function from a set A to a set B is a rule by which each member of the set A is linked to **exactly one member** of the set B.

- The set A (the input set) is called the **domain** of the function.
- The set B is called the **co–domain** (or target set) of the function.
- If a is a member of A and $f(a) = b$ then b is called the **image** of a.
- The set of all images in B is called the **range** (or output set) of the function.

In this chapter you will be dealing with functions where the domain is the set, $\mathbb{R}$, of all real numbers on the x-axis, unless stated otherwise.

Consider the function $f: \mathbb{R} \to \mathbb{R}$ defined by $f(x) = x^2$.

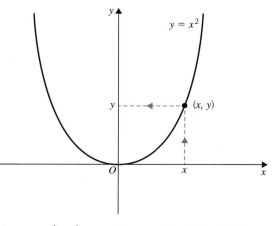

$f: \mathbb{R} \to \mathbb{R}$ tells you that the domain, or input set, is $\mathbb{R}$ and the co-domain, or target set, is also $\mathbb{R}$. The rule $f(x) = x^2$ tells you that every real number, x, in the domain has an image x^2 in the co-domain. This function can be illustrated by the graph $y = x^2$.

The x–axis represents the **domain**, while the y–axis represents the **co–domain**.

A point (x, y) lies on the graph if and only if $y = x^2$, the **image** of x.

The y-coordinates represent the images and so the **range** of the function is the set of all real numbers greater than or equal to zero. This can also be written as $\{y \ \varepsilon \ \mathbb{R}: y \geq 0\}$,

which reads as 'the set of all real numbers y which are greater than or equal to zero'
This is an example of set–builder notation.

Example 4.1

State the range of the functions illustrated by these graphs.

a $f(x) = x^2 - 4x + 3$ **b** $f(x) = \sin x°, \ 0 \le x \le 360$ **c** $f(x) = 2^x$

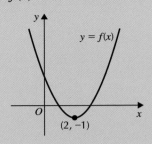
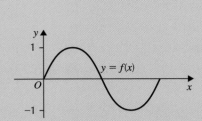
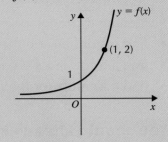

a The range is the set of all real numbers which are greater than or equal to −1 or $\{y \, \varepsilon \, \mathbb{R} : y \ge -1\}$.

b The range is the set of all real numbers between and including −1 and 1 or $\{y \, \varepsilon \, \mathbb{R} : -1 \le y \le 1\}$.

c The range is the set of all real numbers greater than zero or $\{y \, \varepsilon \, \mathbb{R} : y > 0\}$.

> You do not have to use set–builder notation 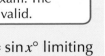 if asked these questions in an exam. The description in words is equally valid.

In part **b** of Example 4.1, a restriction was put on the domain of $f(x) = \sin x°$ limiting x to the real numbers from 0 to 360. That was a restriction of choice, not of necessity, but this is not always the case. Consider these two cases:

- $f(x) = \dfrac{1}{1-x}, \ x \ne 1$
- $f(x) = \sqrt{x}, \ x \ge 0$

In the case of $f(x) = \frac{1}{1-x}$ the value $x = 1$ would not have an image as $= \frac{1}{1-1} = \frac{1}{0}$ which is undefined so $x = 1$ has to be removed from the domain.

In the case of $f(x) = \sqrt{x}$ no value of $x < 0$ would have an image as negative numbers do not have a square root, so all negative values of x have to be removed from the domain.

Example 4.2

Write down any restriction on the domain of each of these functions and hence state the largest suitable domain.

a $f(x) = \dfrac{1}{x + 3}$ **b** $f(x) = \sqrt{x + 5}$ **c** $f(x) = \dfrac{1}{x^2 - 3x + 2}$

d $f(x) = \sqrt{(x + 1)(x - 2)}$ **e** $f(x) = \dfrac{1}{\sqrt{(1 - x)}}$

a $x + 3 \ne 0 \Rightarrow x \ne -3$ •   You cannot have 0 in the denominator.

The largest suitable domain is $\{x \, \varepsilon \, \mathbb{R} : x \ne -3\}$, (or the set of all real numbers except −3).

b $x + 5 \geq 0 \Rightarrow x \geq -5$ •————| You cannot have the square root of a negative number. Ⓢ |

The largest suitable domain is $\{x \, \varepsilon \, \mathbb{R} : x \geq -5\}$.

c $x^2 - 3x + 2 \neq 0 \Rightarrow (x - 1)(x - 2) \neq 0 \Rightarrow x \neq 1$ or 2

The largest suitable domain is $\{x \, \varepsilon \, \mathbb{R} : x \neq 1 \text{ or } 2\}$.

| To see when $(x + 1)(x - 2) \geq 0$, Ⓢ sketch the graph of $y = (x + 1)(x - 2)$. This is a parabola with a minimum turning point cutting the x–axis at -1 and 2. |

d $(x + 1)(x - 2) \geq 0$ •

The largest suitable domain is $\{x \, \varepsilon \, \mathbb{R} : x \leq -1 \text{ or } x \geq 2\}$.

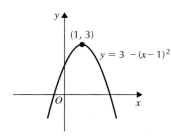

| From the graph you can Ⓟ read that $(x + 1)(x - 2) \geq 0$ $\Rightarrow x \leq -1$ or $x \geq 2$. |

e $1 - x > 0 \Rightarrow x < 1$ •

The largest suitable domain is $\{x \, \varepsilon \, \mathbb{R} : x < 1\}$.

| You cannot have the square root of a Ⓢ negative number, nor can you have 0 on the denominator and so $1 - x$ must be strictly greater than 0. |

Exercise 4A

★ **1** State the range of each of these functions.

a $y = 3 - (x - 1)^2$

b $y = 2\cos x° + 1$; $0 \leq x \leq 360$

c $y = (x - 3)^2 - 2$

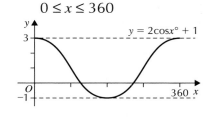

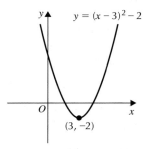

⚙ **2** By sketching the graph, or otherwise, state the range of each of these functions.

a $y = (x - 6)^2 + 5$
b $y = 6 - (x + 3)^2$
c $y = (x - 1)^2 + 15$
d $y = 3\sin x°$
d $y = 4\cos x° - 2$
e $y = 5\sin(x - 30)° + 1$

★ ⚙ **3** State any restriction on the domain of each of these functions and hence write down the largest suitable domain.

a $f(x) = \dfrac{1}{x - 5}$

b $g(x) = \dfrac{5}{x + 9}$

c $f(x) = \sqrt{(x - 8)}$

d $h(x) = \sqrt{5 - 2x}$

e $f(x) = \dfrac{1}{x - 2x^2}$

f $f(x) = \sqrt{(x^2 - 2x - 15)}$

g $h(x) = \sqrt{\left(6 - x - x^2\right)}$

h $g(x) = \dfrac{4}{\sqrt{(7 - x)}}$

i $h(x) = \dfrac{1}{\sqrt{(2x + 3)}}$

j $f(x) = \dfrac{\sqrt{x}}{x^2 - 3x + 2}$

Determining the formula for a composite function

Consider the two functions f such that $f(x) = 2x + 3$ and g such that $g(x) = x^2$ and now see what happens when the two functions are combined one after the other, starting with input x.

$$x \longrightarrow \boxed{2 \times \text{(input)} + 3} \longrightarrow 2x + 3 \longrightarrow \boxed{\text{(input)}^2} \longrightarrow (2x + 3)^2$$

The output for f becomes the input for g and combining f and g in this way creates a third function, h, such that $h(x) = (2x + 3)^2$

$h(x) = g(f(x))$ is said to be the **composition** of the two functions f and g. It is an example of a **composite function**.

Now consider k such that $k(x) = f(g(x))$ with f and g defined as above:

$$x \longrightarrow \boxed{\text{(input)}^2} \longrightarrow x^2 \longrightarrow \boxed{2 \times \text{(input)} + 3} \longrightarrow 2x^2 + 3$$

$k(x)$ is also a composite function with $k(x) = f(g(x)) = 2x^2 + 3$

It is important to recognise that $g(f(x)) \neq f(g(x))$ and so you must always remember to work in the given order.

Given the formulae for two simpler functions f and g, you will be expected to be able to find an expression for the composite functions:

- $h(x) = g(f(x))$, which reads 'g of f of x'
- $k(x) = f(g(x))$ which reads 'f of g of x'

Example 4.3

Let $f(x) = 3x + 4$ and $g(x) = x^2$

a Find an expression for $f(g(x))$

b Find an expression for $g(f(x))$

a $f(g(x)) = f(x^2)$ $g(x) = x^2$ Ⓢ

      $= 3(x^2) + 4$ $f(...) = 3(...) + 4$, with x^2 as the input inside the bracket. Ⓢ

      $= 3x^2 + 4$

b $g(f(x)) = g(3x + 4)$ $f(x) = 3x + 4$ Ⓢ

      $= (3x + 4)^2$ $g(...) = (...)^2$, with $3x + 4$ as the input. Ⓢ

Example 4.4

If $f(x) = x + 1$ and $g(x) = x^2 - 3x + 1$, find an expression for $g(f(x))$ giving your answer in simplest form.

$g(f(x)) = g(x + 1)$

$\qquad = (x + 1)^2 - 3(x + 1) + 1$

$\qquad = (x + 1)^2 - 3(x + 1) + 1$

$\qquad = x^2 + 2x + 1 - 3x - 3 + 1$

$\qquad = x^2 - x - 1$

Example 4.5

$f(x) = \dfrac{x}{x - 1}$ where $x \neq 1$

Find an expression for $f(f(x))$ giving your answer as a fraction in simplest form.

$f(f(x)) = f\left(\dfrac{x}{x - 1}\right)$

$\qquad = \dfrac{\left(\dfrac{x}{(x - 1)}\right)}{\left(\dfrac{x}{(x - 1)}\right) - 1}$ ⎯⎯⎯⎯⎯⎯⎯⎯ $f(\ldots) = \dfrac{(\ldots)}{(\ldots) - 1}$ with $\dfrac{x}{(x-1)}$ as input. Ⓢ

$\qquad = \dfrac{x}{\left(x - (x - 1)\right)}$ ⎯⎯⎯⎯⎯⎯ To simplify the fraction multiply each term of both the numerator and denominator by $(x - 1)$. Ⓢ

$\qquad = \dfrac{x}{1} = x$

Exercise 4B

★ 1 Find an expression for $f(g(x))$ for each of these pairs of functions, giving your answer in simplest form.

    a $f(x) = x + 3$, $g(x) = x^2$ **b** $f(x) = 3x$, $g(x) = x + 4$

    c $f(x) = x - 1$, $g(x) = 3x - 2$ **d** $f(x) = \sin x°$, $g(x) = 4x$

    e $f(x) = 2x - 3$, $g(x) = x + 3$ **f** $f(x) = 3x - 2$, $g(x) = x^2 + 5$

    g $f(x) = x^2 - 2x + 1$, $g(x) = 4 + x$ **h** $f(x) = 1 - 2x^2$, $g(x) = \sin x°$

2 Find an expression for $g(f(x))$ for each of the pairs of functions in Question 1.

3 The functions f and g are defined on the set of real numbers by $f(x) = 3x + k$ and $g(x) = 5x - 1$. Find the value of k for which $f(g(x)) = g(f(x))$.

4 The functions f and g are defined on the set of real numbers by $f(x) = 3x^2 + 2$ and $g(x) = x + 1$ Solve the equation $f(g(x)) - g(f(x)) = 0$.

5 The functions f and g are defined on suitable domains by $f(x) = \frac{1}{x}$ and $g(x) = 3x - 2$.
Find an expression for $h(x) = f(g(x))$, giving any restriction on the domain.

6 **a** Find an expression for $p(x) = q(r(x))$ given $q(x) = \sqrt{x - 3}$ and $r(x) = 3x + 4$.
 b State a suitable domain for $p(x)$.

7 The functions f and g are defined on suitable domains by $f(x) = \frac{1}{x-5}$ and
$g(x) = 3x + 1$.
Find an expression for $h(x) = f(g(x))$, stating any restriction on the domain.

8 A function f is defined on the set of real numbers by $f(x) = \frac{x}{2x-1}$ where $x \neq \frac{1}{2}$
Find an expression for $f(f(x))$, reducing your answer to its simplest form.

9 The functions f, g and h are defined on suitable domains by $f(x) = 1 - x$, $g(x) = \frac{1}{x}$
and $h(x) = \frac{1}{1-x}$
Find an expression for $h(f(g(x)))$, giving your answer in simplest form.

10 Functions $g(x) = 3x + 5$ and $h(x) = \frac{1}{x+2}$ are defined on suitable domains. Find an
expression for $g(h(x))$, giving your answer as a single fraction.

11 The functions f and g are defined on suitable domains by $f(x) = \frac{1}{(x^2-9)}$ and
$g(x) = 2x + 3$.
 a Find an expression for $h(x) = f(g(x))$.
 b State a suitable domain for h.

12 The functions f and g are defined by $f(x) = 3x + 1$ and $g(x) = x^2 + 1$.
 a Show that the equation $g(f(x)) = 0$ has no real roots.
 b If $h(x) = x + k$, find the value of k such that the equation $h(g(f(x)))$ has equal
roots.

13 A tanker is leaking oil at sea forming a circular oil slick 0.04 metres thick. After t
minutes the radius of the oil slick in metres can be
estimated by the function $r(t) = 0.4t^{\frac{1}{3}}$.
Find a formula for the volume, $V\,\text{m}^3$, of the oil slick
as a function of t.

> Volume of a cylinder
> is given by $V = \pi r^2 h$

14 A company estimates that the demand, d, for a new game that they are developing
will be given by the function $d(x) = 4000 - 200x$ where x is the final price, in
pounds, of one game.

They also estimate that the cost, £c, of producing x games is given by the function
$c(x) = 10x + 30\,000$ while the revenue, £r, brought in by selling x games is given by
$r(x) = 20x - 0.005x^2$.

By first calculating $c(d(x))$ and $r(d(x))$ work out a formula for the estimated profit £P.

Finding the inverse of a linear function

Consider the linear functions f and k defined by $f(x) = 3x + 1$ and $g(x) = \frac{1}{3}(x - 1)$.

- $g(f(x)) = \frac{1}{3}((3x + 1) - 1) = \frac{1}{3}(3x) = x$

- $f(g(x)) = 3\left(\frac{1}{3}(x - 1)\right) + 1 = x - 1 + 1 = x$

In a case like this, where a composition of two functions results in the final image of x being back to x itself, each function is said to be the **inverse** of the other. In this case, g is the inverse of f which is denoted by f^{-1} and f is the inverse of g or g^{-1}

Example 4.6

Find the inverse of the linear function, $f(x) = 2x + 1$.

Let $y = 3x - 2$ — Change the subject of the formula to x. Ⓢ

$\Rightarrow 3x - 2 = y$

$\Rightarrow 3x = y + 2$

$\Rightarrow x = \dfrac{(y + 2)}{3}$

$\Rightarrow f^{-1}(y) = \dfrac{1}{3}(y + 2)$

$\Rightarrow f^{-1}(x) = \dfrac{1}{3}(x + 2)$ — It is normal practice to write the formula for a function in terms of x. Ⓒ

Exercise 4C

★ 1 Find the inverse of each of these linear functions.

a $f(x) = 5x - 1$ **b** $f(x) = 3 - 2x$ **c** $h(x) = \dfrac{1}{3}x - 1$

d $g(x) = 6x - 7$ **e** $f(x) = 8 - \dfrac{1}{2}x$ **f** $g(x) = 7 - 5x$

g $f(x) = \dfrac{1}{2}(x + 1)$ **h** $k(x) = \dfrac{1}{4}(x + 1)$

2 **a** Find the inverse of the linear function f such that $f(x) = 7x - 1$.

b Verify that $f^{-1}(f(x)) = x$ and that $f(f^{-1}(x)) = x$.

c Sketch the graphs of $f(x)$ and $f^{-1}(x)$ on the same diagram.

> ⚠ The use of a graphic calculator or other graphing package would be useful in answering Questions 2 and 3

3 For each of these functions:

  i find a formula for the inverse function

  ii draw the graph of both the function and its inverse on the same diagram

  iii draw the line $y = x$ on the diagram.

    a $f(x) = 3x + 2$ **b** $h(x) = 4 - 2x$

    c $g(x) = \frac{1}{4}x - 3$ **d** $f(x) = -3x + 2$

4 Using your answers to Question 3, state a connection between the graph of a function and its inverse.

Extension to include logarithmic and exponential functions

Before starting this section you should have completed Chapters 1 and 3.

Example 4.7

State the range of the function $f(x) = 2^x + 3$.

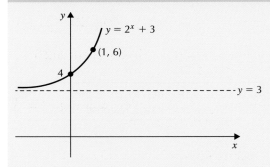

The graph of the exponential function $y = 2^x$ has been slid up 3 and so has boundary line $y = 3$. **S**

The range is $\{y \, \varepsilon \, \mathbb{R} : y > 3\}$.

Example 4.8

State any restriction on the domain of the logarithmic function $f(x) = \log_3(x - 5)$.

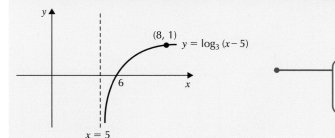

The graph $y = \log_3 x$ has been slid forward 5 and so has boundary line $x = 5$. **S**

Restriction: $x > 5$ and so the largest suitable domain is $\{x \, \varepsilon \, \mathbb{R} : x > 5\}$.

Example 4.9

The functions f and g are defined on suitable domains by $f(x) = \log_3 x$ and $g(x) = 9x^4$.

Find an expression for the function $h(x) = f(g(x))$ expressing your answer in the form $h(x) = A + B\log_3 x$ where A and B are constants.

$$h(x) = f(g(x))$$
$$= f(9x^4)$$
$$= \log_3(9x^4)$$
$$= \log_3 9 + \log_3 x^4$$
$$= 2 + 4\log_3 x \quad\bullet\!\!\!\!\!\!\!\!-\!\!\!-\!\!\!-\!\!\!-\!\!\!-\!\!\!-\!\!\!-\!\!\!-\!\!\!-\!\!\!-\!\!\!-\!\!\!-\!\!\!-\!\!\!-\!\!\!-\!\!\!-\!\!\! \boxed{\log_3 9 = \log_3 3^2 = 2.\ \Ⓟ}$$

Exercise 4D

1 State the range of each of these functions.

     a $f(x) = 3^x - 5$ **b** $g(x) = 4 - 5^x$ **c** $h(x) = 5^{(x-1)} + 2$

2 State the largest possible domain for these functions.

     a $g(x) = \log_2(x + 3)$ **b** $k(x) = \log_7(x - 4)$ **c** $y = \log_3(2x - 5) + 1$

★ ⚙ **3**   If f and g are such that $f(x) = 2^x$ and $g(x) = x + 4$, express $f(g(x))$ in the form $A(2^x)$ where A is a constant.

★ ⚙ **4**   The functions f and g are defined on suitable domains by $f(x) = \log_2 x$ and $g(x) = \dfrac{8}{x^5}$. Express $f(g(x))$ in the form $p + q\log_2 x$ where p and q are constants.

⚙ **5**   The functions f and g are defined on suitable domains by $f(x) = \log_4 2x$ and $g(x) = \dfrac{8}{x^3}$. Express $f(g(x))$ in the form $p + q\log_4 x$ where p and q are constants.

⚙ **6**   The functions h and k are defined on suitable domains as $f(x) = \log_2 x$ and $g(x) = 3x + 3$. Express the function $g(f(x))$ as a single logarithm to base 2.

- I can identify the range of a function. ★ Exercise 4A Q1
- I can identify any restriction on the domain of a function and hence state the largest possible domain. ★ Exercise 4A Q3
- I can find a formula for a composite function $f(g(x))$ where $f(x)$ and $g(x)$ can be algebraic, trigonometric, exponential or logarithmic. ★ Exercise 4B Q1, Exercise 4D Q3, 4
- I can find the formula for the inverse of a linear function. ★ Exercise 4C Q1

For further assessment opportunities, see the Preparation for Assessment for Unit 1 on pages 132–135.

5 Determining vector connections

This chapter will show you how to:

- work with vector properties
- work with vectors defined in terms of the unit vectors $\underset{\sim}{i}$, $\underset{\sim}{j}$ and $\underset{\sim}{k}$
- work with position vectors and coordinates
- calculate the coordinates of an internal division point of a line
- work with the resultant of a vector pathway
- work with parallel vectors and collinearity.

You should already know:

- how to determine the coordinates of a point in three-dimensions from a diagram representing a 3-dimensional object
- how to add and subtract two-dimensional vectors using directed line segments
- how to add and subtract two- or three-dimensional vectors using components
- how to calculate the magnitude of a vector.

Working with vectors

A vector is a quantity with both **magnitude** and **direction** as opposed to a scalar which has magnitude only. Geometrically, a vector can be described by a **directed line segment** which illustrates both its magnitude and direction.

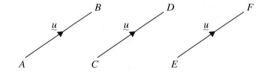

In the directed line segments in the diagram, each line segment has:

- the same magnitude ($AB = CD = EF$; they are equal in length)
- the same direction (AB, CD, and EF are parallel with arrows pointing the same way).

It follows that $\overrightarrow{AB} = \overrightarrow{CD} = \overrightarrow{EF}$.

All three are representatives of the same vector $\underset{\sim}{u}$ which can be visualised as a 'journey' along a straight line path on the two dimensional plane or in three-dimensional space.

$\overrightarrow{BA}$ Is the **negative** of $\overrightarrow{AB}$ and is denoted by $-\underset{\sim}{u}$.

$\underset{\sim}{u}$ and $-\underset{\sim}{u}$ are equal in magnitude but opposite in direction (parallel but with arrows pointing opposite ways).

$\underset{\sim}{u} + (-\underset{\sim}{u}) = \underset{\sim}{0}$, the **zero vector.**

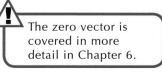

The zero vector is covered in more detail in Chapter 6.

2 dimensions

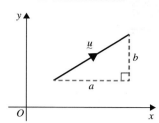

3 dimensions

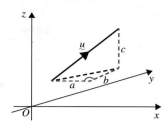

- On the 2–dimensional plane $\underset{\sim}{u}$ is written in component form as:

$$\underset{\sim}{u} = \begin{pmatrix} a \\ b \end{pmatrix}$$

In 3–dimensional space $\underset{\sim}{u}$ is written in component form as:

$$\underset{\sim}{u} = \begin{pmatrix} a \\ b \\ c \end{pmatrix}$$

- The negative of the vector $\underset{\sim}{u}$ is

$$-\underset{\sim}{u} = \begin{pmatrix} -a \\ -b \end{pmatrix}$$

The negative of the vector $\underset{\sim}{u}$ is

$$-\underset{\sim}{u} = \begin{pmatrix} -a \\ -b \\ -c \end{pmatrix}$$

and

$$\underset{\sim}{u} + (-\underset{\sim}{u}) = \underset{\sim}{0} \text{ where } \underset{\sim}{0} = \begin{pmatrix} 0 \\ 0 \end{pmatrix}$$

and

$$\underset{\sim}{u} + (-\underset{\sim}{u}) = \underset{\sim}{0} \text{ where } \underset{\sim}{0} = \begin{pmatrix} 0 \\ 0 \\ 0 \end{pmatrix}$$

- The magnitude or length of the vector $\underset{\sim}{u}$ is given by

$$|\underset{\sim}{u}| = \sqrt{(a^2 + b^2)}$$

The magnitude or length of the vector $\underset{\sim}{u}$ is given by

$$|\underset{\sim}{u}| = \sqrt{(a^2 + b^2 + c^2)}$$

If you visualise a geometric vector as a journey with its components giving the instructions for that journey, it makes sense to think of $\underset{\sim}{u} + \underset{\sim}{v}$, the sum of the two vectors $\underset{\sim}{u}$ and $\underset{\sim}{v}$, as the **resultant** of the vector pathway $\underset{\sim}{u}$ **'followed by'** $\underset{\sim}{v}$

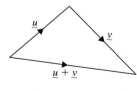

To **add two vectors**, $\underset{\sim}{u}$ and $\underset{\sim}{v}$, choose representatives of each fitting nose to tail to form an unbroken path.

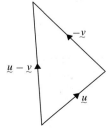

$\underset{\sim}{u} - \underset{\sim}{v} = \underset{\sim}{u} + (-\underset{\sim}{v})$ and so to **subtract two vectors**, $\underset{\sim}{u}$ and $\underset{\sim}{v}$, choose representatives of $\underset{\sim}{u}$ and $(-\underset{\sim}{v})$.

In component terms we **add** and **subtract** vectors as shown:

2 dimensions

If $\underset{\sim}{u} = \begin{pmatrix} a \\ b \end{pmatrix}$ and $\underset{\sim}{v} = \begin{pmatrix} c \\ d \end{pmatrix}$ then:

$$\underset{\sim}{u} \pm \underset{\sim}{v} = \begin{pmatrix} a \pm c \\ b \pm d \end{pmatrix}$$

3 dimensions

If $u = \begin{pmatrix} a \\ b \\ c \end{pmatrix}$ and $v = \begin{pmatrix} d \\ e \\ f \end{pmatrix}$, then:

$$\underset{\sim}{u} \pm \underset{\sim}{v} = \begin{pmatrix} a \pm c \\ b \pm d \\ c \pm e \end{pmatrix}$$

Multiplying a vector by a scalar is simply repeated addition. For example, for any vector $\underset{\sim}{u}$:

$3\underset{\sim}{u} = \underset{\sim}{u} + \underset{\sim}{u} + \underset{\sim}{u}$

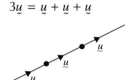

$|3\underset{\sim}{u}| = 3 \times |\underset{\sim}{u}|$

$3\underset{\sim}{u}$ is parallel to $\underset{\sim}{u}$

$3\underset{\sim}{u}$ and $\underset{\sim}{u}$ are in the same direction

$-2\underset{\sim}{u} = -\underset{\sim}{u} + (-\underset{\sim}{u})$

$|-2\underset{\sim}{u}| = 2 \times |\underset{\sim}{u}|$

$-2\underset{\sim}{u}$ is parallel to $\underset{\sim}{u}$

$-2\underset{\sim}{u}$ and $\underset{\sim}{u}$ are in opposite directions

In general, for any scalar $k \neq 0$

- the magnitude of the vector $k\underset{\sim}{u}$ is increased or decreased by a factor of $|k|$
- $k\underset{\sim}{u}$ is parallel to $\underset{\sim}{u}$
- if $k > 0$ the vectors $\underset{\sim}{u}$ and $k\underset{\sim}{u}$ are in the same direction, while for $k < 0$ they are in opposite directions.

Vectors are **equal** if they have the same magnitude and direction:

- **Geometrically**, the directed line segments which represent equal vectors are equal in length; are parallel and have arrows pointing the same way
- **Algebraically**, the corresponding components are equal.

Example 5.1

If $\underset{\sim}{a} = \begin{pmatrix} 2 \\ 1 \\ 4 \end{pmatrix}$, $\underset{\sim}{b} = \begin{pmatrix} -3 \\ 1 \\ -5 \end{pmatrix}$ and $\underset{\sim}{c} = \begin{pmatrix} -2 \\ 2 \\ 1 \end{pmatrix}$

a Calculate the vector $\underset{\sim}{a} + 2\underset{\sim}{b} - 3\underset{\sim}{c}$ in component form.

b Calculate the magnitude of $\underset{\sim}{a} + 2\underset{\sim}{b} - 3\underset{\sim}{c}$.

a $a + 2b - 3c = \begin{pmatrix} 2 \\ 1 \\ 4 \end{pmatrix} + 2\begin{pmatrix} -3 \\ 1 \\ -5 \end{pmatrix} - 3\begin{pmatrix} -2 \\ 2 \\ -1 \end{pmatrix}$

$$= \begin{pmatrix} 2 \\ 1 \\ 4 \end{pmatrix} + \begin{pmatrix} -6 \\ 2 \\ -10 \end{pmatrix} - \begin{pmatrix} -6 \\ 6 \\ -3 \end{pmatrix}$$

$$= \begin{pmatrix} 2 - 6 + 6 \\ 1 + 2 - 6 \\ 4 - 10 + 3 \end{pmatrix}$$

$$= \begin{pmatrix} 2 \\ -3 \\ -3 \end{pmatrix}$$

b The magnitude of $a + 2b - 3c = |a + 2b - 3c|$

$$= \sqrt{\left(2^2 + (-3)^2 + (-3)^2\right)} = \sqrt{22}$$

Example 5.2

$u = \begin{pmatrix} 2x \\ 3 + y \\ 8 \end{pmatrix}$ and $v = \begin{pmatrix} 8 \\ 7 \\ z + 1 \end{pmatrix}$.

If $u = v$, calculate x, y and z.

$u = v \Rightarrow$ corresponding components are equal

$\Rightarrow 2x = 8$, $3 + y = 7$ and $8 = z + 1$

$\Rightarrow x = 4$, $y = 4$, $z = 7$

Expressing a vector in terms of the unit vectors i, j, and k

For any vector v there exists a vector, 1 unit long, which has the same direction as v. This vector is called the **unit vector** in the direction of v.

As the length or magnitude of v is $|v|$, the unit vector in the direction of v is $\dfrac{1}{|v|} v$

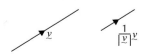

There are three special unit vectors $\underset{\sim}{i}$ $\underset{\sim}{j}$ and $\underset{\sim}{k}$, each representing a 'journey' of 1 unit in the directions of the positive x, y and z-axes respectively.

2 dimensions

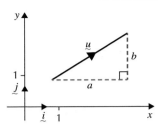

$$\underset{\sim}{i} = \begin{pmatrix} 1 \\ 0 \end{pmatrix} \text{ and } \underset{\sim}{j} = \begin{pmatrix} 0 \\ 1 \end{pmatrix}$$

Every 2-dimensional vector can be expressed in terms of the unit vectors as follows:

$$\underset{\sim}{u} = \begin{pmatrix} a \\ b \end{pmatrix}$$

$$= a\begin{pmatrix} 1 \\ 0 \end{pmatrix} + b\begin{pmatrix} 0 \\ 1 \end{pmatrix}$$

$$= a\underset{\sim}{i} + b\underset{\sim}{j}$$

3 dimensions

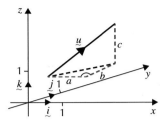

$$\underset{\sim}{i} = \begin{pmatrix} 1 \\ 0 \\ 0 \end{pmatrix} \quad \underset{\sim}{j} = \begin{pmatrix} 0 \\ 1 \\ 0 \end{pmatrix} \text{ and } \underset{\sim}{k} = \begin{pmatrix} 0 \\ 0 \\ 1 \end{pmatrix}$$

Every 3-dimensional vector can be expressed in terms of the unit vectors as follows:

$$\underset{\sim}{u} = \begin{pmatrix} a \\ b \\ c \end{pmatrix}$$

$$= a\begin{pmatrix} 1 \\ 0 \\ 0 \end{pmatrix} + b\begin{pmatrix} 0 \\ 1 \\ 0 \end{pmatrix} + c\begin{pmatrix} 0 \\ 0 \\ 1 \end{pmatrix}$$

$$= a\underset{\sim}{i} + b\underset{\sim}{j} + c\underset{\sim}{k}$$

Example 5.3

If $\underset{\sim}{v} = \begin{pmatrix} 5 \\ 12 \end{pmatrix}$, find the unit vector in the direction of $\underset{\sim}{v}$.

If $\underset{\sim}{v} = \begin{pmatrix} 5 \\ 12 \end{pmatrix}$ then $|\underset{\sim}{v}| = \sqrt{(5^2 + 12^2)} = 13$

so the unit vector in the direction of $\underset{\sim}{v}$ is $\frac{1}{13}\begin{pmatrix} 5 \\ 12 \end{pmatrix} = \begin{pmatrix} \frac{5}{13} \\ \frac{12}{13} \end{pmatrix}$

While there is only one unit vector in the direction of $\underset{\sim}{v}$ there are two unit vectors that are parallel to $\underset{\sim}{v}$.

They are $\begin{pmatrix} \frac{5}{13} \\ \frac{12}{13} \end{pmatrix}$ and $-\begin{pmatrix} \frac{5}{13} \\ \frac{12}{13} \end{pmatrix}$ the unit vector in the opposite direction.

Example 5.4

The following vectors are given in component form. Rewrite them in terms of the unit vectors $\underset{\sim}{i}$ $\underset{\sim}{j}$ and $\underset{\sim}{k}$

a $\underset{\sim}{p} = \begin{pmatrix} 3 \\ -2 \end{pmatrix}$ **b** $\underset{\sim}{q} = \begin{pmatrix} -5 \\ 6 \\ 3 \end{pmatrix}$ **c** $\underset{\sim}{r} = \begin{pmatrix} 1 \\ 0 \\ -4 \end{pmatrix}$

a $\underset{\sim}{p} = \begin{pmatrix} 3 \\ -2 \end{pmatrix} = 3\underset{\sim}{i} - 2\underset{\sim}{j}$ ⬤ $\underset{\sim}{p}$ is a 2-dimensional vector so it is written in terms of $\underset{\sim}{i}$ and $\underset{\sim}{j}$ only.   **S**

b $\underset{\sim}{q} = \begin{pmatrix} -5 \\ 6 \\ 3 \end{pmatrix} = -5\underset{\sim}{i} + 6\underset{\sim}{j} + 3\underset{\sim}{k}$

c $\underset{\sim}{r} = \begin{pmatrix} 1 \\ 0 \\ -4 \end{pmatrix} = \underset{\sim}{i} - 4\underset{\sim}{k}$

Example 5.5

If $\underset{\sim}{p} = 2\underset{\sim}{i} - \underset{\sim}{j} + 3\underset{\sim}{k}$ and $\underset{\sim}{q} = \underset{\sim}{i} - 3\underset{\sim}{j}$

a Calculate $3\underset{\sim}{p} - 2\underset{\sim}{q}$ giving your answer in terms of the unit vectors, $\underset{\sim}{i}$ $\underset{\sim}{j}$ and $\underset{\sim}{k}$

b Calculate the magnitude of $3\underset{\sim}{p} - 2\underset{\sim}{q}$ leaving your answer as a surd.

a $3\underset{\sim}{p} - 2\underset{\sim}{q} = 3\begin{pmatrix} 2 \\ -1 \\ 3 \end{pmatrix} - 2\begin{pmatrix} 1 \\ -3 \\ 0 \end{pmatrix}$ ⬤   It is always easier to work with vectors written in component form.   **S**

$= \begin{pmatrix} 6 \\ -3 \\ 9 \end{pmatrix} - \begin{pmatrix} 2 \\ -6 \\ 0 \end{pmatrix}$

$= \begin{pmatrix} 4 \\ 3 \\ 9 \end{pmatrix}$

$= 4\underset{\sim}{i} + 3\underset{\sim}{j} + 9\underset{\sim}{k}$ ⬤   change back to unit vector form for your answer.   **P**

b $\left| 3\underset{\sim}{i} + 3\underset{\sim}{j} + 9\underset{\sim}{k} \right| = \sqrt{\left(4^2 + 3^2 + 9^2\right)} = \sqrt{106}$

Exercise 5A

★ **1** $m = \begin{pmatrix} 1 \\ 2 \\ -1 \end{pmatrix}$, $n = \begin{pmatrix} 3 \\ 0 \\ -4 \end{pmatrix}$ and $p = \begin{pmatrix} -1 \\ -3 \\ -2 \end{pmatrix}$

    i Calculate the following vectors, giving your answer in component form.

      a $3m - 2n$ **b** $5m - n + 2p$ **c** $2(n - 3p)$

    ii Find the magnitude of each, leaving your answer as a surd.

★ **2** Write each of these column vectors in terms of the unit vectors i j and k.

    a $\begin{pmatrix} 3 \\ 5 \end{pmatrix}$ **b** $\begin{pmatrix} 7 \\ -9 \end{pmatrix}$ **c** $\begin{pmatrix} 3 \\ -2 \\ -6 \end{pmatrix}$ **d** $\begin{pmatrix} 1 \\ 0 \\ -4 \end{pmatrix}$ **e** $\begin{pmatrix} 8 \\ 2 \\ -1 \end{pmatrix}$

3 Write each of these as column vectors.

    a $2i + 5j - k$ **b** $7i - 3j + 9k$ **c** $6i - 5k$ **d** $8j + 5k$

4 $p = 3i - 2j + 4k$, $q = 2i - 3k$ and $r = -2i + 3j - k$

    i Calculate the following vectors, giving your answers in terms of i j and k.

      a $2p - q$ **b** $2 - 5q + r$ **c** $3(r - q) + p$

> ⚠️ Remember to change each vector to component form then change back at the end.

    ii Find the magnitude of each leaving your answer as a surd.

5 The vectors $u = \begin{pmatrix} 3x \\ y \end{pmatrix}$ and $v = \begin{pmatrix} 15 \\ 6 - 2y \end{pmatrix}$ are equal. Find the values of x and y.

6 The vectors $p = (11 - 2x)i + 14j$ and $q = 5i - (3y + 1)j$ are equal. Find the values of x and y.

★ **7** The vectors $u = \begin{pmatrix} 3x \\ y \\ 5 \end{pmatrix}$ and $v = \begin{pmatrix} 12 \\ 9 - 2y \\ z + 3 \end{pmatrix}$ are equal. Find the values of x, y and z.

8 The vectors $p = \begin{pmatrix} 7 - x \\ 13 \\ 2z - 3 \end{pmatrix}$ and $q = \begin{pmatrix} 11 \\ 2y + 5 \\ -7 \end{pmatrix}$ are equal. Find the values of x, y and z.

9 The vectors $m = \begin{pmatrix} 3x + 2y \\ 4 \\ 2z \end{pmatrix}$ and $n = \begin{pmatrix} 5 \\ 2x + y \\ 12 \end{pmatrix}$ are equal. Find the values of x, y and z.

★ **10** Find the unit vector in the direction of each of the following.

    a $\overrightarrow{AB} = \begin{pmatrix} 4 \\ 3 \end{pmatrix}$ **b** $u = \begin{pmatrix} -6 \\ 8 \end{pmatrix}$ **c** $v = \begin{pmatrix} \sqrt{2} \\ \sqrt{2} \end{pmatrix}$

11 Find the two unit vectors parallel to $\underset{\sim}{p} = \begin{pmatrix} 3 \\ -1 \\ 2 \end{pmatrix}$

12 The vector $\begin{pmatrix} \frac{3}{4} \\ \frac{1}{2} \\ z \end{pmatrix}$ is a unit vector. If $z > 0$, calculate the value of z.

13 Find the two possible values for y, given that $\begin{pmatrix} \frac{1}{3} \\ y \\ \frac{1}{2} \end{pmatrix}$ is a unit vector.

14 $\underset{\sim}{u} = 2\underset{\sim}{i} - 2\underset{\sim}{j} + 4\underset{\sim}{k}$ and $\underset{\sim}{v} = \underset{\sim}{i} + 3\underset{\sim}{j} + a\underset{\sim}{k}$. If $|\underset{\sim}{u}| = |\underset{\sim}{v}|$ find the possible values of a.

Working position vectors and coordinates

The position of a point P on the 2-dimensional plane or in 3-dimensional space has up to now been given by its coordinates. As an alternative, the **position vector** $\overrightarrow{OP}$ can be used. The **position vector** $\overrightarrow{OP}$ is the vector which describes the 'journey' from the origin O to the point P.

2 dimensions	**3 dimensions**
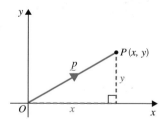	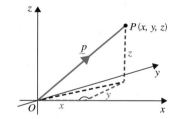
P has coordinates (x, y)	P has coordinates (x, y, z)
P has position vector $\overrightarrow{OP} = \underset{\sim}{p} = \begin{pmatrix} x \\ y \end{pmatrix}$	P has position vector $\overrightarrow{OP} = \underset{\sim}{p} = \begin{pmatrix} x \\ y \\ z \end{pmatrix}$

The **coordinates** give the position of the point P relative to a coordinate grid, while the **position vector** tells you how to get there from the origin. Obviously they each give the same information. If you know one you know the other.

Position vectors can be used derive a formula for the vector $\overrightarrow{AB}$.

Let A and B be points with position vectors $\underset{\sim}{a}$ and $\underset{\sim}{b}$ respectively while O is the origin.

$\overrightarrow{AB} = \overrightarrow{AO} + \overrightarrow{OB}$ — $\overrightarrow{AB}$ is replaced with an alternative 'nose–to–tail' path starting at A and ending at B.

$\quad = -\underset{\sim}{a} + \underset{\sim}{b}$

$\quad = \underset{\sim}{b} - \underset{\sim}{a}$ — $\overrightarrow{AO} = -\overrightarrow{OA}$ **S**

Example 5.6

If A is the point $(5,\ 3)$ and B is the point $(7,\ -1)$ write the vector $\overrightarrow{AB}$ in component form.

Using position vectors:

$\overrightarrow{AB} = \underset{\sim}{b} - \underset{\sim}{a}$

$\phantom{\overrightarrow{AB}} = \begin{pmatrix} 7 \\ -1 \end{pmatrix} - \begin{pmatrix} 5 \\ 3 \end{pmatrix}$

$\phantom{\overrightarrow{AB}} = \begin{pmatrix} 2 \\ -4 \end{pmatrix}$

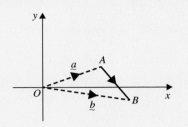

Alternative method

For the vector from $A(5,\ 3)$ to $B(7,\ -1)$, calculate the components as follows:

x-component; to get from 5 to 7 you **add 2**

y-component: to get from 3 to -1 you **subtract 4**

$\overrightarrow{AB} = \begin{pmatrix} 2 \\ -4 \end{pmatrix}$

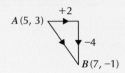

Example 5.7

A is the point $(3,\ -1,\ 2)$ and B is the point $(5,\ 3,\ -2)$

a Write the vector $\overrightarrow{AB}$ in:

   i component form   **ii** in terms of the unit vectors.

b Calculate the magnitude of $\overrightarrow{AB}$.

a i $\overrightarrow{AB} = \underset{\sim}{b} - \underset{\sim}{a}$

$\phantom{\overrightarrow{AB}} = \begin{pmatrix} 5 \\ 3 \\ -2 \end{pmatrix} - \begin{pmatrix} 3 \\ -1 \\ 2 \end{pmatrix}$

$\phantom{\overrightarrow{AB}} = \begin{pmatrix} 2 \\ 4 \\ -4 \end{pmatrix}$

Alternative method

$A(3,\ -1,\ 2)$ and $B(5,\ 3,\ -2)$

$\overrightarrow{AB} = \begin{pmatrix} 2 \\ 4 \\ -4 \end{pmatrix}$

> x-component: to get from 3 to 5 **add 2**
> y-component: to get from -1 to 3 **add 4**
> z-component: to get from 2 to -2 **subtract 4**   **P**

ii $\overrightarrow{AB} = 2\underset{\sim}{i} + 4\underset{\sim}{j} - 4\underset{\sim}{k}$

b $\left|\overrightarrow{AB}\right| = \sqrt{\left(2^2 + 4^2 + (-4)^2\right)} = 6$

As well as being able to calculate the components of the vector $\overrightarrow{AB}$ from the point A to the point B, you must be able to do the opposite For example, if given the coordinates of the point A and the components of the vector $\overrightarrow{AB}$ you must be able to calculate the coordinates of B.

Example 5.8

If A is the point $(2, 3)$ and the vector $\overrightarrow{AB} = \begin{pmatrix} 3 \\ 1 \end{pmatrix}$, calculate the coordinates of B.

A is the point $(2, 3)$.

A to B is described by the vector $\begin{pmatrix} 3 \\ 1 \end{pmatrix}$.

It follows that the point B is $(2 + 3, 3 + 1) = (5, 4)$

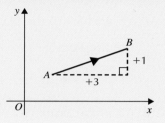

Example 5.9

$ABCD$ is a parallelogram.

A is the point $(2, -1, 4)$, $B(7, 1, 3)$ and $C(-6, 4, 2)$.
Calculate the coordinates of D.

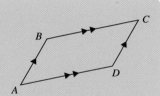

$\overrightarrow{AD} = \overrightarrow{BC}$

$= \begin{pmatrix} -13 \\ 3 \\ -1 \end{pmatrix}$

The vector $\overrightarrow{AD}$ gives the path from A to D.
$\overrightarrow{AD} = \overrightarrow{BC}$ (same length and same direction) Ⓢ

This vector can be calculated using either method. Ⓢ

A is the point $(2, -1, 4)$

A to D is described by the vector $\begin{pmatrix} -13 \\ 3 \\ -1 \end{pmatrix}$

$D = (2 - 13, -1 + 3, 4 - 1)$

$ = (-11, 2, 3)$

Exercise 5B

1 A, B, and C are the points $(-2, 5)$, $(3, 4)$, and $(1, -1)$.

   a Write the vectors $\overrightarrow{CB}$, $\overrightarrow{AB}$ and $\overrightarrow{AC}$ in component form.

   b Calculate the magnitude of each of $\overrightarrow{CB}$, $\overrightarrow{AB}$ and $\overrightarrow{AC}$, leaving your answer in surd form.

2 P, Q, and R are the points $(1, 7)$, $(-4, -5)$ and $(-2, 2)$. Write the vectors $\overrightarrow{PR}$, $\overrightarrow{RQ}$ and $\overrightarrow{QP}$ in terms of the unit vectors $\underset{\sim}{i}$ and $\underset{\sim}{j}$.

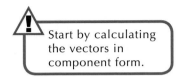

Start by calculating the vectors in component form.

★ 3 A, B, and C are the points $(-2, 5, 3)$, $(3, 4, -1)$ and $(1, -4, 9)$.

   a Write the vectors $\overrightarrow{AB}$, $\overrightarrow{BC}$ and $\overrightarrow{CA}$ in component form.

   b Calculate the magnitude of each of $\overrightarrow{AB}$, $\overrightarrow{BC}$ and $\overrightarrow{CA}$, leaving your answer in surd form.

4 P, Q, and R are the points $(3, 1, 7)$, $(4, -2, 5)$, and $(-5, 3, 2)$

   Write the vectors $\overrightarrow{PR}$, $\overrightarrow{RQ}$ and $\overrightarrow{QP}$ in terms of the unit vectors $\underset{\sim}{i}$ $\underset{\sim}{j}$ and k.

★ 5 Find the coordinates of B in each of the following:

   a A is the point $(-7, 3)$, $\overrightarrow{AB} = \begin{pmatrix} 3 \\ 6 \end{pmatrix}$ b A is the point $(2, 5)$, $\overrightarrow{AB} = \begin{pmatrix} -3 \\ -2 \end{pmatrix}$

   c A is the point $(-7, 3, 5)$, $\overrightarrow{AB} = \begin{pmatrix} 3 \\ -3 \\ 7 \end{pmatrix}$ d A is the point $(2, 5, -4)$, $\overrightarrow{AB} = \begin{pmatrix} -3 \\ -2 \\ 9 \end{pmatrix}$

   e A is the point $(0, 8, 4)$, $\overrightarrow{AB} = \begin{pmatrix} 9 \\ -6 \\ 3 \end{pmatrix}$

6 $ABCD$ is a parallelogram. A is the point $(2, 7)$, B is $(-3, 9)$ and C is $(-1, 4)$. Find the coordinates of D.

7 $PQRS$ is a parallelogram If P is the point $(-3, 2, 5)$, Q is $(5, 3, 9)$ and R is $(6, 5, 8)$ calculate the coordinates of S.

8 The diagram shows a point $P(7, 3, 12)$. If PQR is an isosceles triangle with equal side 13 units long, calculate the coordinates of Q and R.

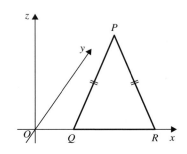

Calculating the coordinates of an internal division point of a line

Consider the following cases of a line *AB* being divided internally by a point *P*.

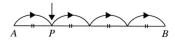

P divides *AB* in the ratio 1:3.

P divides *AB* in the ratio 2:1.

You will be required to find the coordinates of *P* when the coordinates of *A* and *B* are given.

Example 5.10

A is the point $(4, -6, 12)$ and *B* is $(4, 4, -3)$. *P* divides *AB* in the ratio 3:2. Find the coordinates of *P*.

> Draw a straight line to represent the relative positions of *A*, *B* and *P*. Ⓢ

$$\overrightarrow{AP} = \frac{3}{5}\overrightarrow{AB}$$

> *P* is $\frac{3}{5}$ the way along *AB* and so the vector $\overrightarrow{AP}$ from *A* to *P* is $\frac{3}{5}$ of $\overrightarrow{AB}$. Ⓢ

$$= \frac{3}{5}\begin{pmatrix} 0 \\ 10 \\ -15 \end{pmatrix}$$

$$= \begin{pmatrix} 0 \\ 6 \\ -9 \end{pmatrix}$$

A is the point $(4, -6, 12)$ and $\overrightarrow{AP} = \begin{pmatrix} 0 \\ 6 \\ -9 \end{pmatrix}$ so *P* is the point $(4 + 0, -6 + 6, 12 - 9)$

$= (4, 0, 3)$

Example 5.11

C and D are the points $(-4, 3, 6)$ and $(1, 8, -4)$.

Find the coordinates of the point R which divides CD in the ratio 4:1

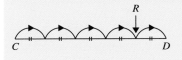

$$\overrightarrow{CR} = \tfrac{4}{5}\,\overrightarrow{CD}$$

$$= \tfrac{4}{5}\begin{pmatrix} 5 \\ 5 \\ -10 \end{pmatrix}$$

$$= \begin{pmatrix} 4 \\ 4 \\ -8 \end{pmatrix}$$

C is the point $(-4, 3, 6)$ and $\overrightarrow{CR} = \begin{pmatrix} 4 \\ 4 \\ -8 \end{pmatrix}$ so R is the point

$(-4 + 4, 3 + 4, 6 - 8) = (0, 7, -2)$

Example 5.12

$PQRS$ is a straight line. P and Q are the points $(-1, 3, 2)$ and $(5, -1, 3)$.

If $PQ = QR = RS$ find the coordinates of S.

$$\overrightarrow{PS} = 3\overrightarrow{PQ} = 3\begin{pmatrix} 6 \\ -4 \\ 1 \end{pmatrix} = \begin{pmatrix} 18 \\ -12 \\ 3 \end{pmatrix}$$

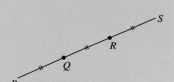

While S is not an interval division point, the same principles apply. The coordinates of P are known, and the vector $\overrightarrow{PS} = 3\overrightarrow{PQ}$ can be calculated.

P is the point $(-1, 3, 2)$ and $\overrightarrow{PS} = \begin{pmatrix} 18 \\ -12 \\ 3 \end{pmatrix}$

$\Rightarrow$ S is the point $(-1 + 18, 3 - 12, 2 + 3) = (17, -9, 5)$

Exercise 5C

★ **1** A and C are the points $(2, 3, -1)$ and $(8, -6, 8)$. If B divides the line AC in the ratio 1:2, find the coordinates of B.

2 P and R are the points $(-2, 5, 7)$ and $(8, 0, 22)$. The point Q divides PR in the ratio 2:3. Find the coordinates of Q.

3 The point Y divides the line XZ in the ratio 2:5.

Given $X(-6, -3, 2)$ and $Y(1, 4, -5)$, find the coordinates of Z.

4 The line *AC* divided in the ratio 2:1.by the point *D*.
If *A* and *C* are the points (3, –1, 6) and (12, –4, 0), find the coordinates of *D*.

5 The line *AC* is divided in the ratio 3:1 by the point *B*. If *A* is the point (–2, 4, 3) and *C* is (2, 0, –5) find the coordinates of *B*.

6 The points *A*, *B*, *C* and *D* lie on a straight line with *AB* = *BC* = *CD*. Given that *A* is the point (3, 5, 4) and *C* is (1, –1, –8), find the coordinates of *B* and *D*.

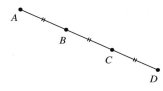

7 The points *P*, *Q*, *R*, *S* and *T* lie on a straight line with *PQ* = *QR* = *RS* = *ST*. Given that *P* is the point (2, 5, 1) and *R* is (6, –1, –3), find the coordinates of *Q*, *S* and *T*.

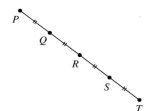

8 Point *A* has coordinates (3, 2, –5) and *B* (1, 4, 1). *B* divides the line *AC* in the ratio 2:3. Find the coordinates of *C*.

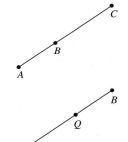

9 *AB* is divided in the ratio 3:2 by the point *Q* (1, 12, –8). Find the coordinates of *A* if *B* is the point (3, 4, 0).

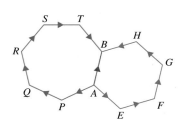

Determining the resultant of vector pathways in 3-dimensions

The vector pathways in the diagram show different routes from *A* to *B*:

$$\overrightarrow{AP} + \overrightarrow{PQ} + \overrightarrow{QR} + \overrightarrow{RS} + \overrightarrow{ST} + \overrightarrow{TB}$$

$$\overrightarrow{AE} + \overrightarrow{EF} + \overrightarrow{FG} + \overrightarrow{GH} + \overrightarrow{HB}$$

In vector terms each is equivalent to the other and to the vector $\overrightarrow{AB}$ which is the shortest path from *A* to *B*.

$\overrightarrow{AB}$ is the **resultant** of each of the two vector pathways and you can get from *A* to *B* by either route. Every other unbroken path made up of directed line segments which start at *A* and end at *B* is also equivalent to $\overrightarrow{AB}$.

Example 5.13

ABCDEFGH is a cuboid.

Find a single vector equivalent to:

$\overrightarrow{AB} + \overrightarrow{AD} + \overrightarrow{AE}$

This question could also be written as:

Find the resultant of $\overrightarrow{AB} + \overrightarrow{AD} + \overrightarrow{AE}$.

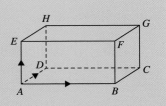

You must find an unbroken pathway equivalent to $\overrightarrow{AB} + \overrightarrow{AD} + \overrightarrow{AE}$

$= \overrightarrow{AB} + \overrightarrow{BC} + \overrightarrow{AE}$

$= \overrightarrow{AB} + \overrightarrow{BC} + \overrightarrow{CG}$

$= \overrightarrow{AG}$

> **S** $\overrightarrow{AD}$ can be replaced by $\overrightarrow{BC}$ as $\overrightarrow{BC} = \overrightarrow{AD}$ (equal length; same direction).

> **S** To complete the path you need a vector equivalent to $\overrightarrow{AE}$ starting at C. That vector is $\overrightarrow{CG}$.

> **C** You now have an unbroken pathway which starts at A and ends at G and so the resultant vector equivalent to $\overrightarrow{AB} + \overrightarrow{AD} + \overrightarrow{AE}$ is $\overrightarrow{AG}$

In the previous example, you were asked to find a resultant vector but you may also be asked to construct a pathway in order to calculate a single vector.

Example 5.14

ABCDE is a pyramid with a rectangular base *ABCD*.

$\overrightarrow{AB} = \begin{pmatrix} 9 \\ 3 \\ 3 \end{pmatrix}$, $\overrightarrow{AD} = \begin{pmatrix} -1 \\ 9 \\ -2 \end{pmatrix}$, $\overrightarrow{AE} = \begin{pmatrix} 2 \\ 6 \\ 7 \end{pmatrix}$

Express the vector $\overrightarrow{CE}$ in component form.

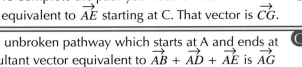

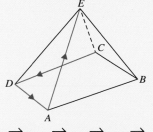

$\overrightarrow{CE} = \overrightarrow{CD} + \overrightarrow{DA} + \overrightarrow{AE}$

$= -\overrightarrow{AB} - \overrightarrow{AD} + \overrightarrow{AE}$

$= -\begin{pmatrix} 9 \\ 3 \\ 3 \end{pmatrix} - \begin{pmatrix} -1 \\ 9 \\ -2 \end{pmatrix} + \begin{pmatrix} 2 \\ 6 \\ 7 \end{pmatrix}$

$= \begin{pmatrix} -9 + 1 + 2 \\ -3 - 9 + 6 \\ -3 + 2 + 7 \end{pmatrix} = \begin{pmatrix} -6 \\ -6 \\ 6 \end{pmatrix}$

> **S** Find an unbroken pathway from C to E. $\overrightarrow{AE}$ must be part of that pathway as it is the only **known** path from the base to E.

> **S** $\overrightarrow{CD} = -\overrightarrow{AB}$ (equal and parallel but opposite direction.) Similarly, $\overrightarrow{DA} = -\overrightarrow{AD}$

$\overrightarrow{CE} = \overrightarrow{CB} + \overrightarrow{BA} + \overrightarrow{AE}$ is another unbroken pathway which starts at C and ends at E.

Example 5.15

PQRST is a pyramid with a rectangular base.
K divides *QR* in the ratio 1:2.

$\overrightarrow{TP}$ represents $-3\underset{\sim}{i} - 5\underset{\sim}{j} - 6\underset{\sim}{k}$

$\overrightarrow{PQ}$ represents $5\underset{\sim}{i} + \underset{\sim}{j} - 2\underset{\sim}{k}$

$\overrightarrow{PS}$ represents $6\underset{\sim}{i} - 3\underset{\sim}{j} + 3\underset{\sim}{k}$

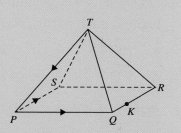

Find the vector $\overrightarrow{TK}$ in terms of the unit vectors $\underset{\sim}{i}$, $\underset{\sim}{j}$, and $\underset{\sim}{k}$.

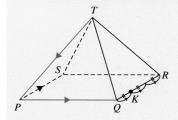

> Find an unbroken vector pathway from **S**
> *T* to *K*. $\overrightarrow{TP}$ must be part of that
> pathway with the next stage starting at *P*.

$\overrightarrow{TK} = \overrightarrow{TP} + \overrightarrow{PQ} + \frac{1}{3}\overrightarrow{QR}$

$= \overrightarrow{TP} + \overrightarrow{PQ} + \frac{1}{3}\overrightarrow{PS}$

$= \begin{pmatrix} -3 + 5 + 2 \\ -5 + 1 - 1 \\ -6 - 2 + 1 \end{pmatrix}$

> Questions and answers may be in terms of **S**
> the unit vectors, but it is always easier to
> work with vectors in component form.

$= \begin{pmatrix} 4 \\ -5 \\ -7 \end{pmatrix}$

$= 4\underset{\sim}{i} - 5\underset{\sim}{j} - 7\underset{\sim}{k}$

Exercise 5D

1 *PQRS* is a parallelogram. Find a single vector to represent

   a $\overrightarrow{RS} + \overrightarrow{RQ}$ **b** $\overrightarrow{RS} - \overrightarrow{RQ}$ **c** $\overrightarrow{PQ} - \overrightarrow{QR} + \overrightarrow{QS}$

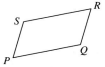

★ **2** *ABCDEF* is a regular hexagon.

   Find the resultant of these vector pathways expressing each
   as a single vector.

   a $\overrightarrow{AB} + \overrightarrow{BC} + \overrightarrow{CD}$ **b** $\overrightarrow{AF} + \overrightarrow{BC} + \overrightarrow{ED}$

   c $\overrightarrow{ED} - \overrightarrow{AF}$ **d** $\overrightarrow{AB} + \overrightarrow{AF} - \overrightarrow{CB}$

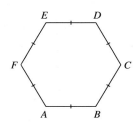

⚙ **3** $\overrightarrow{PQ} = \underset{\sim}{a}$, $\overrightarrow{PT} = \underset{\sim}{b}$ and $\overrightarrow{PS} = \underset{\sim}{c}$

Express each of the following in terms of $\underset{\sim}{a}$, $\underset{\sim}{b}$ and $\underset{\sim}{c}$

a $\overrightarrow{PV}$ b $\overrightarrow{SV}$ c $\overrightarrow{PW}$

d $\overrightarrow{WQ}$ e $\overrightarrow{VT}$ f $\overrightarrow{RP}$

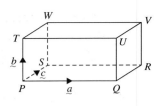

⚙ **4** *PQRSTUVW* is a parallelepiped.

$$\overrightarrow{PQ} = \begin{pmatrix} 3 \\ 1 \\ 0 \end{pmatrix} \quad \overrightarrow{UV} = \begin{pmatrix} 1 \\ 5 \\ 3 \end{pmatrix} \quad \overrightarrow{SW} = \begin{pmatrix} 1 \\ 1 \\ 4 \end{pmatrix}$$

Express these vectors in component form.

a $\overrightarrow{PV}$ b $\overrightarrow{QW}$ c $\overrightarrow{SU}$

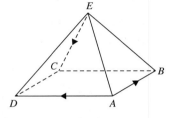

⚙ **5** *ABCDE* is a pyramid with rectangular base *ABCD*.

$$\overrightarrow{AB} = 2\underset{\sim}{i} + 6\underset{\sim}{j} + 2\underset{\sim}{k}$$

$$\overrightarrow{AD} = -4\underset{\sim}{i} + \underset{\sim}{j} - \underset{\sim}{k}$$

$$\overrightarrow{EC} = \underset{\sim}{i} + 2\underset{\sim}{j} - 5\underset{\sim}{k}$$

Express the following vectors in component form:

a $\overrightarrow{EA}$ b $\overrightarrow{BE}$

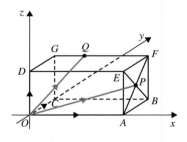

⚙ **6** *OABCDEFG* is a cuboid.

$$\overrightarrow{OA} = \begin{pmatrix} 6 \\ 0 \\ 0 \end{pmatrix} \quad \overrightarrow{OC} = \begin{pmatrix} 0 \\ 3 \\ 0 \end{pmatrix} \quad \overrightarrow{OD} = \begin{pmatrix} 0 \\ 0 \\ 4 \end{pmatrix}$$

Q divides *GF* in the ratio 1:2.

P is the point of intersection of the diagonals of the rectangle *ABFE*.

a Express the position vector $\overrightarrow{OQ}$ of *Q* in component form and hence write down the coordinates of *Q*.

b Express the position vector $\overrightarrow{OP}$ of *P* in component form and hence write down the coordinates of *P*.

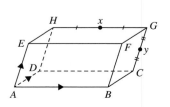

⚙ **7** *ABCDEFGH* is a parallelepiped.

$$\overrightarrow{AB} = 6\underset{\sim}{i} + 2\underset{\sim}{j} + 2\underset{\sim}{k}$$

$$\overrightarrow{AD} = \underset{\sim}{i} + 5\underset{\sim}{j} - 2\underset{\sim}{k}$$

$$\overrightarrow{AE} = 2\underset{\sim}{i} + 2\underset{\sim}{j} + 6\underset{\sim}{k}$$

X is the mid–point of *HG* and *Y* is the mid–point of *CG*.

Find a single vector to represent each of the following, expressing your answer in terms of the unit vectors $\underset{\sim}{i}$, $\underset{\sim}{j}$ and $\underset{\sim}{k}$.

a $\overrightarrow{AG}$ b $\overrightarrow{AX}$ c $\overrightarrow{AY}$ d $\overrightarrow{BX}$ e $\overrightarrow{XY}$

8 *PQRSV* is a pyramid with a rectangular base.

A divides *QR* in the ratio 1:2 while *B* divides *SR* in the ratio 1:3.

$$\overrightarrow{PQ} = \begin{pmatrix} 8 \\ 4 \\ 3 \end{pmatrix} \quad \overrightarrow{PS} = \begin{pmatrix} 3 \\ -6 \\ 3 \end{pmatrix} \quad \overrightarrow{RV} = \begin{pmatrix} -1 \\ -2 \\ 7 \end{pmatrix}$$

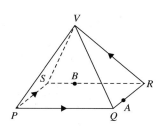

Express in terms of a single vector.

a $\overrightarrow{PA}$ b $\overrightarrow{PB}$ c $\overrightarrow{QV}$ d $\overrightarrow{PV}$ e $\overrightarrow{AB}$

9 *PQRSTUVW* is a cuboid.

A lies three quarters along *WV* and *B* lies two thirds along *UV*.

$$\overrightarrow{PQ} = \begin{pmatrix} 3 \\ 6 \\ 3 \end{pmatrix} \quad \overrightarrow{PS} = \begin{pmatrix} -4 \\ 2 \\ 0 \end{pmatrix} \quad \overrightarrow{PT} = \begin{pmatrix} -6 \\ -12 \\ 30 \end{pmatrix}$$

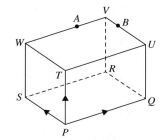

Calculate the components of $\overrightarrow{PA}$ and $\overrightarrow{PB}$

Working with parallel vectors and collinearity

Gradients can be used to show that two lines are parallel in 2 dimensions. This method will not work in 3 dimensions where the concept of gradient does not exist, so an alternative method is required.

For any non–zero scalar *k*, you know that $k\underset{\sim}{u}$ **is parallel to** $\underset{\sim}{u}$.

This means that in 2 or 3 dimensions:

- if $\underset{\sim}{u} \parallel \underset{\sim}{v}$ then there exists a scalar *k* such that $\underset{\sim}{u} = k\underset{\sim}{v}$
- if $\underset{\sim}{u} = k\underset{\sim}{v}$ for some scalar $k \neq 0$ then $\underset{\sim}{u} \parallel \underset{\sim}{v}$

Example 5.16

If the vectors $u = \begin{pmatrix} a \\ 6 \\ c \end{pmatrix}$ and $v = \begin{pmatrix} 3 \\ 2 \\ 4 \end{pmatrix}$ are parallel, calculate the values of a and c.

$u \parallel v \Rightarrow u = kv$ for some scalar $k \neq 0$

$$\Rightarrow \begin{pmatrix} a \\ 6 \\ c \end{pmatrix} = k \begin{pmatrix} 3 \\ 2 \\ 4 \end{pmatrix}$$

$$\begin{pmatrix} a \\ 6 \\ c \end{pmatrix} = 3 \begin{pmatrix} 3 \\ 2 \\ 4 \end{pmatrix}$$

$6 = 3 \times 2$, so $k = 3$ **S**

$\Rightarrow a = 3 \times 3 = 9$ and $c = 3 \times 4 = 12$

Example 5.17

A, B, C and D are the points $(5, 6, -1)$, $(6, 3, 2)$, $(2, 8, -7)$ and $(5, -1, 2)$.
Show that AB and CD are parallel.

$\overrightarrow{AB} = \begin{pmatrix} 3 \\ -9 \\ 9 \end{pmatrix}$ $\overrightarrow{CD} = \begin{pmatrix} 1 \\ -3 \\ 3 \end{pmatrix}$

Find $\overrightarrow{AB}$ and $\overrightarrow{CD}$ in component form. **S**

$\overrightarrow{AB} = 3\overrightarrow{CD}$

Calculate a constant k for which $\overrightarrow{AB} = k\overrightarrow{CD}$. **S**

$\Rightarrow AB \parallel CD$

State the conclusion **C**

Working with collinearity

Points are said to be **collinear** if they lie on the same straight line.

Consider the relative positions of any three points A, B and C. There are only two possibilities:

The points A, B and C are collinear

$AB \parallel BC$

The points A, B and C are not collinear

$AB \not\parallel BC$

- If A, B and C are collinear then AB is parallel to BC.
- If A, B and C are not collinear then AB is not parallel to BC.

Example 5.18

a Show that $A(5, 6, -1)$ $B(6, 3, 2)$ and $C(8, -3, 8)$ are collinear.

b Find the ratio in which B divides AC.

a

$$\overrightarrow{AB} = \begin{pmatrix} 1 \\ -3 \\ 3 \end{pmatrix}, \overrightarrow{BC} = \begin{pmatrix} 2 \\ -6 \\ 6 \end{pmatrix}$$

Express $\overrightarrow{AB}$ and $\overrightarrow{BC}$ in component form

$$\overrightarrow{AB} = \tfrac{1}{2}\overrightarrow{BC} \Rightarrow AB \parallel BC$$

$\overrightarrow{AB} = k\,\overrightarrow{BC}$. Make the formal conclusion that $AB \parallel BC$.

$AB \parallel BC$

$AB \parallel BC$ and so as B is a common point, A, B and C are collinear

Final justification.

b

$$\overrightarrow{AB} = \tfrac{1}{2}\overrightarrow{BC}$$ and so B divides AC in the ratio 1:2.

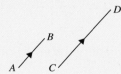

Part a is a formal proof so it is important that all three stages are written out fully and in order. In the Higher exam, you will be expected to state explicitly that B is a **common point** although when dealing with only 3 points this is not, strictly speaking, necessary as the commonality of B is implicit. The reason is that, when dealing with more than three points, there is the possibility that two vectors can be parallel but the points not collinear as you can see in the diagram below where:

$AB \parallel CD$, but A, B, C, and D are not collinear as AB and CD do not share a common point.

Example 5.19

a Show that $A(-1, 4, 2)$, $B(5, 1, 11)$ and $C(7, 0, 14)$ are collinear

b Find the ratio in which B divides AC.

$$\overrightarrow{AB} = \begin{pmatrix} 2 \\ -1 \\ 3 \end{pmatrix} \quad \overrightarrow{BC} = \begin{pmatrix} 6 \\ -3 \\ 9 \end{pmatrix}$$

$$\overrightarrow{AB} = \tfrac{1}{3}\overrightarrow{BC} \Rightarrow AB \parallel BC$$

$AB \parallel BC$ and so as B is a common point, A, B and C are collinear

b $\overrightarrow{AB} = \tfrac{1}{3}\overrightarrow{BC}$ and so B divides AC in the ratio 1:3.

Exercise 5E

★ 1 Each of these pairs of vectors is parallel. Find the value of t in each case.

a $\quad p = \begin{pmatrix} 2 \\ -1 \\ 3 \end{pmatrix}$ and $q = \begin{pmatrix} 6 \\ -3 \\ t-1 \end{pmatrix}$ b $\quad -2\underset{\sim}{i} + 3\underset{\sim}{j} - 5\underset{\sim}{k}$ and $t\underset{\sim}{i} - 6\underset{\sim}{j} + 10\underset{\sim}{k}$

2 Vectors $\underset{\sim}{p} = \begin{pmatrix} 2 \\ h+1 \\ \frac{3}{2} \end{pmatrix}$ and $\underset{\sim}{q} = \begin{pmatrix} 6 \\ -3 \\ \frac{k+1}{2} \end{pmatrix}$ are parallel. Find the values of h and k.

3 Vectors $\begin{pmatrix} -8 \\ c-1 \\ 12 \end{pmatrix}$ and $\begin{pmatrix} 4 \\ -3 \\ 2d+4 \end{pmatrix}$ are parallel. Find the values of c and d.

★ 4 Show that each of the following pairs of vectors is parallel.

a $\quad \begin{pmatrix} 6 \\ 12 \\ 15 \end{pmatrix}$ and $\begin{pmatrix} 2 \\ 4 \\ 5 \end{pmatrix}$ b $\quad \begin{pmatrix} 3 \\ -2 \\ 1 \end{pmatrix}$ and $\begin{pmatrix} 6 \\ -4 \\ 3 \end{pmatrix}$

c $\quad 4\underset{\sim}{i} + 6\underset{\sim}{j} - 8\underset{\sim}{k}$ and $10\underset{\sim}{i} + 15\underset{\sim}{j} - 20\underset{\sim}{k}$

d $\quad 15\underset{\sim}{i} - 21\underset{\sim}{j} + 3\underset{\sim}{k}$ and $-10\underset{\sim}{i} + 14\underset{\sim}{j} - 2\underset{\sim}{k}$

⚙ 5 The vectors $\overrightarrow{AB}$ and $\overrightarrow{CD}$ are parallel. $\overrightarrow{CD}$ is twice the length of $\overrightarrow{AB}$.
A, B and C have coordinates $(-2, 3, 5)$, $(5, 3, 1)$ and $(9, -2, 1)$ respectively.
Find the coordinates of D.

6 Which of the following groups of points are collinear?

a $\quad A(2, -1, 3)$, $B(1, 4, 1)$ and $C(-1, 14, -3)$

b $\quad P(3, 0, 2)$, $Q(1, 3, 3)$ and $R(-5, 12, 5)$

c $\quad X(2, 3, 5)$, $Y(5, -2, 11)$ and $Z(11, -10, 22)$

d $\quad D(1, 0, 3)$, $E(3, -4, 1)$ and $F(6, -10, -2)$

★ 7 a Show that the points $A(3, -2, 5)$, $B(5, 3, 2)$ and $C(9, 13, -4)$ are collinear.
b Find the ratio in which B divides AC.

8 a Show that the points $P(2, 5, 3)$, $Q(3, 1, 8)$ and $R(6, -11, 23)$ are collinear.
b Find the ratio in which Q divides PR.

9 a Show that the points $M(2, 5, 3)$, $N(4, 1, 7)$ and $P(5, -1, 9)$ are collinear.

b Find the ratio in which N divides MP.

10 a Show that the points $A(3, 1, 7)$, $B(5, 5, 9)$ and $C(8, 11, 12)$ are collinear.

b Find the ratio in which B divides AC.

11 Show that the points $E(5, -2, 7)$, $F(3, 4, 5)$ and $G(0, 12, 2)$ are **not** collinear.

12 P, Q and R have coordinates $(-2, 1, 7)$, $(1, 2, 5)$ and $(7, 4, 1)$ respectively.

a Show that P, Q and R are collinear

b If $\overrightarrow{PS} = 4\overrightarrow{PQ}$, find the coordinates of S.

13 A ship leaves port P with coordinates $(1, 3)$ at 13:25 and sets off in a straight line path towards a lighthouse situated at $B(21, 51)$.

After 20 minutes the boat is at the point $A(6, 15)$.

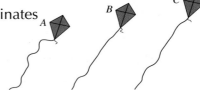

a Show that the boat is heading directly to the lighthouse.

b If the boat continues to sail at the same speed, when will it reach the lighthouse?

14 Relative to a nearby phone mast, 3 kites have coordinates $A(-2, -6, -2)$, $B(1, -9, 4)$ and $C(3, -11, 8)$.

Prove that A, B and C are collinear.

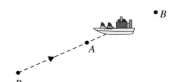

15 At a given point in time an aeroplane is at the point $A(-3, -5, 2)$. One minute later it is at the point $B(-1, -11, 7)$

a If it continues on a straight line path, reaching the point C five minutes after it leaves B, find the coordinates of C.

b Does the aeroplane pass through the point $D(5, -19, 22)$ en route? Justify your answer.

- I can work with vector properties. ★ Exercise 5A Q1, 7, 10
- I can express vectors in terms of the unit vectors $\underset{\sim}{i}$, $\underset{\sim}{j}$ and $\underset{\sim}{k}$. ★ Exercise 5A Q2
- I can work with position vectors and coordinates. ★ Exercise 5B Q3, 5
- I can calculate the coordinates of an internal division point of a line. ★ Exercise 5C Q1
- I can work with the resultant of a vector pathway. ★ Exercise 5D Q2
- I can work with parallel vectors and collinearity. ★ Exercise 5E Q1, 4, 7

For further assessment opportunities, see the Preparation for Assessment for Unit 1 on page 132–135.

6 Working with vectors

This chapter will show you how to:

- work with the zero vector
- calculate the angle between two vectors using the scalar product
- work with perpendicular vectors
- work with the algebraic properties of the scalar product.

You should already know:

- how to determine the coordinates of a point in three-dimensions
- how to add and subtract two-dimensional vectors using directed line segments
- how to add and subtract two- or three-dimensional vectors using components
- how to calculate the magnitude of a vector
- how to determine the vector joining two points
- how to express a vector in terms of the unit vectors $\underline{i}$, $\underline{j}$ and $\underline{k}$
- how to determine the coordinates of the internal division point of a line.

Working with the zero vector

Each of the vector pathways shown below starts at X and ends at X. The resultant effect each time is equivalent to staying exactly where you started.

$$\overrightarrow{XA} + \overrightarrow{AB} + \overrightarrow{BC} + \overrightarrow{CX}$$

$$= \overrightarrow{XD} + \overrightarrow{DE} + \overrightarrow{EF} + \overrightarrow{FG} + \overrightarrow{GX}$$

$$= \overrightarrow{XH} + \overrightarrow{HI} + \overrightarrow{IJ} + \overrightarrow{JK} + \overrightarrow{KX}$$

$$= \underline{0} \text{ the zero vector}$$

The zero vector is $\begin{pmatrix} 0 \\ 0 \end{pmatrix}$ in 2-dimensions and $\begin{pmatrix} 0 \\ 0 \\ 0 \end{pmatrix}$ in 3-dimensions.

The zero vector can be used in solving practical problems involving equilibrium. A system is in equilibrium if the sum of the forces acting on it is $\underline{0}$.

Example 6.1

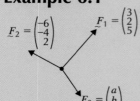
$F_2 = \begin{pmatrix} -6 \\ -4 \\ 2 \end{pmatrix}$ $F_1 = \begin{pmatrix} 3 \\ 2 \\ 5 \end{pmatrix}$ $F_3 = \begin{pmatrix} a \\ b \\ c \end{pmatrix}$

Three forces, measured in newtons, are acting on a point as shown in the diagram.

If the point is stationary, calculate a, b and c.

The point is stationary, so the system is in equilibrium and the sum of the forces must be the zero vector.

Explain your strategy with reason before starting the calculation. **C**

$F_1 + F_2 + F_3 = 0$

$$\Rightarrow \begin{pmatrix} -6 \\ -4 \\ 2 \end{pmatrix} + \begin{pmatrix} 3 \\ 2 \\ 5 \end{pmatrix} + \begin{pmatrix} a \\ b \\ c \end{pmatrix} = \begin{pmatrix} 0 \\ 0 \\ 0 \end{pmatrix}$$

$$\Rightarrow \begin{pmatrix} -3 + a \\ -2 + b \\ 7 + c \end{pmatrix} = \begin{pmatrix} 0 \\ 0 \\ 0 \end{pmatrix}$$

$\Rightarrow a = 3, b = 2, c = -7$

Exercise 6A

★ **1** If each of these sets of forces is in equilibrium, calculate a, b and c. Each force is measured in newtons.

a $F_1 = \begin{pmatrix} 6 \\ 4 \end{pmatrix}$, $F_2 = \begin{pmatrix} -2 \\ -5 \end{pmatrix}$, $F_3 = \begin{pmatrix} a \\ b \end{pmatrix}$

b $F_1 = \begin{pmatrix} 4 \\ 3 \\ -1 \end{pmatrix}$, $F_2 = \begin{pmatrix} 1 \\ -5 \\ 2 \end{pmatrix}$, $F_3 = \begin{pmatrix} a \\ b \\ c \end{pmatrix}$

c $F_1 = 5\underset{\sim}{i} + a\underset{\sim}{j} + \underset{\sim}{k}$, $F_2 = b\underset{\sim}{i} - 6\underset{\sim}{j} - \underset{\sim}{k}$, $F_3 = -3\underset{\sim}{i} + 2\underset{\sim}{j} + c\underset{\sim}{k}$

2 The system of forces shown in the diagram is in equilibrium.

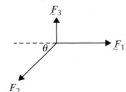

$F_1 = \begin{pmatrix} 3 \\ 0 \end{pmatrix}$, $F_2 = \begin{pmatrix} 2\sqrt{3}\cos\theta \\ 2\sqrt{3}\sin\theta \end{pmatrix}$ and $F_3 = \begin{pmatrix} 0 \\ x \end{pmatrix}$

a Calculate θ.

b Calculate x.

3 Four children are arguing over a sledge each trying to pull it away from the others. Anna is pulling with a force of $\begin{pmatrix} 87 \\ -35 \end{pmatrix}$N and Beth, Connor and David with forces of $\begin{pmatrix} -148 \\ -71 \end{pmatrix}$N, $\begin{pmatrix} 61 \\ 0 \end{pmatrix}$N, and $\begin{pmatrix} 0 \\ 106 \end{pmatrix}$N respectively.

Calculate the resultant force acting on the sledge and explain what it means.

4 a A child is taking two dogs for a walk. If the dogs exert horizontal forces $F_1 = \begin{pmatrix} 40 \\ 0 \end{pmatrix}$N and $F_2 = \begin{pmatrix} -25 \\ 37 \end{pmatrix}$N on the boy, calculate the force which he must exert to keep them still.

Give your answer in component form.

b Find the magnitude of the force to the nearest whole number.

5 When a kite is flying it is acted on by four forces: its weight w; the tension t in the string; the drag d which acts in the direction of the wind and the lift l.

The vectors for weight, lift and drag are defined, in newtons (N), by:

$$w = \begin{pmatrix} 0 \\ -5 \end{pmatrix}, \quad l = \begin{pmatrix} 0 \\ 7 \end{pmatrix}, \quad d = \begin{pmatrix} 15 \\ 0 \end{pmatrix}.$$

Calculate the magnitude of the tension t in the string if the kite is stationary.

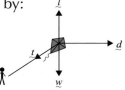

6 A box is resting on a smooth sloping plane as shown in the diagram.

When a force p acting up the plane and a horizontal force q are applied, the box is at rest.

The other forces acting on the box are the weight w and the reaction r between the box and the plane. All the forces are measured in newtons.

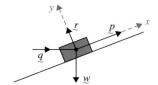

The forces p, r and w are given by:

$$p = \begin{pmatrix} 7.6 \\ 0 \end{pmatrix}, \quad r = \begin{pmatrix} 0 \\ 40.9 \end{pmatrix} \text{ and } w = \begin{pmatrix} -20 \\ -34.6 \end{pmatrix}$$

a Calculate the components of the vector q.

b Calculate the magnitudes of the force q to the nearest newton.

7 A 100 kg traffic light is supported by a system of three cables attached to rings at A, B and C.

The forces in action are shown in the diagram with all forces measured in newtons.

$$F_w = \begin{pmatrix} 0 \\ 0 \\ -980 \end{pmatrix} N, \quad F_B = \begin{pmatrix} -314 \\ -235 \\ 230 \end{pmatrix} N, \quad F_C = \begin{pmatrix} 784 \\ 0 \\ 460 \end{pmatrix} N$$

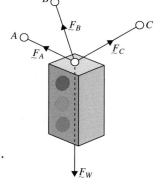

As the traffic light is stationary, the forces are in equilibrium.

a Calculate the force F_A in component form.

b Calculate the magnitude of the force F_A to the nearest newton.

8 A hot air balloon is tethered to the ground by three ropes as shown in the diagram. There is also a vertical lift of 800 newtons.

Calculate x, y and z and hence the magnitude of the force F_B to the nearest newton.

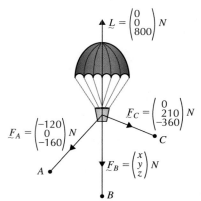

Working with the scalar product of two vectors

Vectors can be added, subtracted or multipled by a scalar. A vector cannot be multiplied by another vector - it makes no sense to multiply directions.

There is a quantity, however, defined as the **scalar product** or **dot product** of two vectors, which involves the multiplication of the magnitudes of the two vectors and the cosine of the angle between them.

If a and b are two non-zero vectors with θ, the angle between their positive directions, then the scalar product is defined as:

$$a \bullet b = |a||b|\cos\theta$$

If either a or $b = 0$ then $a \bullet b = 0$

The angle θ is the angle between the positive directions of the vectors a and b.

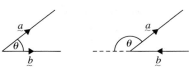

The scalar product is a scalar quantity, and is **not** a vector.

Example 6.2

Calculate $a \bullet b$ for these pairs of vectors.

a

The magnitudes of a and b are 2 and 3 units respectively.

b

The magnitude of a is 3 units and that of b is 5 units.

a

> Check both arrows are pointing out from the vertex. **P**

$$a \bullet b = |a||b|\cos\theta°$$
$$= 2 \times 3 \times \cos 30°$$
$$= 2 \times 3 \times \frac{\sqrt{3}}{2}$$
$$= 3\sqrt{3}$$

b

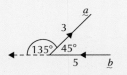

> The arrows are **not** both pointing out from the vertex so extend the vector b to show that the angle between a and b is 135°. **P** **S**

$$a \bullet b = |a||b|\cos\theta°$$
$$= 3 \times 5 \times \cos 135°$$
$$= 3 \times 5 \times (-\cos 45°)$$
$$= 3 \times 5 \times \left(\frac{-1}{\sqrt{2}}\right)$$
$$= \frac{-15}{\sqrt{2}} \text{ or } \frac{-15\sqrt{2}}{2}$$

While the **geometric** definition of the scalar product is given in terms of the lengths of a and b and the angle between them, there is another form. The scalar product can also be defined **algebraically** in terms of the components of a and b. The algebraic definition can be derived from the geometric definition as follows:

Let 0 be the origin and let the points A and B have position vectors a and b respectively:

$$a = \begin{pmatrix} a_1 \\ a_2 \\ a_3 \end{pmatrix} \text{ and } b = \begin{pmatrix} b_1 \\ b_2 \\ b_3 \end{pmatrix}$$

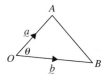

It follows that $\overrightarrow{AB} = \begin{pmatrix} b_1 - a_1 \\ b_2 - a_2 \\ b_3 - a_3 \end{pmatrix}$

The magnitudes of a, b and $\overrightarrow{AB}$ are therefore:

$$\sqrt{\left(a_1^2 + a_2^2 + a_3^2\right)}, \ \sqrt{\left(b_1^2 + b_2^2 + b_3^2\right)} \text{ and } \sqrt{\left((b_1 - a_1)^2 + (b_2 - a_2)^2 + (b_3 - a_3)^2\right)}$$

respectively.

Applying the cosine rule to triangle AOC then gives:

$$\cos\theta = \frac{|a|^2 + |b|^2 - |\overrightarrow{AB}|^2}{2|a||b|}$$

$$\Rightarrow 2|a||b|\cos\theta = |a|^2 + |b|^2 - |\overrightarrow{AB}|^2$$

$$\Rightarrow 2a \bullet b = (a_1^2 + a_2^2 + a_3^2) + (b_1^2 + b_2^2 + b_3^2) - ((b_1 - a_1)^2 + (b_2 - a_2)^2 + (b_3 - a_3)^2)$$

$$= a_1^2 + a_2^2 + a_3^2 + b_1^2 + b_2^2 + b_3^2 - b_1^2 + 2a_1b_1 - a_1^2 - b_2^2 + 2a_2b_2 - a_2^2 - b_3^2 + 2a_3b_3 - a_3^2$$

$$= 2(a_1b_1 + a_2b_2 + a_3b_3)$$

$$\Rightarrow a \bullet b = a_1b_1 + a_2b_2 + a_3b_3$$

So the algebraic definition of the scalar product, given in terms of the components of the vectors a and b, is:

$$\text{If } a = \begin{pmatrix} a_1 \\ a_2 \\ a_3 \end{pmatrix} \text{ and } b = \begin{pmatrix} b_1 \\ b_2 \\ b_3 \end{pmatrix} \text{ then } a \bullet b = a_1b_1 + a_2b_2 + a_3b_3$$

Example 6.3

Calculate the scalar product for each of these pairs of vectors.

a $\quad a = \begin{pmatrix} 3 \\ 2 \\ 1 \end{pmatrix}, b = \begin{pmatrix} -1 \\ 4 \\ 6 \end{pmatrix}$

b $\quad p = i + j - 2k, \ q = j = 6k$

a $\underset{\sim}{a} \bullet \underset{\sim}{b} = \begin{pmatrix} 3 \\ 2 \\ 1 \end{pmatrix} \bullet \begin{pmatrix} -1 \\ 4 \\ 6 \end{pmatrix}$ ⟶ Components are given, so use the algebraic form of the definition. **P**

$= 3 \times (-1) + (2 \times 4) + (1 \times 6)$

$= -3 + 8 + 6$

$= 11$

b $\underset{\sim}{p} \bullet \underset{\sim}{q} = \begin{pmatrix} 1 \\ 1 \\ -2 \end{pmatrix} \bullet \begin{pmatrix} 0 \\ 1 \\ 6 \end{pmatrix}$ ⟶ Rewrite $\underset{\sim}{p}$ and $\underset{\sim}{q}$ in component form. **P**

$= 1 \times 0 + 1 \times 1 + (-2) \times 6$

$= 0 + 1 - 12$

$= -11$

Example 6.4

If A is the point $(3, -1, 2)$, B $(4, 5, -3)$ and C $(0, -3, 6)$.

Calculate $\overrightarrow{AB} \bullet \overrightarrow{AC}$.

$\overrightarrow{AB} = \begin{pmatrix} 1 \\ 6 \\ -5 \end{pmatrix}$, $\overrightarrow{AC} = \begin{pmatrix} -3 \\ -2 \\ 4 \end{pmatrix}$ ⟶ Find $\overrightarrow{AB}$ and $\overrightarrow{AC}$ in component form. **P**

$\overrightarrow{AB} \bullet \overrightarrow{AC} = \begin{pmatrix} 1 \\ 6 \\ -5 \end{pmatrix} \bullet \begin{pmatrix} -3 \\ -2 \\ 4 \end{pmatrix}$ ⟶ Use algebraic formula. **P**

$= 1 \times (-3) + 6 \times (-2) + (-5) \times 4$

$= -3 - 12 - 20$

$= -35$

Exercise 6B

★ **1** Calculate the exact value of $\underset{\sim}{a} \bullet \underset{\sim}{b}$ in these pairs of vectors.

a

$|\underset{\sim}{a}| = 3$, $|\underset{\sim}{b}| = 2$

b

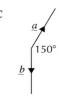

$|\underset{\sim}{a}| = 1$, $|\underset{\sim}{b}| = 3$

c

$|\underset{\sim}{a}| = 5$, $|\underset{\sim}{b}| = 3$

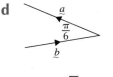
d

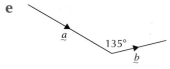
e

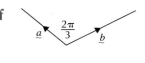
f

$|\underset{\sim}{a}| = \sqrt{5}, |\underset{\sim}{b}| = 2$ $\qquad$ $|\underset{\sim}{a}| = 4, |\underset{\sim}{b}| = \sqrt{2}$ $\qquad$ $|\underset{\sim}{a}| = 2, |\underset{\sim}{b}| = 3$

★ **2** Calculate the scalar product for each pair of vectors.

a $\underset{\sim}{u} = 3\underset{\sim}{i} - 2\underset{\sim}{j} + \underset{\sim}{k}, \underset{\sim}{v} = 2\underset{\sim}{i} + 5\underset{\sim}{j}$ $\qquad$ **b** $\underset{\sim}{p} = 4\underset{\sim}{i} - 6\underset{\sim}{j} - 4\underset{\sim}{k}, \underset{\sim}{q} = -3\underset{\sim}{i} - 2\underset{\sim}{j} + \underset{\sim}{k}$

c $\underset{\sim}{a} = 6\underset{\sim}{i} - \sqrt{3}\underset{\sim}{j} + \underset{\sim}{k}, \underset{\sim}{b} = 2\underset{\sim}{i} + \sqrt{2}\underset{\sim}{j} - 4\underset{\sim}{k}$ $\qquad$ **d** $\underset{\sim}{u} = \begin{pmatrix} 2 \\ -4 \\ 1 \end{pmatrix}, \underset{\sim}{v} = \begin{pmatrix} 2 \\ -3 \\ 6 \end{pmatrix}$

e $\underset{\sim}{v} = 2\underset{\sim}{j} - \sqrt{5}\underset{\sim}{k}, \underset{\sim}{w} = \underset{\sim}{i} + 3\underset{\sim}{j} + 2\sqrt{3}\underset{\sim}{k}$ $\qquad$ **f** $\underset{\sim}{u} = \begin{pmatrix} 2\sqrt{3} \\ 4 \\ -3 \end{pmatrix}, \underset{\sim}{v} = \begin{pmatrix} \sqrt{6} \\ -2 \\ -3 \end{pmatrix}$

★ **3** Given $A\,(-2, 4, 3)$, $B\,(3, 6, 0)$ and $C\,(5, -3, 2)$, calculate:

a $\overrightarrow{AB} \bullet \overrightarrow{AC}$ $\qquad\qquad$ **b** $\overrightarrow{BC} \bullet \overrightarrow{AB}$

4 Given $P\,(3, 0, 1)$, $Q\,(4, 2, -7)$ and $R\,(-3, 5, 3)$, calculate:

a $\overrightarrow{QR} \bullet \overrightarrow{PQ}$ $\qquad\qquad$ **b** $\overrightarrow{RP} \bullet \overrightarrow{QP}$

5 $\underset{\sim}{u} = \begin{pmatrix} 2 \\ -1 \\ p \end{pmatrix}, \underset{\sim}{v} = \begin{pmatrix} 2p \\ 3 \\ 1 \end{pmatrix}$

Find the value of p given that $\underset{\sim}{u} \bullet \underset{\sim}{v} = 22$.

6 $\underset{\sim}{a} = \begin{pmatrix} t \\ 5t \\ -8 \end{pmatrix}, \underset{\sim}{b} = \begin{pmatrix} 3 \\ 2 \\ -4 \end{pmatrix}.$

Find the value of t given that $\underset{\sim}{a} \bullet \underset{\sim}{b} = 6$.

7 $\underset{\sim}{a} = \underset{\sim}{i} - 3\underset{\sim}{j} + 4\underset{\sim}{k}, \underset{\sim}{b} = 2\underset{\sim}{i} - 5\underset{\sim}{k}, C = 4\underset{\sim}{i} - 3\underset{\sim}{j} + \underset{\sim}{k}.$
Calculate:

a $\underset{\sim}{a} \bullet (\underset{\sim}{b} + \underset{\sim}{c})$ $\qquad\qquad$ **b** $\underset{\sim}{b} \bullet (\underset{\sim}{a} - \underset{\sim}{c})$ $\qquad\qquad$ **c** $\underset{\sim}{a} \bullet (\underset{\sim}{a} + \underset{\sim}{b} + \underset{\sim}{c})$

Using the scalar product to calculate the angle between two vectors

There are two definitions for the scalar product:

• the geometric definition: $\underset{\sim}{a} \bullet \underset{\sim}{b} = |\underset{\sim}{a}||\underset{\sim}{b}|\cos\theta$
• the algebraic or component definition: $\underset{\sim}{a} \bullet \underset{\sim}{b} = a_1b_1 + a_2b_2 + a_3b_3$

These two definitions can be combined to find the angle between two vectors:

If θ is the angle between $\underset{\sim}{a}$ and $\underset{\sim}{b}$ then $\cos\theta$ can be derived from the geometric definition:

$$\underset{\sim}{a} \bullet \underset{\sim}{b} = |a||b|\cos\theta$$

$$\Rightarrow \cos\theta = \frac{\underset{\sim}{a} \bullet \underset{\sim}{b}}{|\underset{\sim}{a}||\underset{\sim}{b}|}$$

The component form of the definition for $\underset{\sim}{a} \bullet \underset{\sim}{b}$ can then be substituted to give the formula for $\cos\theta$:

$$\cos\theta = \frac{a_1b_1 + a_2b_2 + a_3b_3}{|\underset{\sim}{a}||\underset{\sim}{b}|}$$

Example 6.5

Calculate the angle between the vectors $\underset{\sim}{u} = \begin{pmatrix} 4 \\ 2 \\ 5 \end{pmatrix}$ and $\underset{\sim}{v} = \begin{pmatrix} 3 \\ 1 \\ 2 \end{pmatrix}$.

$$\underset{\sim}{u} \bullet \underset{\sim}{v} = |\underset{\sim}{u}||\underset{\sim}{v}|\cos\theta$$

$$\Rightarrow \cos\theta = \frac{\underset{\sim}{u} \bullet \underset{\sim}{v}}{|\underset{\sim}{u}||\underset{\sim}{v}|}$$

> Start with the geometric formula and rearrange to make $\cos\theta$ the subject. **S**

$$\underset{\sim}{u} \bullet \underset{\sim}{v} = \begin{pmatrix} 4 \\ 2 \\ 5 \end{pmatrix} \bullet \begin{pmatrix} 3 \\ 1 \\ 2 \end{pmatrix} = 12 + 2 + 10 = 24$$

$$|\underset{\sim}{u}| = \sqrt{(4^2 + 2^2 + 5^2)} = \sqrt{45}$$

$$|\underset{\sim}{v}| = \sqrt{(3^2 + 1^2 + 2^2)} = \sqrt{14}$$

> To use the formula you must first calculate $\underset{\sim}{u} \bullet \underset{\sim}{v}$, $|\underset{\sim}{u}|$ and $|\underset{\sim}{v}|$ using components. **S P**

$$\cos\theta = \left(\frac{24}{\sqrt{45} \times \sqrt{14}}\right)$$

> Substitute into the formula. **S P**

$$\theta = \cos^{-1}\left(\frac{24}{\sqrt{45} \times \sqrt{14}}\right)$$

$$\approx 17.0°$$

> θ is the angle between two vectors and can only be acute or obtuse. If $\cos\theta$ is positive, θ is acute. If $\cos\theta$ is negative, θ is obtuse. **C**

Example 6.6

If the points A, B and C have coordinates A (2, 1, 5), B (3, −2, 4) and C (−3, 4, 1), calculate the angle $\angle BAC$.

> Sketch an angle and label it BAC. The size of the angle chosen does not matter because the sketch is simply used to ensure that the vectors used are pointing away from the vertex. The vectors pointing away from the vertex are $\overrightarrow{AB}$ and $\overrightarrow{AC}$. **S**

(continued)

$$\vec{AB} \bullet \vec{AC} = \left|\vec{AB}\right|\left|\vec{AC}\right|\cos\theta$$

$$\Rightarrow \cos\theta = \frac{\vec{AB} \bullet \vec{AC}}{\left|\vec{AB}\right|\left|\vec{AC}\right|}$$

Rearrange the geometric definition. **S**

$$\vec{AB} = \begin{pmatrix} 1 \\ -3 \\ -1 \end{pmatrix} \quad \vec{AC} = \begin{pmatrix} -5 \\ 3 \\ -4 \end{pmatrix}$$

Write $\vec{AB}$ and $\vec{AC}$ in component form. **P**

$$\vec{AB} \bullet \vec{AC} = \begin{pmatrix} 1 \\ -3 \\ -1 \end{pmatrix} \bullet \begin{pmatrix} -5 \\ 3 \\ -4 \end{pmatrix} = 5 - 9 + 4 = -10$$

$$\left|\vec{AB}\right| = \sqrt{\left(1^2 + (-3)^2 + (-1)^2\right)} = \sqrt{11}$$

$$\left|\vec{AC}\right| = \sqrt{\left((-5)^2 + 3^2 + (-4)^2\right)} = \sqrt{50}$$

Calculate $\vec{AB} \bullet \vec{AC}$, $\left|\vec{AB}\right|$ and $\left|\vec{AC}\right|$ for use in the formula. **P**

$$\cos\theta = \frac{-10}{\sqrt{11} \times \sqrt{50}}$$

Substitute in the formula for $\cos\theta$. **P**

$$\theta = 180° - \cos^{-1}\left(\frac{10}{\sqrt{11} \times \sqrt{50}}\right)$$

Cos θ is negative, so θ must be obtuse. **C**

$$= 115.2°$$

Example 6.7

DOABC is a pyramid.

A is the point (12, 0, 0), *B* (12, 6, 0) and *D* (6, 3, 9).

F divides *DB* in the ratio 1:2.

a Find the coordinates of *F*.

b Find the size of angle *BOF*.

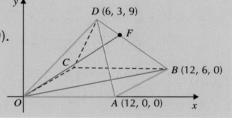

a

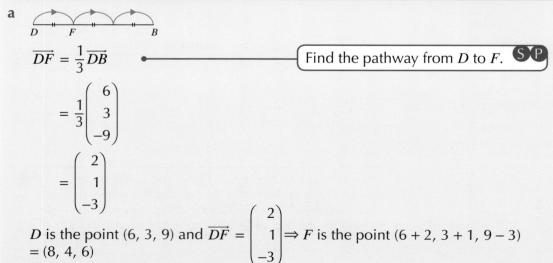

$$\vec{DF} = \frac{1}{3}\vec{DB}$$

Find the pathway from *D* to *F*. **S** **P**

$$= \frac{1}{3}\begin{pmatrix} 6 \\ 3 \\ -9 \end{pmatrix}$$

$$= \begin{pmatrix} 2 \\ 1 \\ -3 \end{pmatrix}$$

D is the point (6, 3, 9) and $\vec{DF} = \begin{pmatrix} 2 \\ 1 \\ -3 \end{pmatrix} \Rightarrow$ *F* is the point (6 + 2, 3 + 1, 9 − 3)
= (8, 4, 6)

b

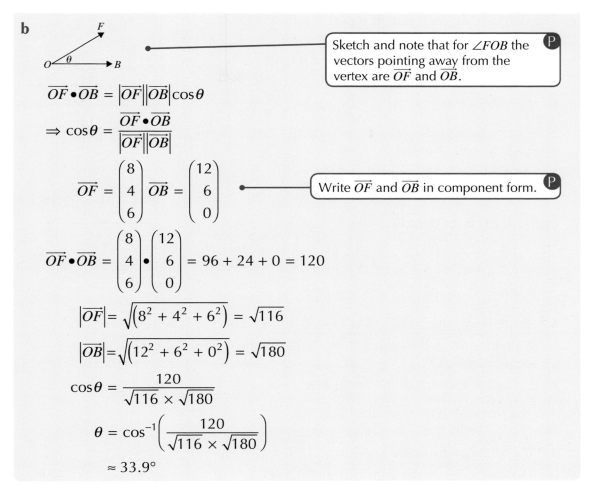

Sketch and note that for ∠FOB the vectors pointing away from the vertex are $\overrightarrow{OF}$ and $\overrightarrow{OB}$. **P**

$$\overrightarrow{OF} \bullet \overrightarrow{OB} = \left|\overrightarrow{OF}\right|\left|\overrightarrow{OB}\right|\cos\theta$$

$$\Rightarrow \cos\theta = \frac{\overrightarrow{OF} \bullet \overrightarrow{OB}}{\left|\overrightarrow{OF}\right|\left|\overrightarrow{OB}\right|}$$

$$\overrightarrow{OF} = \begin{pmatrix} 8 \\ 4 \\ 6 \end{pmatrix} \quad \overrightarrow{OB} = \begin{pmatrix} 12 \\ 6 \\ 0 \end{pmatrix}$$

Write $\overrightarrow{OF}$ and $\overrightarrow{OB}$ in component form. **P**

$$\overrightarrow{OF} \bullet \overrightarrow{OB} = \begin{pmatrix} 8 \\ 4 \\ 6 \end{pmatrix} \bullet \begin{pmatrix} 12 \\ 6 \\ 0 \end{pmatrix} = 96 + 24 + 0 = 120$$

$$\left|\overrightarrow{OF}\right| = \sqrt{\left(8^2 + 4^2 + 6^2\right)} = \sqrt{116}$$

$$\left|\overrightarrow{OB}\right| = \sqrt{\left(12^2 + 6^2 + 0^2\right)} = \sqrt{180}$$

$$\cos\theta = \frac{120}{\sqrt{116} \times \sqrt{180}}$$

$$\theta = \cos^{-1}\left(\frac{120}{\sqrt{116} \times \sqrt{180}}\right)$$

$$\approx 33.9°$$

Exercise 6C

★ **1** Calculate the size of angle θ between these pairs of vectors.

   a $a = 5i + 3j - k$, $b = i - 2j - 3k$

   b $u = 3i - 5j$, $v = 2i + 4j + 5k$

   c $p = 3i + 2j + k$, $q = 5i - 3j - 4k$

★ **2** Use the coordinates of A, B and C to calculate the angles.

   a A (−2, 3, 5), B (4, 0, 7) and C (2, 2, 1) calculate angle *ABC*.

   The vectors must be chosen so that they are pointing away from the vertex of the angle. **!**

   b A (2, 2, 3), B (−5, 3, 0) and C (3, 2, −1) calculate angle *BAC*.

   c A (−1, 0, 0), B (3, 2, −5) and C (−4, 1, 3) calculate angle *BCA*.

3 Given points P (3, 2, 0), Q (5, −6, 1) and R (4, 0, −9), find the size of angle *QRP*.

4 *OA* is 6 units long, *OC* is 3 units and *OD* is 5 units.

   a Write down the coordinates of E, B and G.

   b Hence calculate the size of angle *BEG*.

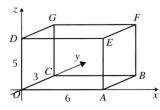

5 The diagram shows a square based pyramid with height 9 units.

The square *OABC* has a side length of 8 units.

a Write down the coordinates of *B*.

b Calculate the angle *ADB*.

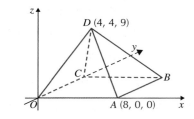

6 A rectangular based pyramid has vertices *O*, *P*, *Q* (8, 6, 0), *R* and *S* (4, 3, 9).

a Find the coordinates of *P* and *R*.

b Calculate the size of angle *PSR*.

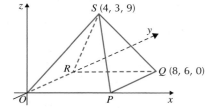

7 *OABCDEFG* is a cuboid.

$$\overrightarrow{OA} = \begin{pmatrix} 8 \\ 0 \\ 0 \end{pmatrix}, \overrightarrow{OC} = \begin{pmatrix} 0 \\ 5 \\ 0 \end{pmatrix} \text{ and } \overrightarrow{OD} = \begin{pmatrix} 0 \\ 0 \\ 6 \end{pmatrix}$$

a Write down the coordinates of *A*, *B* and *F*.

b *P* divides *BF* in the ratio 1:2. Write down the coordinates of *P*.

c Calculate the size of the angle *BOP*.

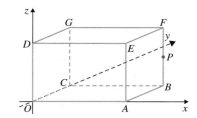

8 *A* is the point with coordinates (5, 2, 1), *B* is (7, 7, 3) and *C* is (−1, −2, 8).

a Calculate angle *ABC*.

b Hence calculate the area of triangle *ABC*.

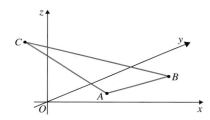

9 *ABCDEFGH* is a cuboid.

$$\overrightarrow{AB} = \begin{pmatrix} 5 \\ 15 \\ -10 \end{pmatrix}, \overrightarrow{AD} = \begin{pmatrix} 4 \\ -2 \\ 2 \end{pmatrix} \text{ and } \overrightarrow{AE} = \begin{pmatrix} -6 \\ 3 \\ -9 \end{pmatrix}$$

X is the mid-point of *CG* and *Y* divides *HG* in the ratio 2:3.

a Calculate the components of *AY* and *AX*.

b Calculate the angle *YAX*.

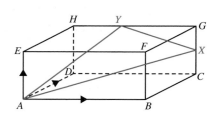

10 *OABCDEFG* is a cuboid.

The base *OABC* lies in the *x*-*y* plane.

F is the point (8, 9, 15).

P divides *AE* in the ratio 1:2

Q divides *EF* in the ratio 3:2

R divides *DE* in the ratio 1:3

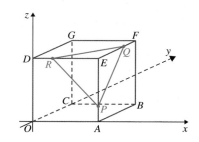

a Write down the coordinates of *A*, *D* and *E*.

b Calculate the coordinates of *P*, *Q* and *R*.

c Find the size of angle *RQP*.

d Hence find the area of triangle *PQR*.

11 Relative to a suitable set of axes, a fighter jet, *J*, with coordinates $(3, -5, h)$ is at a point h km above sea level. An aircraft carrier is stationary at the point *A* $(9, 4, 0)$.

The distance between the jet and aircraft carrier is $\sqrt{181}$ km.

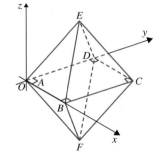

a Calculate the height, *h*, of the jet above sea level.

The jet and a submarine, *S*, which has coordinates $(2, 5, -3)$ both head towards the aircraft carrier.

b Calculate the angle *JAS*.

12 A rocket is designed to be propelled by three engines exerting forces of

$$\begin{pmatrix} 5 \\ 12 \\ 9 \end{pmatrix}, \begin{pmatrix} -3 \\ 4 \\ 16 \end{pmatrix} \text{ and } \begin{pmatrix} 4 \\ -5 \\ 8 \end{pmatrix} \text{ meganewtons respectively.}$$

a Calculate the resultant force acting on the rocket.

b If the third engine fails to start on take-off, calculate the angle between the actual and intended paths of the rocket.

13 A crystal is in the shape of a **regular** octahedron *ABCDEF*.

Coordinate axes are chosen with *A* at the origin and *B* and *D* on the *x*- and *y*-axes respectively.

If *C* is the point $(4, 6, 0)$, and the distance from *E* to *F* is 14 units, calculate the total surface area of the crystal.

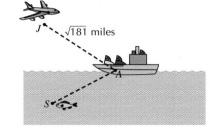

Working with perpendicular vectors

Gradients can be used to identify perpendicular lines in 2 dimensions only. There is another method using the scalar product which can be used in **both** 2 and 3 dimensions.

The geometric formula for the scalar product for two non-zero vectors *a* and *b* gives the following two-way relationship:

If $a \bullet b = 0$ then $|a||b|\cos\theta = 0$
$\Rightarrow \cos\theta = 0$ as a and b are non-zero
$\Rightarrow \theta = 90°$
$\Rightarrow a$ is perpendicular to b

If a is perpendicular to b then $\theta = 90°$
$\Rightarrow \cos\theta = 0$
$\Rightarrow |a||b|\cos\theta = 0$
$\Rightarrow a \bullet b = 0$

In the same way it can be shown that $a \bullet b > 0$ corresponds to an acute angle θ, while $a \bullet b < 0$ corresponds to an obtuse angle θ.

Example 6.8

$\underset{\sim}{u} = 3\underset{\sim}{i} + 2\underset{\sim}{j}$, $\underset{\sim}{v} = 2\underset{\sim}{i} - 3\underset{\sim}{j} + 4\underset{\sim}{k}$.

Show that $\underset{\sim}{u}$ and $\underset{\sim}{v}$ are perpendicular.

$$\underset{\sim}{u} \bullet v = \begin{pmatrix} 3 \\ 2 \\ 0 \end{pmatrix} \bullet \begin{pmatrix} 2 \\ -3 \\ 4 \end{pmatrix}$$

$\underset{\sim}{u} \perp \underset{\sim}{v}$ if $\underset{\sim}{u} \bullet v = 0$ so calculate $\underset{\sim}{u} \bullet v$, check the result and comment. **S P C**

$$= 3 \times 2 + 2 \times (-3) + 0 \times 4$$
$$= 6 - 6 + 0$$
$$= 0$$

$\underset{\sim}{u} \bullet v = 0 \Rightarrow \underset{\sim}{u}$ and $\underset{\sim}{v}$ are perpendicular.

This statement, with the reasoning to explain that the vectors are perpendicular, is an important and necessary part of the solution. **C**

Example 6.9

A is the point $(-1, 2, 0)$, B is $(1, 5, 4)$ and C is $(2, -1, 8)$.

Show that AB is perpendicular to BC.

$$\overrightarrow{AB} = \begin{pmatrix} 2 \\ 3 \\ 4 \end{pmatrix} \quad \overrightarrow{BC} = \begin{pmatrix} 1 \\ -6 \\ 4 \end{pmatrix}$$

$AB \perp BC$ if $\overrightarrow{AB} \bullet \overrightarrow{BC} = 0$ so: **S P C**
- write $\overrightarrow{AB}$ and $\overrightarrow{BC}$ in component form
- calculate $\overrightarrow{AB} \bullet \overrightarrow{BC}$
- check and comment.

$$\overrightarrow{AB} \bullet \overrightarrow{BC} = \begin{pmatrix} 2 \\ 3 \\ 4 \end{pmatrix} \bullet \begin{pmatrix} 1 \\ -6 \\ 4 \end{pmatrix}$$

$$= 2 - 18 + 16$$
$$= 0$$

$\overrightarrow{AB} \bullet \overrightarrow{BC} = 0 \Rightarrow AB$ is perpendicular to BC.

Example 6.10

$\underset{\sim}{a} = \begin{pmatrix} 1 \\ 2 \\ k \end{pmatrix}$ and $\underset{\sim}{b} = \begin{pmatrix} -1 \\ 4 \\ 3 \end{pmatrix}$ are perpendicular.

Calculate k.

$\underset{\sim}{a} \perp \underset{\sim}{b}$ and so $\underset{\sim}{a} \bullet \underset{\sim}{b} = 0$

The vectors are perpendicular. State that $\underset{\sim}{a} \bullet \underset{\sim}{b} = 0$ with the reason and go on to solve equation for k. **S C**

$$\Rightarrow \begin{pmatrix} 1 \\ 2 \\ k \end{pmatrix} \bullet \begin{pmatrix} -1 \\ 4 \\ 3 \end{pmatrix} = 0$$

$\Rightarrow -1 + 8 + 3k = 0$

$\Rightarrow 7 + 3k = 0$

$\Rightarrow k = \dfrac{-7}{3}$

Exercise 6D

1 Show that each of these pairs of vectors is perpendicular.

   a $\underset{\sim}{u} = 3\underset{\sim}{i} + 2\underset{\sim}{j} - \underset{\sim}{k}$, $\underset{\sim}{v} = 5\underset{\sim}{i} - 3\underset{\sim}{j} + 9\underset{\sim}{k}$ **b** $\underset{\sim}{p} = 2\underset{\sim}{i} + 3\underset{\sim}{k}$, $\underset{\sim}{q} = 3\underset{\sim}{i} - 5\underset{\sim}{j} - 2\underset{\sim}{k}$

   c $\underset{\sim}{a} = 2\underset{\sim}{i} + 8\underset{\sim}{j} + 7\underset{\sim}{k}$, $\underset{\sim}{b} = 6\underset{\sim}{i} - 5\underset{\sim}{j} + 4\underset{\sim}{k}$ **d** $\underset{\sim}{u} = \begin{pmatrix} -1 \\ -3 \\ -2 \end{pmatrix}$, $\underset{\sim}{v} = \begin{pmatrix} 8 \\ -10 \\ 11 \end{pmatrix}$

   e $\underset{\sim}{r} = \begin{pmatrix} 4 \\ -3 \\ -5 \end{pmatrix}$, $\underset{\sim}{s} = \begin{pmatrix} -1 \\ 2 \\ -2 \end{pmatrix}$ **f** $\underset{\sim}{c} = \begin{pmatrix} 6 \\ 8 \\ 9 \end{pmatrix}$, $\underset{\sim}{d} = \begin{pmatrix} 5 \\ 3 \\ -6 \end{pmatrix}$

2 State, with reason, if the angle between these pairs of vectors is acute or obtuse.

   a $\underset{\sim}{a} = 3\underset{\sim}{i} + 2\underset{\sim}{j} - 4\underset{\sim}{k}$ and $\underset{\sim}{b} = 2\underset{\sim}{i} + 4\underset{\sim}{j} + \underset{\sim}{k}$ **b** $\underset{\sim}{q} = \begin{pmatrix} 2 \\ 3 \\ -5 \end{pmatrix}$ and $\underset{\sim}{r} = \begin{pmatrix} 2 \\ -8 \\ -1 \end{pmatrix}$

   c $\underset{\sim}{u} = -2\underset{\sim}{i} + 5\underset{\sim}{j} - 2\underset{\sim}{k}$ and $\underset{\sim}{v} = 4\underset{\sim}{i} + \underset{\sim}{j} + 6\underset{\sim}{k}$ **d** $\overrightarrow{AB} = \begin{pmatrix} 3 \\ -2 \\ -4 \end{pmatrix}$ and $\overrightarrow{AC} = \begin{pmatrix} -3 \\ -4 \\ -1 \end{pmatrix}$

★ 3 $P\,(2, -3, 5)$, $Q\,(2, 2, 8)$ and $R\,(3, -1, 13)$ are vertices of the triangle PQR. Show that the triangle is right-angled at Q.

★ 4 Show that the triangle with vertices $A\,(7, 5, -5)$, $B\,(7, 4, -3)$ and $C\,(6, 2, -4)$ is right-angled.

★ 5 Vectors $\underset{\sim}{u} = \begin{pmatrix} 2 \\ y \\ -1 \end{pmatrix}$ and $\underset{\sim}{v} = \begin{pmatrix} 3 \\ -4 \\ 4 \end{pmatrix}$ are perpendicular. Find the value of y.

6 Vectors $\underset{\sim}{a} = \begin{pmatrix} p \\ 3 \\ 2 \end{pmatrix}$ and $\underset{\sim}{b} = \begin{pmatrix} 2 \\ -1 \\ 2p \end{pmatrix}$ are perpendicular. Find the value of p.

7 Vectors $\underset{\sim}{p} = \begin{pmatrix} 2x \\ -7 \\ 3 \end{pmatrix}$ and $\underset{\sim}{q} = \begin{pmatrix} x \\ x \\ -5 \end{pmatrix}$ are perpendicular. Find the possible values for x.

8 A, B, C and D are the points $(-3, 1, -1)$, $(-1, 4, 5)$, $(-2, 0, 3)$ and $(-1, 2, 8)$ respectively.

    a Find the components of $\overrightarrow{AB}$ and $\overrightarrow{CD}$.

    The vector $\begin{pmatrix} 1 \\ p \\ q \end{pmatrix}$ is perpendicular to both $\overrightarrow{AB}$ and $\overrightarrow{CD}$.

    b Find the values of p and q.

9 $\underset{\sim}{a} = \begin{pmatrix} 1 \\ k \\ -1 \end{pmatrix}$, $\underset{\sim}{b} = \begin{pmatrix} -3 \\ 3 \\ 3 \end{pmatrix}$

    Given that $\underset{\sim}{a} + \underset{\sim}{b}$ is perpendicular to $\underset{\sim}{a} - \underset{\sim}{b}$, find the possible values of k.

10 A is the point $(-3, 0, 1)$ and B is the point $(5, 2, 7)$. If a point C lies on the y-axis and angle ACB is a right angle, find the possible coordinates of C.

Working with the properties of the scalar product

There are two main properties of the scalar product:

1 $\underset{\sim}{a} \bullet \underset{\sim}{b} = \underset{\sim}{b} \bullet \underset{\sim}{a}$ The order of $\underset{\sim}{a}$ and $\underset{\sim}{b}$ does not matter.

2 $\underset{\sim}{a} \bullet (\underset{\sim}{b} + \underset{\sim}{c}) = \underset{\sim}{a} \bullet \underset{\sim}{b} + \underset{\sim}{a} \bullet \underset{\sim}{c}$ You can multiply out brackets in exactly the same way as for normal multiplication.

It is also useful to note that:

3 $\underset{\sim}{a} \bullet \underset{\sim}{a} = |a|^2$ The angle between any vector and itself is 0 and so
$\underset{\sim}{a} \bullet \underset{\sim}{a} = = |\underset{\sim}{a}||\underset{\sim}{a}|\cos\theta = |\underset{\sim}{a}||\underset{\sim}{a}| \times 1 = |\underset{\sim}{a}|^2$

Example 6.11

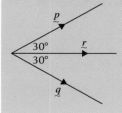

If $\left|\underset{\sim}{p}\right| = \left|\underset{\sim}{r}\right| = 3$ and $|q| = 4$

Calculate $\underset{\sim}{p} \bullet (\underset{\sim}{r} + \underset{\sim}{q})$

$\underset{\sim}{p} \bullet (\underset{\sim}{r} + \underset{\sim}{q}) = \underset{\sim}{p} \bullet \underset{\sim}{r} + \underset{\sim}{p} \bullet \underset{\sim}{q}$ **(P)** Multiply out the brackets.

$= \left|\underset{\sim}{p}\right|\left|\underset{\sim}{r}\right|\cos 30° + \left|\underset{\sim}{p}\right|\left|\underset{\sim}{q}\right|\cos 60°$ **(P)** Use the geometric formula. The angle between p and r is 30° and angle between $\underset{\sim}{p}$ and $\underset{\sim}{q}$ is 60°.

$= \left(3 \times 3 \times \frac{\sqrt{3}}{2}\right) + \left(3 \times 4 \times \frac{1}{2}\right)$

$= \frac{9\sqrt{3}}{2} + 6$

Example 6.12

The diagram shows a right-angled isosceles triangle with equal sides 2 units long.

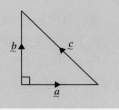

a Find the value of $\underset{\sim}{a} \bullet (\underset{\sim}{a} + \underset{\sim}{b} + \underset{\sim}{c})$

b Comment on your answer.

$$\underset{\sim}{a} \bullet (\underset{\sim}{a} + \underset{\sim}{b} + \underset{\sim}{c}) = \underset{\sim}{a} \bullet \underset{\sim}{a} + \underset{\sim}{a} \bullet \underset{\sim}{b} + \underset{\sim}{a} \bullet \underset{\sim}{c}$$

> Multiply out the brackets and **S P** then in order to use the geometric definition, find $|\underset{\sim}{a}|$, $|\underset{\sim}{b}|$ and $|\underset{\sim}{c}|$ and the angles between their respective directions.

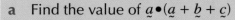

$\sqrt{(2^2 + 2^2)} = \sqrt{8}$

$$\underset{\sim}{a} \bullet (\underset{\sim}{a} + \underset{\sim}{b} + \underset{\sim}{c}) = |\underset{\sim}{a}||\underset{\sim}{a}|\cos 0° + |\underset{\sim}{a}||\underset{\sim}{b}|\cos 90° + |\underset{\sim}{a}||\underset{\sim}{c}|\cos 135°$$

$$= (2 \times 2 \times 1) + (2 \times 2 \times 0) + \left(2 \times \sqrt{8} \times \left(\frac{-1}{\sqrt{2}}\right)\right)$$

$$= 4 + 0 - 4$$

$$= 0$$

> **P**
> $2 \times \sqrt{8} \times \left(\dfrac{-1}{\sqrt{2}}\right) = -2 \times \sqrt{4} = -4$

b As neither $\underset{\sim}{a}$ nor $\underset{\sim}{a} + \underset{\sim}{b} + \underset{\sim}{c} = \underset{\sim}{0}$ it follows that $\underset{\sim}{a}$ is perpendicular to $\underset{\sim}{a} + \underset{\sim}{b} + \underset{\sim}{c}$.

Exercise 6E

1 Using the vectors $\underset{\sim}{a} = \begin{pmatrix} a_1 \\ a_2 \\ a_3 \end{pmatrix}$ and $\underset{\sim}{b} = \begin{pmatrix} b_1 \\ b_2 \\ b_3 \end{pmatrix}$ prove the two properties:

    1 $\underset{\sim}{a} \bullet \underset{\sim}{b} = \underset{\sim}{b} \bullet \underset{\sim}{a}$ **2** $\underset{\sim}{a} \bullet (\underset{\sim}{b} + \underset{\sim}{c}) = \underset{\sim}{a} \bullet \underset{\sim}{b} + \underset{\sim}{a} \bullet \underset{\sim}{c}$

★ 2 If $|\underset{\sim}{a}| = 5$ and $|\underset{\sim}{b}| = 3$ evaluate

    a $\underset{\sim}{a} \bullet (\underset{\sim}{a} + \underset{\sim}{b})$ **b** $(\underset{\sim}{a} + \underset{\sim}{b}) \bullet (\underset{\sim}{a} - \underset{\sim}{b})$ **c** $(\underset{\sim}{a} + 2\underset{\sim}{b}) \bullet (\underset{\sim}{a} - 3\underset{\sim}{b})$

3 $|\underset{\sim}{u}| = |\underset{\sim}{v}| = 5$ units. Calculate:

    a $\underset{\sim}{u} \bullet (\underset{\sim}{u} + \underset{\sim}{v})$ **b** $2\underset{\sim}{u} \bullet (\underset{\sim}{u} + 3\underset{\sim}{v})$

4 $|\underset{\sim}{p}| = 2$, and $|\underset{\sim}{q}| = 5$ while $\underset{\sim}{p} \bullet (\underset{\sim}{p} + \underset{\sim}{q}) = 9$.
Calculate the size of θ.

★ ⚙ **5** *ABC* is an equilateral triangle with sides 3 units long.

$\overrightarrow{AB}$, $\overrightarrow{AC}$ and $\overrightarrow{BC}$ are represented by the vectors $\underset{\sim}{u}$, $\underset{\sim}{v}$ and $\underset{\sim}{w}$ respectively.

Calculate $\underset{\sim}{u} \bullet (\underset{\sim}{u} + \underset{\sim}{v} + \underset{\sim}{w})$

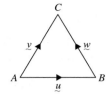

⚙ **6** *PQRSX* is a square-based pyramid with 8 equal edges, each of length 5 units.

Calculate:

   a $\underset{\sim}{a} \bullet (\underset{\sim}{a} + \underset{\sim}{b} + \underset{\sim}{c})$ **b** $\underset{\sim}{b} \bullet (\underset{\sim}{a} + \underset{\sim}{c})$

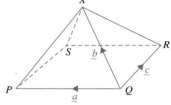

⚙ **7** $\underset{\sim}{a} = 2\underset{\sim}{u} + 3\underset{\sim}{v}$ where $\underset{\sim}{u}$ and $\underset{\sim}{v}$ are vectors of unit length inclined at an angle of 60° to each other. By first calculating $\underset{\sim}{a} \bullet \underset{\sim}{a}$ find the length of the vector $\underset{\sim}{a}$.

⚙ **8** In triangle *ABC*, $\overrightarrow{AB} = \underset{\sim}{u}$ and $\overrightarrow{BC} = \underset{\sim}{v}$.

If $(\underset{\sim}{u} + \underset{\sim}{v}) \bullet (\underset{\sim}{u} + \underset{\sim}{v}) = \underset{\sim}{u} \bullet \underset{\sim}{u} + \underset{\sim}{v} \bullet \underset{\sim}{v}$ show that triangle *ABC* is right angled at *B*.

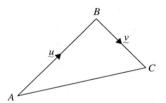

⚙ **9** **a**   Given that $\underset{\sim}{a} = \begin{pmatrix} 1 \\ 3 \\ 0 \end{pmatrix}$ and $\underset{\sim}{a} \bullet (\underset{\sim}{a} + \underset{\sim}{b}) = 7$, what is the value of $\underset{\sim}{a} \bullet \underset{\sim}{b}$?

   b   What does your answer to part **a** tell you about the angle between $\underset{\sim}{a}$ and $\underset{\sim}{b}$?

⚙ **10** In the trapezium *PQRS* the lengths of the sides *PQ* and *PS* are 5 and 4 units respectively. The vectors $\underset{\sim}{a}$, $\underset{\sim}{b}$ and $\underset{\sim}{c}$ are as shown on the diagram.

Evaluate $\underset{\sim}{a} \bullet (\underset{\sim}{b} + \underset{\sim}{c})$ and $\underset{\sim}{c} \bullet (\underset{\sim}{a} - \underset{\sim}{b})$.

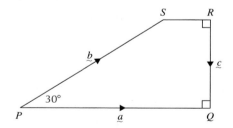

⚙ **11** *AC* is a diameter of a circle with centre O and *B* is any point on the circumference, so that $\overrightarrow{OA} = \underset{\sim}{a}$ and $\overrightarrow{OB} = \underset{\sim}{b}$.

By expressing the vectors $\overrightarrow{BA}$ and $\overrightarrow{BC}$ in terms of $\underset{\sim}{a}$ and $\underset{\sim}{b}$ show that ∠*ABC* is right-angled.

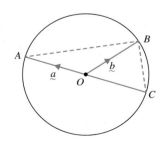

🏔 Challenge

In Chapter 2 the addition formulae were introduce without proof:

- $\sin(A+B) = \sin A \cos B + \cos A \sin B$
- $\cos(A+B) = \cos A \cos B - \cos A \sin B$
- $\sin(A-B) = \sin A \cos B - \cos A \sin B$
- $\cos(A-B) = \cos A \cos B + \cos A \sin B$

They can now be proved using vectors.

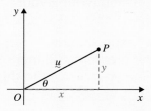

Consider a unit vector $\underset{\sim}{u} = \begin{pmatrix} x \\ y \end{pmatrix}$ which makes an angle of $\theta°$ with the positive x-axis then

- horizontal displacement $= x = \cos\theta$
- vertical displacement $= y = \sin\theta$

so P has coordinates $P(\cos\theta°, \sin\theta°)$

Let $\underset{\sim}{a}$ and $\underset{\sim}{b}$ be unit vectors which make angles of $\theta°$ and $\alpha°$ respectively with the x-axis as shown, meaning that P is the point $(\cos\theta°, \sin\theta°)$ and Q is the point $(\cos\alpha°, \sin\alpha°)$.

$\angle POQ = \theta - \alpha$

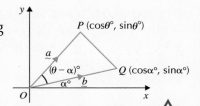

1 a Write the vector $\overrightarrow{PQ}$ in component form and hence calculate the length of PQ.

b Using the cosine rule show that
$\cos(\theta - \alpha)° = \cos\theta°\cos\alpha° + \sin\theta°\sin\alpha°$.

> ⚠ Remember $|\underset{\sim}{a}| = |\underset{\sim}{b}| = 1$ as a and B are unit vectors.

2 By letting $\cos(\theta + \alpha)° = \cos(\theta - (-\alpha))°$ find a similar formula for $\cos(\theta + \alpha)°$.

3 By letting $\sin(\theta + \alpha)° = \cos(90 - (\theta + \alpha))° = \cos((90 - \theta) - \alpha)°$ find a formula for $\sin(\theta + \alpha)°$.

4 Finally derive a formula for $\sin(\theta - \alpha)°$.

- I can work with the zero vector. ★ Exercise 6A Q1
- I can calculate the scalar product of two vectors using either the geometric or algebraic form of the definition. ★ Exercise 6B Q1, 2, 3
- I can calculate the angle between two vectors using the scalar product. ★ Exercise 6C Q1, 2
- I can work with perpendicular vectors. ★ Exercise 6D Q3, 4
- I can work with the algebraic properties of the scalar product. ★ Exercise 6E Q2, 5

For further assessment opportunities, see the Preparation for Assessment for Unit 1 on pages 132–135.

Unit 1 Expressions and Functions

The questions in this section cover the minimum competence for the content of the course in Unit 1. They are a good preparation for your unit assessment. In an assessment you will get full credit only if your solution includes the appropriate steps of process and accuracy. So make sure you show your thinking when writing your answers.

Remember that reasoning questions marked with the symbol ⚙ expect you to interpret the situation, select an appropriate strategy to solve the problem and then clearly explain your solution. If a context is given you must relate your answer to this context.

The use of a calculator is allowed.

Applying algebraic skills to logarithms and exponentials (Chapter 1)

1 Write $64 = 2^6$ as a logarithm.

2 Write $\log_5 125 = 3$ in index form.

3 Simplify $\log_e 6w + \log_e 2t$.

4 Express $\log_3 p^6 - \log_3 p^2$ as $a \log_3 p$

5 Express $\log_{10} \dfrac{3r}{5}$ as $\log_{10} b - \log_{10} c$

6 Simplify $\log_2 \dfrac{1}{8}$

7 Solve $\log_2(x - 2) = 5$

8 Solve $\log_8 16 + \log_8 4 = x - 4$

9 Solve $3^{x+2} = 81$

Applying trigonometric skills to manipulating expressions (Chapter 2)

10 Express $5 \cos x + 12 \sin x$ in the form $k \cos (x - a)°$ where $k > 0$ and $0 \le a \le 360$.

11 Express $4 \sin x + 3 \cos x$ in the form $k \sin (x + a)°$ where $k > 0$ and $0 \le a \le 360$.

12 Express $2 \cos x + 7 \sin x$ in the form $k \cos (x + a)$ where $k > 0$ and $0 \le a < 2\pi$.

13 Express $3 \sin x + 5 \cos x$ in the form $k \sin (x - a)$ where $k > 0$ and $0 \le a < 2\pi$.

14 Prove that $\dfrac{\sin(A + B)}{\cos A \cos B} = \tan A + \tan B$.

15 In the diagram $ABCD$ is a rectangle. Find the exact value of $\sin(x + y)$.

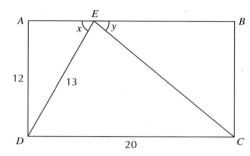

Applying algebraic and trigonometric skills to functions (Chapters 3 and 4)

16 Sketch the following graphs, clearly showing the maximum and minimum values and the x- and y-intercepts.

a $y = 3 \sin (2x)$, for $0 \leq x \leq 360$

b $y = k \cos \left(x + \dfrac{\pi}{2} \right)$, where $k > 0$ and $0 \leq x < 2\pi$

17 The diagram shows the graph of the function $y = f(x)$. The points A, B and C have coordinates $(0, 0)$, $(2, 4)$ and $(3, 0)$ respectively.

a Sketch the graph of the function $y = f(x - 3) + 3$

b Sketch the graph of the function $y = \frac{1}{2} f(4x)$

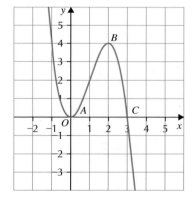

18 The diagram shows the graph of the function $y = a^x$. Find the value of a.

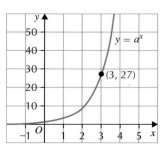

19 The diagram shows the graph of $y = a \sin(bx + c)$.
What are the values of a, b and c?

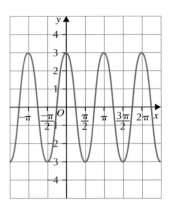

20 The diagram shows the graph of the function
$y = \log_a(x + b)$. Determine the values of a and b.

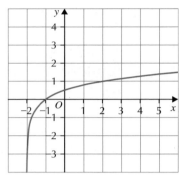

21 The functions f and g are defined by:

$f(x) = \sqrt{x - 2}$ where $x \geq 2$ and $g(x) = x^2 + 3$ where $x \geq 0$

a Find an expression for $g(f(x))$.

b State the domain of $g(f(x))$.

c State the range of $g(f(x))$.

22 The function $h(x)$ is defined as $f(g(x))$.

a If $f(x) = 3x^2$ and $g(x) = \sin x$ find $h(x)$.

b State the range of the function h.

23 The functions f and g are defined by:

$f(x) = 2x - 3$ and $g(x) = \frac{1}{2}(x + 3)$

a Find $f(g(x))$.

b What does your answer to part **a** tell you about the relationship between the functions f and g?

24 Find the inverse function of $\dfrac{2}{x + 7}$ where $x \neq -7$.

Applying geometric skills to vectors (Chapters 5 and 6)

25 *OABCD* is a square based pyramid with coordinates as shown. Find the vector $\overrightarrow{DB}$.

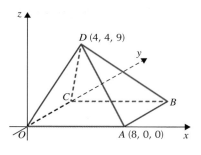

26 The aeroplanes in the diagram have been directed to remain at an angle of no less than 30° relative to each other.

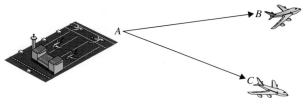

The coordinates of *A*, *B* and *C* are $(2, -5, 0)$, $(3, -3, 8)$ and $(-1, -7, 10)$.
Calculate if the pilots have been successful.

27 Three wind turbines are to be built at points *A*, *B* and *C*, placed on a straight line from *O*. A surveyor measures the positions from *O* before building work starts. He measures the position vectors as: $\overrightarrow{OA} = 3i + 8j - k$; $\overrightarrow{OB} = 6i + 4j - 5k$; $\overrightarrow{OC} = 12i - 4j + 23k$.

a Are the turbines marked correctly in a straight line? You must justify your answer.

b Find the ratio *AB*:*BC*

28 The points *S*, *T* and *U* lie on a straight line. If *T* divides the line in the ratio 5:2 find the coordinates of *T*.

7 Solving algebraic equations

This chapter will show you how to:

- factorise a **cubic polynomial** with a unitary x^3 coefficient
- factorise a cubic or quartic polynomial in which the highest powered term has a non-unitary coefficient
- use the **remainder theorem**
- use the remainder theorem and the **factor theorem** to determine unknown coefficients in a polynomial
- determine the points of intersection of the graph of a polynomial with the x- and y-axes
- determine the function of a polynomial from its graph
- solve a cubic polynomial equation with a unitary coefficient
- solve a polynomial equation of degree ≤ 4 in which the highest powered term has a non-unitary coefficient
- find the coordinates of the point(s) of intersection of a straight line and a curve or of two curves
- use the discriminant to determine an unknown coefficient in a quadratic equation given the nature of the roots

You should already know:

- how to factorise an algebraic expression using the Highest Common Factor (HCF)
- how to evaluate $f(x)$ at $x = h$
- how to complete the square
- how to solve quadratic equations by factorising, completing the square and using the quadratic formula
- how to use the discriminant to determine the nature of the roots of a quadratic equation

Introduction to polynomials

A polynomial is a sum or difference of algebraic terms such as:

- a quadratic $3x^2 - 5$
- a cubic $2x^3 - 3x^2 + x + 1$
- a quartic $x^4 - x^2 + 3$

All indices must be **whole** numbers.

You should also be able to identify the **coefficient** of an algebraic term. A coefficient is a number (or sometimes a letter) multiplying a variable. For example, in the term $3x^2$ the coefficient of x^2 is 3 and in ax^3, the coefficient of x^3 is a.

The **degree** of a polynomial is the value of the highest power:

- $3x^2 - 5$ is of degree 2, and the coefficient of x^2 is 3
- $x^5 - 4x^3 + x^2 - 4$ is of degree 5, and the coefficient of x^3 is -4
- $4 - 3x + x^7$ is of degree 7.

Often, a polynomial will contain a **constant** term, which is a number without a letter. In the quadratic $3x^2 - 5$, the constant is -5.

You will be expected to factorise polynomials of degree ≤ 4 in this course.

Factorising is an extremely useful skill and is essential in curve sketching. Previously, you have factorised to identify the roots of a quadratic. We will extend these factorising skills to higher degree polynomials, in particular, cubics and quartics.

Example 7.1

$f(x) = x^2 - 4x + 3$

Where does the curve $y = f(x)$ cut the x-axis?

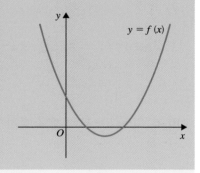

On the x-axis, $y = 0 \therefore f(x) = 0$

$x^2 - 4x + 3 = 0$

$(x - 1)(x - 3) = 0$

$x - 1 = 0$ and $x - 3 = 0$

$x = 1$ and $x = 3$

The curve crosses the x-axis at $x = 1$ and $x = 3$

Note: $f(1) = 0$ and $f(3) = 0$.

The factor theorem

In Example 1.1 we used the fact that $f(x) = 0$ on the y-axis to identify the factors and roots of the quadratic expression.

In general:

If $f(a) = 0$ then $x - a$ is a factor of $f(x)$.

The converse is also true:

If $x - a$ is a factor of $f(x)$ then $f(a) = 0$.

Example 7.2

Is $x - 1$ a factor of $f(x) = x^3 - 3x^2 + 3x - 1$?

$f(x) = x^3 - 3x^2 + 3x - 1$ ———— If $x - 1$ is a factor of $f(x)$, $f(1)$ should $= 0$.

$f(1) = 1^3 - 3(1)^2 + 3(1) - 1$

$\quad = 1 - 3 + 3 - 1$

$\quad = 0$

Since $f(1) = 0$, $x - 1$ is a factor of $f(x)$ ——— This line of communication is essential. Ⓒ

Example 7.3

Is $x + 2$ a factor of $f(x) = x^3 - 2x^2 - x + 2$?

$f(x) = x^3 - 2x^2 - x + 2$

$f(-2) = (-2)^3 - 2(-2)^2 - (-2) + 2$ ———— Note that $x + 2 = x - (-2)$. Ⓟ

$\quad = -8 - 8 + 2 + 2$

$\quad = -12$

Since $f(-2) \neq 0$, $x + 2$ is not a factor of $f(x)$

Exercise 7A

1 Which of the following expressions represent a polynomial?

   a $4x + 1$
   b $3x^4 - 5x^2 + 2x$
   c $\sqrt{x^2 + 2}$

   d $5 - a + a^3$
   e $a^{\frac{3}{2}} + 2a - 1$
   f $x(x + 2)(x + 3)$

2 State the degree of each polynomial:

   a $x^3 + 2x^2 - 3x + 4$
   b $3 - 4x$
   c $6x^5 - x^7$

   d $4t^5 + 3t^4 - 3t + 1$
   e $x(4x + 3)$

3 Is $x - 3$ a factor of $f(x) = x^2 - 7x + 12$?

4 Is $x + 4$ a factor of $f(x) = x^3 + 2x^2 - 2x$?

5 Is $x + 2$ a factor of $f(x) = x^3 - 6x^2 + 11x - 6$?

6 Is $x - 1$ a factor of $f(x) = x^3 + 8x^2 - x - 8$?

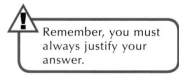

Remember, you must always justify your answer.

7 Determine whether or not $x + 2$ is a factor of $f(x) = 2x^3 + 3x^2 - 2x - 1$

8 Determine whether or not $x - 1$ is a factor of $f(x) = 2x^5 + 3x^2 + 2x + 1$

9 Determine whether or not $x - \frac{1}{2}$ is a factor of $f(x) = 4x^3 + x^2 - x - \frac{1}{4}$

Factorising a cubic polynomial

A cubic polynomial is one which can be expressed in the standard form:

$ax^3 + bx^2 + cx + d$ where $a \neq 0$

This is a polynomial of degree 3, expressed in decreasing powers of x ($cx = cx^1$ and $d = dx^0$).

If $a = 1$ then we describe the expression as having a **unitary coefficient** of x^3.

To factorise a cubic polynomial, we need to express it as a product of irreducible factors, usually $(x - a)(x - b)(x - c)$. Sometimes we may obtain an irreducible quadratic factor and so the cubic polynomial will factorise to $(x - a)(x^2 + bx + c)$

Being able to factorise a polynomial is an important skill which will be required again when you start solving polynomial equations, curve sketching (Chapter 10), and when identifying the points of intersection of two functions (Chapters 12 and 17). Three methods for factorising polynomials can be used:

- factorising by long division
- factorising by synthetic division
- factorising by inspection

You will have opportunities to practise all three methods. Discuss with your teacher the most appropriate method for you.

Factorising by long division

If you go on to study further mathematics at Advanced Higher level or beyond, factorising by long division will be an essential skill as you will be required to divide more complex polynomial expressions, particularly in work on curve sketching.

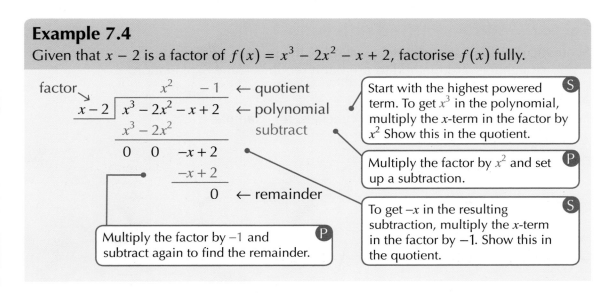

Example 7.4

Given that $x - 2$ is a factor of $f(x) = x^3 - 2x^2 - x + 2$, factorise $f(x)$ fully.

factor

$$x^2 \qquad -1 \qquad \leftarrow \text{quotient}$$
$$x - 2 \;\overline{)\; x^3 - 2x^2 - x + 2} \qquad \leftarrow \text{polynomial}$$
$$\underline{x^3 - 2x^2} \qquad\qquad \text{subtract}$$
$$0 \quad 0 \quad -x + 2$$
$$\underline{-x + 2}$$
$$0 \qquad \leftarrow \text{remainder}$$

Start with the highest powered term. To get x^3 in the polynomial, multiply the x-term in the factor by x^2 Show this in the quotient. **S**

Multiply the factor by x^2 and set up a subtraction. **P**

To get $-x$ in the resulting subtraction, multiply the x-term in the factor by -1. Show this in the quotient. **S**

Multiply the factor by -1 and subtract again to find the remainder. **P**

(continued)

The remainder is 0. This confirms that $x - 2$ is a factor.

The **quotient** $x^2 - 1$ is the result of the division.

So we obtain:

$$f(x) = (x - 2)(x^2 - 1)$$

$$= (x - 2)(x - 1)(x + 1)$$

Factorise the quadratic using the difference of two squares. **P**

Example 7.5

Given that $x + 1$ is a factor of $f(x) = x^3 + 3x^2 - 13x - 15$, factorise $f(x)$ fully.

$$\begin{array}{r} x^2 + 2x - 15 \\ x + 1 \overline{\smash{\big)}\ x^3 + 3x^2 - 13x - 15} \\ \underline{x^3 + x^2} \\ 0 \quad 2x^2 - 13x - 15 \\ \underline{2x^2 + 2x} \\ 0 \quad -15x - 15 \\ \underline{-15x - 15} \\ 0 \end{array}$$

Multiply the x-term in the factor by x^2 to get x^3 in the polynomial. Show x^2 in the quotient. **S**

Multiply the factor by x^2 and set up a subtraction **P**

To get $2x^2$ in the resulting subtraction, multiply the x-term in the factor by $2x$. Show this in the quotient. **S**

Multiply the factor by the $2x$ and set up a subtraction **P**

Multiply the factor by -15 **P**

So:

$$f(x) = (x + 1)(x^2 + 2x - 15)$$

$$= (x + 1)(x + 5)(x - 3)$$

Factorising by synthetic division

Synthetic division uses 'nesting' to simplify the process of dividing, or factorising. It is a very straightforward method to use but it requires some explanation.

Nested form

Polynomials can be rewritten using brackets, for example:

$$f(x) = ax^3 + bx^2 + cx + d$$

$$= x\left[ax^2 + bx + c\right] + d$$

$$= x\left[x(ax + b) + c\right] + d$$

which we can rearrange to

$$= \left[(ax + b)x + c\right]x + d$$

It can be useful to have a polynomial expressed in this form in order to evaluate $f(x)$ at a point. We can also use this method to factorise polynomials.

Example 7.6

$f(x) = x^3 + 7x^2 + 14x + 8$ evaluate $f(3)$.

If $f(x) = x^3 + 7x^2 + 14x + 8$, we can rewrite $f(x)$ as

$f(x) = x\left[x^2 + 7x + 14\right] + 8$

$\qquad = x\left[x(x + 7) + 14\right] + 8$

$\qquad = \left[(x + 7)x + 14\right]x + 8$

$f(3) = \left[(3 + 7) \times 3 + 14\right] \times 3 + 8$

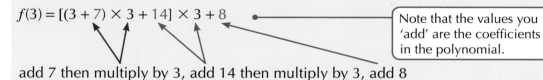

> Note that the values you 'add' are the coefficients in the polynomial. **C**

add 7 then multiply by 3, add 14 then multiply by 3, add 8

$= \left[10 \times 3 + 14\right] \times 3 + 8$

$= 44 \times 3 + 8$

$= 132 + 8$

$= 140$

We can write this more easily using a table, starting with the degree of the polynomial down to the constant term:

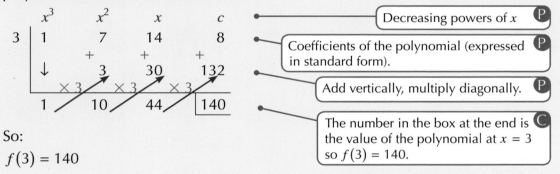

> Decreasing powers of x **P**

> Coefficients of the polynomial (expressed in standard form). **P**

> Add vertically, multiply diagonally. **P**

> The number in the box at the end is the value of the polynomial at $x = 3$ so $f(3) = 140$. **C**

So:

$f(3) = 140$

Example 7.7

Given that $x - 2$ is a factor of $f(x) = x^3 - 2x^2 - x + 2$, factorise $f(x)$ fully.

Looking back at the factor theorem, we know that if $x - a$ is a factor of $f(x)$ then $f(a) = 0$.

If $x - 2$ is a factor then $f(2) = 0$

$$
\begin{array}{c|cccc}
 & x^3 & x^2 & x & c \\
\hline
2 & 1 & -2 & -1 & 2 \\
 & \downarrow & 2 & 0 & -2 \\
\hline
 & 1 & 0 & -1 & \boxed{0} \\
 & x^2 & x & c &
\end{array}
$$

> You must always communicate the solution fully. **C**

since $f(2) = 0$, $x - 2$ is a factor and $x = 2$ is a root

(continued)

The values at the bottom to the left of the box give the coefficients of the remaining polynomial factor, or the **quotient** (the result of the division). This polynomial is one degree lower than the one with which we started.

So:

$$f(x) = x^3 - 2x^2 - x + 2$$
$$= (x - 2)(x^2 + 0x - 1)$$

1st factor   Quadratic factor with coefficients from the bottom row of the table

$$= (x - 2)(x^2 - 1)$$
$$= (x - 2)(x - 1)(x + 1)$$

Example 7.8

Given that $x + 1$ is a factor of $f(x) = x^3 + 3x^2 - 13x - 15$, factorise $f(x)$ fully.

If $x + 1$ is a factor, then $f(-1) = 0$

	x^3	x^2	x	c
-1	1	3	-13	-15
$\downarrow$		-1	-2	15
	1	2	-15	0
	x^2	x	c	

since $f(-1) = 0$, $x + 1$ is a factor.

So:

$$f(x) = x^3 + 3x^2 - 13x - 15$$

$$= (x + 1)(x^2 + 2x - 15)$$

$$= (x + 1)(x + 5)(x - 3)$$

Example 7.9

Show that $x - 2$ is a factor of $f(x) = x^3 - 7x + 6$ and factorise $f(x)$ fully.

If $x - 2$ is a factor, then $f(2) = 0$

	x^3	x^2	x	c
2	1	0	-7	6
$\downarrow$		2	4	-6
	1	2	-3	0
	x^2	x	c	

There is no x^2 term in the polynomial, so use a 0 in the x^2 column. **P**

since $f(2) = 0$, $x - 1$ is a factor.

So:
$$f(x) = x^3 - 7x + 6$$
$$= (x-2)(x^2 + 2x - 3)$$
$$= (x-2)(x+3)(x-1)$$

Factorising by inspection

Inspection works on the same principle as long division but we write it more concisely:

Example 7.10

Given that $x - 2$ is a factor of $f(x) = x^3 - 2x^2 - x + 2$, factorise $f(x)$ fully.

$$f(x) = x^3 - 2x^2 - x + 2$$
$$= (x-2)(x^2 \ldots\ldots$$

$x^3 \quad -2x^2$

> **P** Find the term needed to give x^3 when multiplied by x: x^2. So x^2 goes into the second bracket. Multiplying through gives $x^3 - 2x^2$. We now need to determine the remaining terms.

$$= (x-2)(x^2 - 1)$$

$-x \quad +2$

So:
$$f(x) = (x-2)(x-1)(x+1)$$

Example 7.11

Given that $x + 1$ is a factor of $f(x) = x^3 + 3x^2 - 13x - 15$, factorise $f(x)$ fully.

$$f(x) = x^3 + 3x^2 - 13x - 15$$
$$= (x+1)(x^2 \ldots$$

$x^3 \quad +x^2$

> **P** We require x^2 in the second bracket to obtain x^3.

$$= (x+1)(x^2 + 2x \ldots$$

$x^3 \quad +x^2 \quad +2x^2 \quad +2x$

> **P** Multiplying by x^2 only gives $+1x^2$ so multiply the x in the factor by $2x$ in order to give a total of $3x^2$.

(continued)

$$= (x + 1)(x^2 + 2x - 15)$$

Finally, we need to ensure that we have $-13x$ **P**

$$x^3 + x^2 + 2x^2 + 2x - 15x - 15 = x^3 + 3x^2 - 13x - 15$$

So:

$$f(x) = (x + 1)(x + 5)(x - 3)$$

Exercise 7B

1 Show that $x - 3$ is a factor of $f(x) = x^3 - 8x^2 + 19x - 12$ and factorise fully.

2 Show that $x + 4$ is a factor of $f(x) = x^3 + 4x^2 - x - 4$ and factorise fully.

★ 3 Show that $x - 2$ is a factor of $f(x) = x^3 - 6x^2 + 11x - 6$ and factorise fully.

4 Show that $x - 1$ is a factor of $f(x) = x^3 + 8x^2 - x - 8$ and factorise fully.

5 Show that $x + 3$ is a factor of $f(x) = x^3 - 13x - 12$ and factorise fully.

6 Show that $x + 2$ is a factor of $f(x) = x^4 + 5x^3 + 2x^2 - 8x$ and express $f(x)$ in factorised form.

7 Prove that $x - 3$ is a factor of $f(x) = x^3 - 2x^2 - 9x + 18$ and find all of the factors of $f(x)$.

8 Prove that $x + 5$ is a factor of $f(x) = x^4 + x^3 - 16x^2 + 20x$ and express $f(x)$ in factorised form.

Identifying factors and factorising polynomials

So far, we have been given a factor of each polynomial to help us to begin factorising. This will not always be the case. When we factorise polynomials without being given a factor, we always start by looking at factors of the constant.

Example 7.12
Factorise $f(x) = x^3 - 7x - 6$.

$$f(x) = x^3 - 7x - 6$$

Look at factors of 6 and use trial and error to identify **S** roots and factors of $f(x)$: ± 1, ± 2, ± 3, ± 6

$$f(1) = 1^3 - 7(1) - 6 = 1 - 7 - 6 = -12.$$
$$f(1) \neq 0, \text{ so } x - 1 \text{ is not a factor.}$$

Use the factor theorem to **P** determine a factor.

$$f(-1) = (-1)^3 - 7(-1) - 6 = -1 + 7 - 6 = 0.$$
$$f(-1) = 0, \text{ so } x + 1 \text{ is a factor.}$$

Use synthetic division to find factors. Use $f(-1)$:

$$
\begin{array}{c|cccc}
 & x^3 & x^2 & x & c \\
-1 & 1 & 0 & -7 & -6 \\
 & \downarrow & -1 & 1 & 6 \\
\hline
 & 1 & -1 & -6 & 0 \\
 & x^2 & x & c &
\end{array}
$$

since $f(-1) = 0$, $x + 1$ is a factor.

$$
\begin{aligned}
f(x) &= x^3 - 7x - 6 \\
&= (x + 1)(x^2 - x - 6) \\
&= (x + 1)(x + 2)(x - 3)
\end{aligned}
$$

Use inspection to find factors:

$$
\begin{aligned}
f(x) &= x^3 - 7x - 6 \\
&= (x + 1)(x^2 - x - 6)
\end{aligned}
$$

$$x^3 \quad +x^2 \quad -x^2 \quad -x \quad -6x \quad -6$$

$$= (x + 1)(x + 2)(x - 3)$$

Exercise 7C

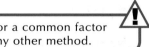
Always check for a common factor before trying any other method.

1 Fully factorise each expression.

   a $f(x) = x^3 - x^2 - x + 1$ **b** $f(x) = x^3 + x^2 - x - 1$

   c $f(x) = x^3 + 3x^2 - x - 3$ **d** $f(x) = x^3 + 3x^2 - 9x + 5$

   e $f(x) = x^3 - 2x^2 - 4x + 8$ **f** $f(x) = x^3 - 6x^2 + 11x - 6$

2 Express each polynomial as a product of its factors.

   a $f(x) = x^3 + 2x^2 - 9x - 18$ **b** $f(x) = x^3 + 5x^2 + 2x - 8$

   c $f(x) = x^3 - 4x^2 + x + 6$ **d** $f(x) = x^3 - 13x + 12$

   e $f(x) = x^3 + 2x^2 - 19x - 20$ **f** $f(x) = x^3 - 19x + 30$

3 Express $f(x)$ in factorised form.

   a $f(x) = x^3 - 4x^2 + 3x$ **b** $f(x) = x^3 - x$

   c $f(x) = x^4 - 3x^3 - 6x^2 + 8x$ **d** $f(x) = x^4 - 2x^2 + 1$

Factorising a cubic or quartic polynomial

So far all of the cubic polynomials we have tackled have had an x^3 term with a coefficient of 1. This will not always be the case. We will now consider polynomials of degree ≤ 4 with non-unitary coefficients. Later in this chapter, and in Chapter 10, we will factorise polynomials to interpret graphs or to solve problems in context. The same techniques can also be applied to higher degree polynomials.

Example 7.13

Factorise $f(x) = 2x^3 + x^2 - 2x - 1$

$f(x) = 2x^3 + x^2 - 2x - 1$ •————— Determine a factor of $f(x)$.

$f(1) = 2(1)^3 + (1)^2 - 2(1) - 1 = 0 \therefore$ since $f(1) = 0$, then $x - 1$ is a factor.

By synthetic division

$f(x) = 2x^3 + x^2 - 2x - 1$

Using $f(x)$:

$$
\begin{array}{c|cccc}
 & x^3 & x^2 & x & c \\
\hline
1 & 2 & 1 & -2 & -1 \\
 & \downarrow & 2 & 3 & 1 \\
\hline
 & 2 & 3 & 1 & 0 \\
\end{array}
$$

since $f(1) = 0$, $x - 1$ is a factor.

$f(x) = 2x^3 + x^2 - 2x - 1$

$\quad = (x - 1)(2x^2 + 3x + 1)$

$\quad = (x - 1)(2x + 1)(x + 1)$

By inspection

$f(x) = 2x^3 + x^2 - 2x - 1$

$\quad = (x - 1)(2x^2 \ldots$

$2x^3 \; -2x^2$

$\quad = (x - 1)(2x^2 + 3x \ldots$

$+3x^2 \; -3x$

$\quad = (x - 1)(2x^2 + 3x + 1)$

$+x \; -1$

$\quad = (x - 1)(2x + 1)(x + 1)$ •————— As the polynomial is non-unitary, we obtain at least one non-unitary factor.

Example 7.14

$f(x) = 2x^4 + x^3 - 6x^2 + x + 2$

Determine a factor of $f(x)$.

$f(x) = 2x^4 + x^3 - 6x^2 + x + 2$

$f(1) = 2(1)^4 + (1)^3 - 6(1)^2 + 1 + 2 = 0$ ∴ since $f(1) = 0$, $x - 1$ is a factor.

By synthetic division

$f(x) = 2x^4 + x^3 - 6x^2 + x + 2$

Using $f(1)$:

$$
\begin{array}{c|ccccc}
 & x^4 & x^3 & x^2 & x & c \\
1 & 2 & 1 & -6 & 1 & 2 \\
 & \downarrow & 2 & 3 & -3 & -2 \\
\hline
 & 2 & 3 & -3 & -2 & \boxed{0} \\
 & x^3 & x^2 & x & c &
\end{array}
$$ since $f(1) = 0$, $x - 1$ is a factor.

$f(x) = (x - 1)(2x^3 + 3x^2 - 3x - 2)$

Try $f(1)$:

$2(1)^3 + 3(1)^2 - 3(1) - 2 = 0$ so $x - 1$ is a factor again

$$
\begin{array}{c|cccc}
 & x^3 & x^2 & x & c \\
1 & 2 & 3 & -3 & -2 \\
 & \downarrow & 2 & 5 & 2 \\
\hline
 & 2 & 5 & 2 & \boxed{0}
\end{array}
$$

Repeat the synthetic division for the cubic factor. (P)

since $f(1) = 0$, $x - 1$ is a factor.

$f(x) = (x - 1)(x - 1)(2x^2 + 5x + 2)$

$ = (x - 1)(x - 1)(2x + 1)(x + 2)$

By inspection

$f(x) = 2x^4 + x^3 - 6x^2 + x + 2$

$ = (x - 1)(2x^3 \ldots$

$ = (x - 1)(2x^3 + 3x^2 \ldots$

$ = (x - 1)(2x^3 + 3x^2 - 3x \ldots$

$ = (x - 1)\left(2x^3 + 3x^2 - 3x - 2\right)$

Find a factor of the cubic, i.e. substitute values into $2x^3 + 3x^2 - 3x - 2$ to find another factor. (S)

Try $x = 1$:

$ = (x - 1)\left[(x - 1)\left(2x^2 + 5x + 2\right)\right]$

$ = (x - 1)(x - 1)(2x + 1)(x + 2)$

Example 7.15

Factorise $f(x) = x^4 - 7x^2 - 18$

Determine a factor of $f(x)$:

$f(x) = x^4 - 7x^2 - 18$ ● ───────────────────── | Try factors of 18. | S

Try $f(3)$:

$f(3) = 3^4 - 7(3)^2 - 18 = 81 - 63 - 18 = 0$ ∴ since $f(3) = 0$, $x - 3$ is a factor.

By synthetic division

$f(x) = x^4 - 7x^2 - 18$

Using $f(3)$:

$$
\begin{array}{c|ccccc}
 & x^4 & x^3 & x^2 & x & c \\
\hline
3 & 1 & 0 & -7 & 0 & -18 \\
 & \downarrow & 3 & 9 & 6 & 18 \\
\hline
 & 1 & 3 & 2 & 6 & \boxed{0} \\
 & x^3 & x^2 & x & c
\end{array}
$$

since $f(3) = 0$, $x - 3$ is a factor.

$f(x) = (x - 3)(x^3 + 3x^2 + 2x + 6)$

Try $f(-3)$:

$$
\begin{array}{c|cccc}
 & x^3 & x^2 & x & c \\
\hline
-3 & 1 & 3 & 2 & 6 \\
 & \downarrow & -3 & 0 & -6 \\
\hline
 & 1 & 0 & 2 & \boxed{0}
\end{array}
$$

● ───── | Repeat the synthetic division for the cubic factor. | P

since $f(-3) = 0$, $x + 3$ is a factor.

$f(x) = (x - 3)(x + 3)\left(x^2 + 2\right)$ ● ───── | You must always justify that the quadratic factor is irreducible by calculating the discriminant or completing the square to prove that the factor is > 0 for all values of x. | C

Discriminant

For $x^2 + 2$, $a = 1$, $b = 0$, $c = 2$

$b^2 - 4ac = 0^2 - 4(1)(2)$

$\qquad = 0 - 8$

$\qquad < 0$ ∴ no real roots

Completing the square

$x^2 + 2 > 0$ for all values of x

Squared term   Plus a constant > 0

By inspection

$f(x) = x^4 - 7x^2 - 18$

$\quad = (x - 3)(x^3 \ldots$

$\quad = (x - 3)(x^3 + 3x^2 \ldots$

$\quad = (x - 3)(x^3 + 3x^2 + 2x \ldots$

$\quad = (x - 3)(x^3 + 3x^2 + 2x + 6)$

> You can substitute into either the original quartic polynomial or the cubic factor. In this example the original quartic had fewer terms to deal with so is perhaps easier to use. **P**

Try $f(-3)$: $f(-3) = (-3)^4 - 7(-3)^2 - 18 = 81 - 63 - 18 = 0 \therefore x + 3$ is a factor.

$\quad = (x - 3)(x + 3)(x^2 + 2)$

$f(x)$ is now fully factorised as $(x^2 + 2)$ is an irreducible quadratic.

> You must always justify that the quadratic factor is irreducible by calculating the discriminant or completing the square to prove that the factor is > 0 for all values of x. **C**

Discriminant

For $x^2 + 2$, $a = 1$, $b = 0$, $c = 2$

$b^2 - 4ac = 0^2 - 4(1)(2)$

$\quad\quad\quad = 0 - 8$

$\quad\quad\quad < 0 \therefore$ no real roots

Completing the square

$x^2 + 2 > 0$ for all values of x

Squared term   Plus a constant > 0

Exercise 7D

1 **a** Show that $2x - 1$ is a factor of $f(x) = 2x^3 + 11x^2 + 4x - 5$.

   b Hence, or otherwise, express $f(x)$ in factorised form.

2 Fully factorise each cubic:

   a $f(x) = 2x^3 + x^2 - 2x - 1$ **b** $f(x) = 3x^3 - 8x^2 + 3x + 2$

   c $f(x) = 2x^3 - 11x^2 + 17x - 6$ **d** $f(x) = 4x^3 + 3x^2 - 4x - 3$

   e $f(x) = 5x^3 - 3x^2 - 32x - 12$ **f** $f(x) = 4x^3 - 3x^2 - 49x - 12$

   g $f(x) = 6x^3 + x^2 - 5x - 2$ **h** $f(x) = 6x^3 + 11x^2 - 3x - 2$

   i $f(x) = x^3 + x^2 + x + 6$ **j** $f(x) = 2x^3 - 5x^2 + 6x - 3$

3 Fully factorise each quartic:

   a $f(x) = x^4 - 8x^2 + 16$ **b** $f(x) = x^4 - 2x^3 + 2x - 1$

   c $f(x) = x^4 - x^3 - 3x^2 + x + 2$ **d** $f(x) = x^4 - 5x^3 + x^2 + 21x - 18$

★ 4 Fully factorise each polynomial:

a $f(x) = 3x^4 - 3x^2$

b $f(x) = x^4 - 6x^3 + 11x^2 - 6x$

c $f(x) = x^4 - 5x^2 + 4$

d $f(x) = x^4 - 3x^3 - 7x^2 + 27x - 18$

e $f(x) = x^4 + 6x^3 + 4x^2 - 6x - 5$

f $f(x) = 2x^4 - 9x^3 + 5x^2 - 3x - 4$

g $f(x) = x^4 - x^2 - 12$

h $f(x) = 4x^4 + 4x^3 - 9x^2 - x + 2$

i $f(x) = 2x^4 - 2x^3 - 36x^2 + 32x + 64$

j $f(x) = x^4 - 2x^3 - 6x^2 + 6x + 9$

5 A company carves candles from cuboids of wax with whole number dimensions. The volume of a cuboid, in cm^3, is given by the function $V(x) = x^3 - 10x^2 + 13x + 24$ and the height of the cuboid is $x + 1\,cm$. Determine expressions for the other two dimensions of the cuboid.

6 The volume, in cm^3, of a square-based cuboid is modelled by the function $V(x) = x^3 + x^2 - 8x - 12$.

When the height of the cuboid is $x - 3\,cm$, calculate the difference between the breadth and the height.

7 As part of a fundraising project, pupils are selling chocolates in square-based pyramid boxes with a volume, in cm^3, given by the function $V(x) = \frac{1}{3}(3x^3 - 17x^2 + 21x + 9)$.

a Given that the height of the pyramid is $3x + 1\,cm$, determine an expression for the length of the base.

b Given that x is a whole number, state the minimum value that x can take and calculate the volume for this value of x.

Calculating unknown coefficients using the factor theorem

We can use the factor theorem to determine unknown coefficients in a polynomial.

Example 7.16

If $x + 2$ is a factor of $f(x) = x^3 + 2x^2 + kx - 2$, determine the value of k.

If $x + 2$ is a factor then $f(-2) = 0$

$$f(x) = x^3 + 2x^2 + kx - 2$$

$$f(-2) = (-2)^3 + 2(-2)^2 + k(-2) - 2 = 0$$

$$-8 + 8 - 2k - 2 = 0$$

$$-2k = 2$$

$$k = -1$$

Example 7.17

$(x - 1)$ and $(x + 3)$ are both factors of $f(x) = 2x^3 + bx^2 - 4x + d$. Determine the values of b and d and fully factorise $f(x)$

$f(x) = 2x^3 + bx^2 - 4x + d$

If $x - 1$ is a factor, then $f(1) = 0$

$2(1)^3 + b(1)^2 - 4(1) + d = 0$

$2 + b - 4 + d = 0$

$b + d = 2$

If $x + 3$ is a factor, then $f(-3) = 0$

$2(-3)^3 + b(-3)^2 - 4(-3) + d = 0$

$-54 + 9b + 12 + d = 0$

$9b + d = 42$

$9b + d = 42$

$b + d = 2$

$8b = 40$ ●————— Use simultaneous equations to calculate b and d. Ⓟ

$b = 5$

$d = -3$

$f(x) = 2x^3 + 5x^2 - 4x - 3$ ●

$= (x - 1)(2x^2 + 7x + 3)$

$= (x - 1)(2x + 1)(x + 3)$

You can use your preferred method to factorise at the end. Ⓟ

Exercise 7E

★ 1 $x - 1$ is a factor of $f(x) = x^3 + px^2 + 2x - 8$ Determine the value of p.

2 Find q if $x - 2$ is a factor of $f(x) = x^3 - 9x^2 + 24x - q$

3 $f(x) = 2kx^3 + kx^2 - 2x - 1$ has $x - 1$ as a factor. Determine the value of k and fully factorise $f(x)$.

4 Determine the value of a given that $x + 3$ is a factor of $f(x) = x^4 + ax^3 - 7x^2 - 43x + 42$. Hence, factorise $f(x)$.

5 Given that $2x + 3$ is a factor of $f(x) = 2x^3 + kx^2 - 2x - 3$, identify the value of k and factorise $f(x)$.

6 $x + 1$ and $x - 2$ are factors of $f(x) = ax^3 - 3x^2 - 3x + b$. Determine the values of a and b.

7 Given that $x + 3$ and $x - 2$ are factors of $f(x) = x^4 + 2x^3 - 7x^2 + px + q$ determine the values of p and q and factorise $f(x)$ fully.

The remainder theorem

We have been using the factor theorem throughout this chapter to identify factors of polynomials and this is an extension of the remainder theorem.

Consider the division $(x^3 + x^2 + x + 1) \div (x + 2)$

This can be performed using long division:

$$\begin{array}{r}
x^2 - x + 3 \quad \longleftarrow \text{Quotient} \\
x + 2 \enclose{longdiv}{x^3 + x^2 + x + 1} \quad \longleftarrow \text{Polynomial} \\
\end{array}$$

Divisor$\longrightarrow$

$$
\begin{array}{r}
x^3 + 2x^2 \\
\hline
-x^2 + x + 1 \\
-x^2 - 2x \\
\hline
3x + 1 \\
3x + 6 \\
\hline
-5 \quad \longleftarrow \text{Remainder}
\end{array}
$$

Solving gives a quotient of $x^2 - x + 3$, with a remainder of -5.

So:

$$(x^3 + x^2 + x + 1) \div (x + 2) = x^2 - x + 3 - \frac{5}{x + 2}$$

or:

$$x^3 + x^2 + x + 1 = (x + 2)(x^2 - x + 3) - 5$$

divisor   quotient   remainder

If we substitute $x = -2$ into the original polynomial:

$$
\begin{aligned}
f(-2) &= (-2)^3 + (-2)^2 + (-2) + 1 \\
&= -8 + 4 - 2 + 1 \\
&= -5
\end{aligned}
$$

Note that $f(-2)$ is equal to the remainder when we divide the polynomial by $x + 2$.

In general, when we divide a polynomial we obtain a quotient and a remainder:

$$f(x) \div (x - a) = q(x) + \frac{R}{x - a} \qquad \text{remainder}$$

quotient

which can be written as $f(x) = (x - a)q(x) + R$

So, when $x = a$, $f(a) = (a - a)q(x) + R$

$$
\begin{aligned}
&= 0 \times q(x) + R \\
&= R
\end{aligned}
$$

Which gives us the remainder theorem:

If a polynomial $f(x)$ is divided by $x - a$, the remainder is $f(a)$

This extends to division by any linear factor:

If a polynomial $f(x)$ is divided by $ax - b$, the remainder is $f\left(\frac{b}{a}\right)$

Example 7.18
Determine the remainder when $f(x) = x^3 + 2x^2 - x + 2$ is divided by $x - 1$.

Dividing $f(x)$ by $x - 1$ produces a remainder of $f(1)$.
$f(1) = 1 + 2 - 1 + 2 = 4 \therefore$ remainder $= 4$.

Example 7.19
Calculate the remainder when $f(x) = x^3 + x^2 - x + 2$ is divided by $2x + 1$.

Remainder $= f\left(-\frac{1}{2}\right)$ $\qquad$ $\left(\text{consider } 2x + 1 = 0 \Rightarrow x = -\frac{1}{2}\right)$

$f\left(-\frac{1}{2}\right) = \left(-\frac{1}{2}\right)^3 + \left(-\frac{1}{2}\right)^2 - \left(-\frac{1}{2}\right) + 2$

$\qquad = -\frac{1}{8} + \frac{1}{4} + \frac{1}{2} + 2$

$\qquad = 2\frac{5}{8}$

Example 7.20
Determine the quotient and remainder for $(x^3 + 3x^2 - x + 5) \div (x + 2)$

$(x^3 + 3x^2 - x + 5) \div (x + 2)$

By long division

$$
\begin{array}{r}
x^2 + x - 3 \\
x + 2 \enclose{longdiv}{x^3 + 3x^2 - x + 5} \\
\underline{x^3 + 2x^2} \\
x^2 - x \\
\underline{x^2 + 2x} \\
-3x + 5 \\
\underline{-3x - 6} \\
11
\end{array}
$$

Quotient $= x^2 + x - 3$
Remainder $= 11$

By synthetic division

$$
\begin{array}{c|cccc}
-2 & 1 & 3 & -1 & 5 \\
& \downarrow & -2 & -2 & 6 \\
\hline
& 1 & 1 & -3 & \boxed{11}
\end{array}
$$

Quotient $= x^2 + x - 3$
Remainder $= 11$

Example 7.21

When $f(x) = ax^3 - 2x^2 + x - 1$ is divided by $x - 2$, the remainder is 17. Determine the value of a.

On dividing $f(x) = ax^3 - 2x^2 + x - 1$ by $x - 2$, the remainder is 17, so $f(2) = 17$.

$f(2) = a(2)^3 - 2(2)^2 + 2 - 1$

$\therefore 17 = 8a - 8 + 2 - 1$

$24 = 8a$

$a = 3$

Exercise 7F

★ 1 Calculate the remainder when:

a $f(x) = x^3 - x^2 + x + 1$ is divided by $x - 1$

b $f(x) = 2x^3 + x^2 - 3x - 4$ is divided by $x + 2$

c $f(x) = x^4 + x^2 - 5$ is divided by $x - 3$

d $f(x) = 2x^3 - 8x + 3$ is divided by $2x - 1$

e $f(x) = 9x^3 - 6x + 1$ is divided by $3x + 1$

★ 2 Determine the value of a given that:

a $f(x) = x^3 + ax^2 - 3x + 1$ divided by $x + 1$ has a remainder of 5

b $f(x) = x^3 - x^2 + ax - 3$ divided by $x - 3$ has a remainder of -6

c $f(x) = ax^3 + 6x - 1$ divided by $2x - 1$ has a remainder of 3.

3 $f(x) = ax^3 - 2x^2 + 5x + b$ has a remainder of 5 when it is divided by $x - 1$ and a remainder of -11 when it is divided by $x + 1$. Determine the values of a and b.

4 $f(x) = x^4 + ax^3 - 7x^2 + bx + 6$ has a factor $x - 1$ and a remainder of -12 when divided by $x - 2$. Determine a and b.

★ 5 Determine the quotient and remainder for each polynomial division:

a $(x^3 + 5x^2 - 3x + 1) \div (x - 2)$ b $(x^3 + 2x^2 - 3x + 4) \div (x + 3)$

c $(2x^3 - x^2 + 3x + 6) \div (x + 1)$ d $(8x^3 + 4x^2 + 1) \div (2x - 1)$

Introduction to polynomial equations

Polynomials are often used to model real-life situations, such as modelling population growth, profit, production costs, designing rollercoasters or efficient packaging or simulating the flight of a basketball or an arrow.

The skills developed on factorising polynomials can be used to solve polynomial equations, interpret graphs and solve real-life problems.

Determining the points of intersection of the graph of a polynomial with the x- and y-axes

Factorisation is used to find the roots of a quadratic function. The roots are the points at which the graph of the function intersects the x-axis (i.e. $y = 0$). The same basic skills are used for higher degree polynomials in this chapter, and are then used with differentiation to sketch the graphs of polynomials, labelling the intersections with the axes and any stationary points (see Chapter 10).

The diagram (right) shows the intersections with the x-axis ($y = 0$) and the y-axis ($x = 0$), and the stationary points of a cubic function.

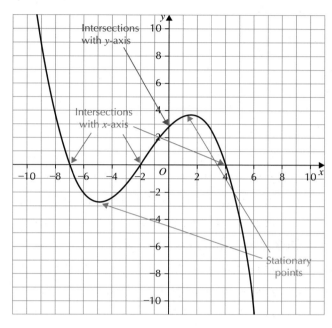

When examining the graphs of polynomial functions, the use of graphic calculators can be extremely useful. You can use the table function to help you find the points at which $f(x) = 0$ or you can draw the graph and use the graph to find the 'zeros'.

If we look at the graph of $y = x^3 + x^2 - 4x - 4$ (below), it is clear that the graph cuts the x-axis at $x = -2$, $x = -1$ and $x = 2$.

This information can be found algebraically by factorising:

$y = x^3 + x^2 - 4x - 4$

Let $x = -1$, $y = (-1)^3 + (-1)^2 - 4(-1) - 4 = 0$

Since $y = 0$ when $x = -1$, $x + 1$ is a factor

$y = x^3 + x^2 - 4x - 4$

$ = (x + 1)(x^2 - 4)$

$ = (x + 1)(x + 2)(x - 2)$

> ⚠️ Use one of the three methods from this chapter.

Now, on the x-axis, $y = 0$, so

$(x + 1)(x + 2)(x - 2) = 0$

$x + 1 = 0$, $x + 2 = 0$, $x - 2 = 0$

$x = -1$, $\quad x = -2$ $\quad x = 2$

So the graph intersects the x-axis at $x = -1$, $x = -2$ and $x = 2$.

$x = -1$, $x = -2$ and $x = 2$ are the roots of the polynomial.

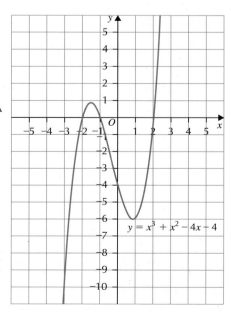

$y = x^3 + x^2 - 4x - 4$

It is also clear from the graph that it cuts the y-axis at $y = -4$. Again this is obtained from the function:

On the y-axis, $x = 0$,

$y = x^3 + x^2 - 4x - 4$

$\quad = 0^3 + 0^2 - 4(0) - 4$

$\quad = -4$

Example 7.22

Determine the coordinates of the points at which the graph of $y = x^3 - 3x + 2$ intersects the x- and y-axes

$y = x^3 - 3x + 2$

On the y-axis, $x = 0$ ∴ $y = 0 - 0 + 2 = 2$. The graph meets the y-axis at $(0,2)$.

On the x-axis, $y = 0$ ∴ $x^3 - 3x + 2 = 0$ ●——— To factorise this equation, look at factors of 2: ± 1, ± 2 ⓢ

Let $x = 1$: $1^3 - 3(1) + 2 = 0$

Since $y = 0$ when $x = 1$, $x - 1$ is a factor. ●——— Use the factor theorem to justify your factor. (See Chapter 1) ⓒ

$x^3 - 3x + 2 = 0$

$(x - 1)(x^2 + x - 2) = 0$

$(x - 1)(x - 1)(x + 2) = 0$

∴ $x - 1 = 0$, $x + 2 = 0$

	x^3	x^2	x	c
1	1	0	-3	2
	↓	1	1	-2
	1	1	-2	$\boxed{0}$

since $f(1) = 0$, $x - 1$ is a factor and we obtain $(x - 1)(x^2 + x - 2)$

So $x = 1$ (a repeated root) and $x = -2$. The graph meets the x-axis at $(-2,0)$ and $(1,0)$.

From inspection of the graph of $y = x^3 - 3x + 2$ as featured in Example 7.1, it can be seen that the point $(1,0)$ is also a stationary point. In this example, two identical factors of $(x - 1)$ were found, which results in a turning point on the x-axis.

A double factor will always result in a turning point on the x-axis.

You could use a graphic calculator to plot the function then calculate the 'zeros' or use the table function to determine when the graph is equal to zero.

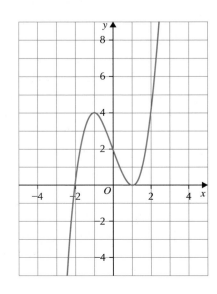

Exercise 7G

★ 1 Determine the points at which each graph meets the x- and y-axes.

a $y = x^3 - 5x^2 + 8x - 4$

b $y = x^3 + 3x^2 - 13x - 15$

c $y = x^3 + x^2 - 25x - 25$

d $y = x^3 + 12x^2 + 35x + 24$

e $y = -x^3 + 5x^2 + 12x - 36$

f $y = 2x^3 - 5x^2 + x + 2$

g $y = x^4 - 5x^2 + 4$

2 A new business is set up and for the first 8 months of operation, the profits can be modelled by the function $P(t) = t^3 - 8t^2 + 15t$ where t is the number of months. For which months did the business make zero profit?

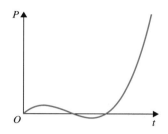

3 Part of a run-away train ride at a fairground can be modelled by the function $h(t) = -2t^4 + 20t^3 - 50t^2 + 72$ where $h(t)$ is the height above ground in metres at time t, seconds.

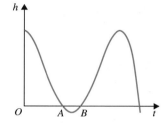

a Determine the coordinates of A and B when the train goes underground (A) and re-emerges (B).

b At what height did the ride begin?

Determining the function of a polynomial from its graph

Exercise 7G involved factorising a polynomial expression to identify the roots (the points of intersection with the x-axis). This process can be reversed in order to identify the function of the polynomial from the information given in the graph.

The values of the roots are used to express the polynomial in factorised form:
$y = k(x - a)(x - b)(x - c)$ where a, b and c are the roots and k is a numerical common factor which affects the amplitude of the function.

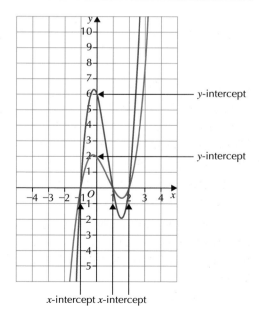

The diagram shows the graphs of $y = (x - 1)(x + 1)(x - 2)$ (in red) and $y = 3(x - 1)(x + 1)(x - 2)$ (in blue). Note that both of these graphs cross the x-axis at $x = -1$, $x = 1$ and $x = 2$ but that the amplitudes are different. The first graph crossed the y-axis at $y = 2$ and the second crosses at $y = 6$, that is 3 times as high as the first graph.

Example 7.23

Determine an expression for $f(x)$ from the graph of $y = f(x)$.

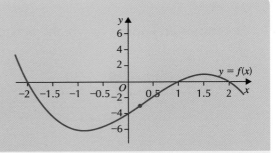

$x = -2$, $x = 1$ and $x = 2$

$(x + 2)$, $(x - 1)$, $(x - 2)$

So, $y = k(x + 2)(x - 1)(x - 2)$

$-4 = k(0 + 2)(0 - 1)(0 - 2)$

$-4 = 4k$

$k = -1$

So we obtain:

$f(x) = -(x + 2)(x - 1)(x - 2)$.

> First, identify the roots from the graph. **P**

> Use the roots to determine the factors. **P**

> Don't forget k, which represents the possibility **P** of a numerical common factor.

> Use the y-intercept to work out the value of k. **P** When $x = 0, y = -4$. Substitute these values into the expression.

Example 7.24

Determine an expression for $f(x)$ from the graph of $y = f(x)$.

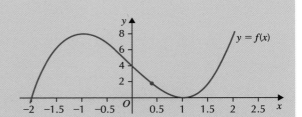

Roots:

$x = -2$, $x = 1$

Factors:

$k(x + 2)(x - 1)(x - 1)$

Expression:

$y = k(x + 2)(x - 1)(x - 1)$

$4 = k(2)(-1)(-1)$

$4 = 2k$

$k = 2$

$\therefore f(x) = 2(x + 2)(x - 1)(x - 1)$

$= 2(x + 2)(x - 1)^2$

> $x = 1$ is a repeated root as it is a turning point **P** on the x-axis.

> Use the y-intercept: $y = 4$ when $x = 0$ so we **P** can replace the values of x and y.

Exercise 7H

★ **1** Determine an expression for $f(x)$ from the graph of $y = f(x)$:

a

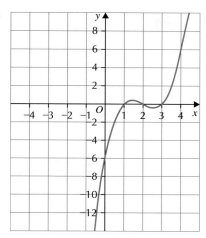

b

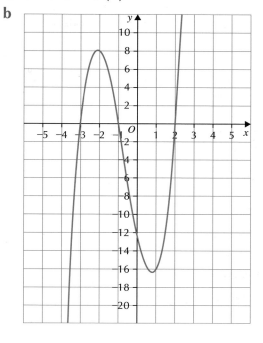

c

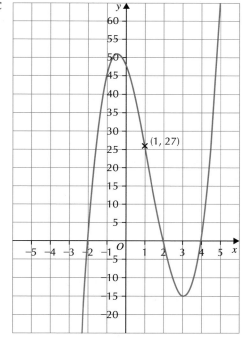

d

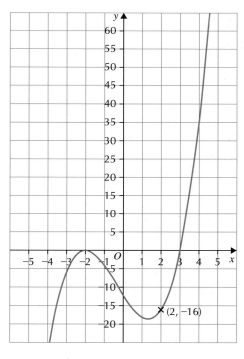

e

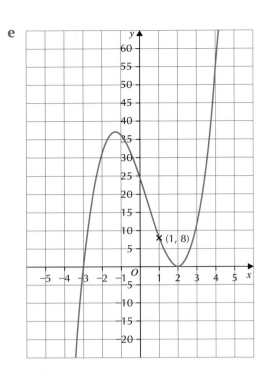

f

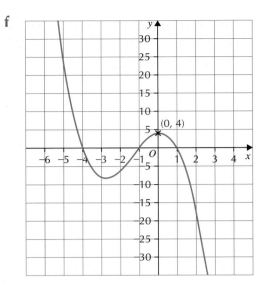

g

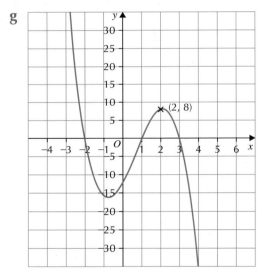

h

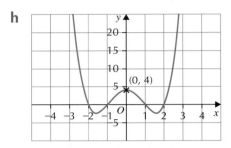

i

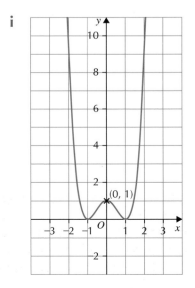

2   Nina is developing a new logo for her company, part of which is modelled in the graph (right).

Determine the equation of the 'N'.

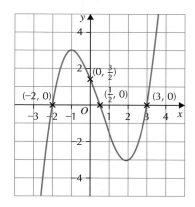

3   Part of a rollercoaster ride is modelled by the graph of $y = f(x)$ (right).

Determine the function, $f(x)$.

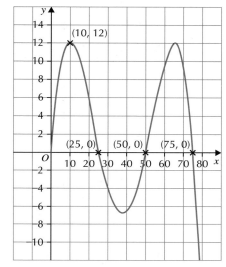

🛠 4   Determine an expression for $f(x)$ from the graph of $y = f(x)$ and calculate the value of a.

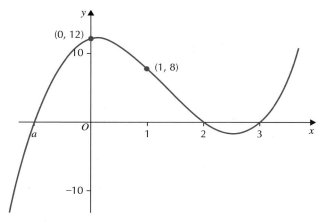

Solving a polynomial equation of degree ≤4

Solving a polynomial equation is a straightforward process once the skills of factorising a polynomial are known. In a similar fashion to solutions for quadratic equations, one side of the equation is made equal to zero before factorising and solving. We can also use polynomial equations to identify the points of intersection of a straight line and a curve or of two curves.

Example 7.25

Solve $x^3 - 2x^2 - 4x + 8 = 0$ This equation is already equal to zero. **C**

$x^3 - 2x^2 - 4x + 8 = 0$

Let $x = 2$

$2^3 - 2(2)^2 - 4(2) + 8 = 8 - 8 - 8 + 8 = 0$ Find a factor of 8 which gives 0 when substituted into the expression. **S**

Since we obtain 0, $x - 2$ is a factor

$x^3 - 2x^2 - 4x + 8 = 0$

$(x - 2)(x^2 - 4) = 0$

$(x - 2)(x - 2)(x + 2) = 0$

$x - 2 = 0, \ x + 2 = 0$

	x^3	x^2	x	c
2	1	−2	−4	8
	↓	2	0	−8
	1	0	−4	0

Since $f(2) = 0$, $x - 2$ is a factor and we obtain $(x - 2)(x^2 - 4) = 0$

$\therefore x = 2$ (repeated root) and $x = -2$.

Example 7.26

Solve $x^3 + 3x^2 + 5x + 5 = 2x^2 + 15x - 3$

$x^3 + 3x^2 + 5x + 5 = 2x^2 + 15x - 3$ Collect like terms so that one side of the equation is equal to zero. **S**

$x^3 + x^2 - 10x + 8 = 0$ Find a factor of the polynomial by looking at factors of 8. **S**

Let $x = 1$

$1^3 + 1^2 - 10(1) + 8 = 1 + 1 - 10 + 8 = 0 \ \therefore x - 1$ is a factor

$x^3 + x^2 - 10x + 8 = 0$

$(x - 1)(x^2 + 2x - 8) = 0$

$(x - 1)(x - 2)(x + 4) = 0$

$x - 1 = 0, \ x - 2 = 0, \ x + 4 = 0$

$\therefore x = 1, \ x = 2, \ x = -4$

	x^3	x^2	x	c
1	1	1	−10	8
	↓	1	2	−8
	1	2	−8	0

Example 7.27

Solve $2x^4 - 3x^3 - x^2 + 3x - 1 = 0$

$2x^4 - 3x^3 - x^2 + 3x - 1 = 0$

Let $x = 1$:

$2(1)^4 - 3(1)^3 - 1^2 + 3(1) - 1 = 0 \ \therefore x - 1$ is a factor.

$$2x^4 - 3x^3 - x^2 + 3x - 1 = 0$$
$$(x - 1)(2x^3 - x^2 - 2x + 1) = 0$$

	x^4	x^3	x^2	x	c
1	2	−3	−1	3	−1
	↓	2	−1	−2	1
	2	−1	−2	1	$\boxed{0}$

Let $x = 1$ in the cubic factor:
$2(1) - 1 - 2 + 1 = 0$ ∴ $x - 1$ is a factor again
$$(x - 1)(2x^3 - x^2 - 2x + 1) = 0$$
$$(x - 1)(x - 1)(2x^2 + x - 1) = 0$$
$$(x - 1)(x - 1)(2x - 1)(x + 1) = 0$$
$$x - 1 = 0 \qquad 2x - 1 = 0 \qquad x + 1 = 0$$
$x = 1$ (repeated root) $x = \frac{1}{2}$ and $x = -1$

	x^3	x^2	x	c
1	2	−1	−2	1
	↓	2	1	−1
	2	1	−1	$\boxed{0}$

Exercise 7I

1 Solve each cubic equation.

   a $x^3 + 3x^2 - 6x - 8 = 0$ **b** $x^3 + 7x^2 + 14x + 8 = 0$

   c $x^3 - 7x^2 + 7x + 15 = 0$ **d** $x^3 - 12x + 16 = 0$

2 Solve each polynomial equation.

   a $3x^3 + 9x^2 - 3x - 9 = 0$ **b** $x^4 - 19x^2 - 30x = 0$

   c $2x^3 + 3x^2 - 9x - 10 = 0$ **d** $x^4 - 4x^3 - x^2 + 16x - 12 = 0$

   e $2x^4 - x^3 - 14x^2 - 5x + 6 = 0$ **f** $6x^4 + x^3 - 25x^2 - 4x + 4 = 0$

★ 3 Solve these polynomial equations.

   a $x^3 + 6x^2 - 12x + 3 = 15 - 17x$ **b** $2x^3 + 7x^2 - 6x + 1 = x^3 - 10x + 13$

   c $x^3 - 8 = 8x^2 - x - 50$ **d** $2x^3 - 3x^2 - 13x - 5 = 1 + x^2 - x^3$

   e $x^4 + 2x^3 - 21x^2 + 5x + 8 = x^3 + 6x - 12$

   f $2x^4 - 8x^3 + 2x = 1 - 3x^2 - 2x^4$

4 Given that the equation $2x^3 + 3x^2 + ax - 2 = 0$ has a root at $x = 1$, determine the value of a and solve for x.

5 **a** Given that $x + 3$ is a factor of $f(x) = x^3 + kx^2 + 3x + 9$, determine the value of k.

   b Prove that $x = -3$ is the only root of $f(x) = 0$.

6 A porpoise is jumping in and out of the water. Its height, in metres, above the surface of the water after t seconds can be modelled by the function $h(t) = -t^4 + 18t^3 - 107t^2 + 234t - 144$.

   a Show that the porpoise breaks the surface of the water after 1 second.

   b At $t = 3$ seconds the porpoise re-enters the water. Fully factorise $h(t)$.

   c State the value of t at which the porpoise next breaks the surface of the water.

7 Hazel has just set up a new business selling special occasion cakes, cupcakes and other treats. Her profit, in £ after t months, over the first year can be modelled by the function $P(t) = 2t^3 - 16t^2 + 43t - 44$.

   a   After how many months does she begin to make a profit?

   b   Prove that she continues to make a profit for the remainder of the year.

   c   Calculate her profit for the 10th month.

8 Hazel sells boxes of 6 cupcakes in cuboidal boxes measuring 18 cm by 12 cm by 8 cm. In order to run a special offer, she wants to increase the length, breadth and height of these boxes by the same amount, x cm, to produce a box with a volume of 3465 cm³.

   a   Determine an expression for the volume of the box in terms of x.

   b   Show that $x^3 + 38x^2 + 456x - 1737 = 0$.

   c   Show algebraically that there is only one possible value for x and hence determine the dimensions of the new box.

9 Valentine's cookies are packed in gift boxes in the shape of a square based cuboid. The height of the box is 4 cm greater than its width and the volume of the box is 128 cm³. Determine the dimensions of the box.

10 Solve $x^4 - 2x^3 - 4x^2 + 2x + 3 > 0$

Determining the points of intersection of the graphs of a straight line and a curve or of two curves

Graphs are widely used to model real life problems and finding the points of intersection of two functions is a very useful skill. You should already be familiar with finding the point of intersection of two straight lines, a straight line and a parabola (or even two parabolae) so we now want to be able to find the points of intersection of higher degree polynomials.

Example 7.28

Determine the coordinates of the points at which the line $y = 2x + 5$ intersects the graph of $y = x^3 - 17x - 25$.

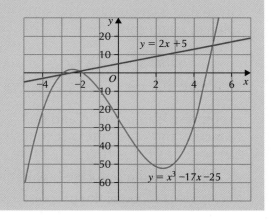

$x^3 - 17x - 25 = 2x + 5$

> To find the point of intersection of two polynomials, set them equal to each other. Ⓢ

$x^3 - 19x - 30 = 0$

> Collect like terms so that one side of the equation is equal to zero. Ⓢ

Let $x = -2$,

$(-2)^3 - 19(-2) - 30 = -8 + 38 - 30 = 0$

> Find a factor of the polynomial by looking at factors of 30. Ⓢ

Since $y = 0$ when $x = -2$, $x + 2$ is a factor.

$x^3 - 19x - 30 = 0$

$(x + 2)(x^2 - 2x - 15) = 0$

$(x + 2)(x + 3)(x - 5) = 0$

$x = -2$, $x = -3$ and $x = 5$

$$
\begin{array}{c|cccc}
 & x^3 & x^2 & x & c \\
-2 & 1 & 0 & -19 & -30 \\
 & \downarrow & -2 & 4 & 30 \\
\hline
 & 1 & -2 & -15 & \boxed{0}
\end{array}
$$

at $x = -2$, $y = 2x + 5 = 2(-2) + 5 = 1$

at $x = -3$, $y = 2(-3) + 5 = -1$

at $x = 5$, $y = 2(5) + 5 = 15$

> Calculate the y-coordinate for each value of x by substituting into the equation of the straight line. You could also substitute the values of x into $y = x^3 - 17x - 25$ but it is usually easier to use the simpler of the two functions. Ⓢ

So the points of intersection are $(-2,1)$, $(-3,-1)$ and $(5,15)$.

> You could plot the two functions on a graphic calculator and use the 'intersect' function to determine, or check, the points of intersection. Ⓒ

Exercise 7J

1   Identify the coordinates of the points of intersection of $y = f(x)$ and $y = g(x)$ from the graph.

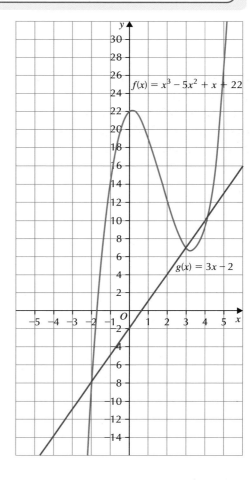

$f(x) = x^3 - 5x^2 + x + 22$

$g(x) = 3x - 2$

2 Determine the coordinates of the points of intersection of each pair of graphs algebraically.

a $y = x^3 - 25$ and $y = 11 - 7x^2$

b $y = x^3 + 9x^2 + 3x - 5$ and $y = 16 - 8x$

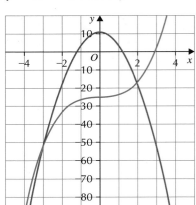

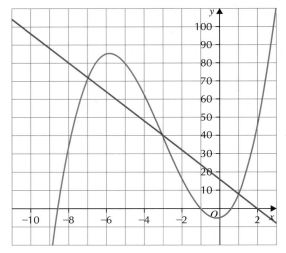

c $y = x^3 + 4x^2 - 3x + 1$ and $y = 5x^2 + 6x - 8$

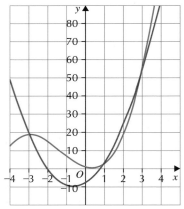

★ **3** Find the points of intersection of each pair of lines.

a $y = x^3 - 10x^2 + 19x + 10$ and $y = 2x - 18$

b $y = 3x^3 + 3x^2 + x - 5$ and $y = x^3 - 2x^2 + 2x + 1$

c $y = x^4 - 25x + 24$ and $y = 10x^3 - 35x^2 + 25x$

d $y = 4x^4 + x^3 - x^2 + 1$ and $y = x^3 + 4x^2$

e $y = 2x^3 + x^2 + x + 10$ and $y = 5 + 16x + 2x^2 - x^3$

Using the discriminant to determine an unknown coefficient in a quadratic equation

The discriminant is used to determine the nature of the roots of a quadratic equation.

For $ax^2 + bx + c = 0$

- if $b^2 - 4ac = 0$ the roots are real and equal
- if $b^2 - 4ac > 0$ there are 2 real, distinct roots

- if $b^2 - 4ac < 0$ there are no real roots
- if $b^2 - 4ac$ is a square number the roots are rational (therefore the expression can be factorised).

We have seen that the discriminant can be used to prove that a polynomial has an irreducible quadratic factor. The properties of the discriminant can now be used to determine the value of an unknown coefficient in a quadratic expression.

Example 7.29

The equation $3x^2 + kx + 3 = 0$ has equal roots. Determine the value of k.

If $3x^2 + kx + 3 = 0$ has equal roots then $b^2 - 4ac = 0$
$a = 3, b = k, c = 3$

$$b^2 - 4ac = k^2 - 4(3)(3) = 0$$
$$k^2 - 36 = 0$$
$$(k - 6)(k + 6) = 0$$
$$k = \pm 6$$

Example 7.30

Determine the range of values of k for which $x^2 + kx - k = 0$ has:

a real roots **b** no real roots.

a If $x^2 + kx - k = 0$ has real roots then $b^2 - 4ac \geq 0$
$$k^2 - 4(1)(-k) \geq 0$$
$$k^2 + 4k \geq 0$$

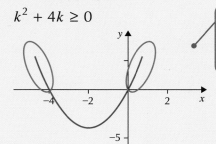

> To solve a quadratic inequality, sketch the graph of $y = k(k + 4) \geq 0$. Note that the parabola has roots at $k = -4$ and $k = 0$ and is u-shaped as it has a positive k^2 term. **C**

So $k(k + 4) \geq 0$, when $k \leq -4$ and $k \geq 0$ so $x^2 + kx - k = 0$ has real roots when $k \leq -4$ and $k \geq 0$ as shown by the sections circled on the graph.

Alternatively, you could identify the roots then use a table of signs to determine when the function is positive or negative

values either side of the roots   roots

k	-6	-4	-1	0	1
k	$-$	$-$	$-$	0	$+$
$(k + 4)$	$-$	0	$+$	$+$	$+$
$k(k + 4)$	$+$	0	$-$	0	$+$

> This line tells you whether the function is positive or negative overall. **C**

(continued)

This alternative method is used again in Chapter 10 to investigate stationary points.

b $x^2 + kx - k = 0$ has no real roots when $-4 < k < 0$.

Note the use of the 'strictly less than' symbol here.

Example 7.31

Show that the roots of the equation $x^2 + kx - 2k - 4 = 0$ are always real.

For real roots, $b^2 - 4ac \geq 0$

$$b^2 - 4ac = k^2 - 4(1)(-2k - 4)$$
$$= k^2 + 8k + 16$$
$$= (k + 4)^2$$

Since $(k + 4)^2$ is a squared term it is always greater than or equal to zero
∴ $b^2 - 4ac \geq 0$ for all values of k so the roots are always real.

Exercise 7K

1 Determine the value(s) of k such that the equation has equal roots.

a $x^2 - kx + 16 = 0$

b $x^2 - 12x + k - 4 = 0$

c $x^2 - 2kx + 5k = 0$

d $kx^2 - 4x + k - 3 = 0$

e $(2k + 3)x^2 + (k + 3)x + 1 = 0$

★ 2 **a** Determine the value of $k, k > 0$, such that $(2k + 1)x^2 + 12x + k = 0$ has equal roots.

b Determine the roots for this value of k.

★ 3 Solve

a $x^2 - 2x - 3 > 0$

b $2x^2 + 5x - 3 \leq 0$

c $4 + 3x - x^2 > 0$

d $16 + 6x - x^2 < 0$

e $9x^2 - 1 \geq 0$

f $-6x^2 + 5x - 1 \leq 0$

4 Given that each equation has real roots, determine the range of possible values for k.

a $3x^2 - 4x + 1 + k = 0$

b $kx^2 - 6x + 3 = 0$

c $(k + 1)x^2 - 4kx + k = 0$

★ 5 Determine the range of values of k for which $x^2 + 6x + 9 + k = 0$ has no real roots.

6 Determine the range of values for k such that $kx^2 - 4x + 3 + k = 0$ has real roots.

7 Show that the roots of the equation $2kx^2 + (3k - 2)x - 2 + k = 0$ are real for all values of k.

8 Given that $f(x) = 2x^2 - kx + 5$ and $g(x) = x^2 - 3kx + 5 - 4k$, determine the values of k such that $f(x) = g(x)$ has real and distinct roots.

9 a The graph of $y = 3x^3 + ax^2 - 19x + 6$ has a root at $x = 2$. Determine the value of a.

   b Prove that this graph meets the curve $y = x^2 - 25x + 16$ at one point only and determine the coordinates of this point.

10 a The graphs of $y = x^3 + x^2 + ax - 1$ and $y = x^2 + x + b$ both have a root at $x = 1$, determine the values of a and b.

   b Show that the two graphs in part **a** have only one rational point of intersection and determine its coordinates.

- I can fully factorise a polynomial given a factor. ★ Exercise 7B Q3
- I can fully factorise a polynomial of degree ≤4. ★ Exercise 7D Q4
- I can determine an unknown coefficient in a polynomial given a factor. ★ Exercise 7E Q1
- I can determine the remainder when a polynomial function is divided by a linear function. ★ Exercise 7F Q1
- I can use the remainder theorem to determine an unknown coefficient in a polynomial expression. ★ Exercise 7F Q2
- I can find the quotient and remainder when dividing a polynomial expression by a linear expression. ★ Exercise 7F Q5
- I can determine the points at which the graph of a polynomial function intersects the x- and y-axes. ★ Exercise 7G Q1
- I can determine the equation of a polynomial function from its graph. ★ Exercise 7H Q1
- I can solve a polynomial equation of degree ≤ 4. ★ Exercise 7I Q3
- I can determine the points of intersection of the graphs of two polynomial functions. ★ Exercise 7J Q3
- I can use the properties of the discriminant to determine an unknown coefficient in a quadratic equation. ★ Exercise 7K Q2, Q5
- I can solve quadratic inequalities. ★ Exercise 7K Q3

For further assessment opportunities, see the Preparation for Assessment for Unit 2 on pages 291–292.

8 Solving trigonometric equations

This chapter will show you how to:

- solve linear and quadratic trigonometric equations, including those involving compound angles, working in degrees or radians
- solve equations using standard trigonometric identities working in degrees or radians
- solve equations involving the wave function $f(x) = a\cos x + b\sin x$ working in degrees or radians
- solve equations using further trigonometric identities working in degrees or radians.

You should already know:

- how to solve linear trigonometric equations where the angle is given in degrees
- exact value ratios for sine, cosine and tangent
- radian equivalents of $30°$; $45°$; $60°$; $90°$ and be able to convert their related angles to radians
- the identity $\cos^2 A + \sin^2 A = 1$
- the addition and double angle formulae
- how to convert $a\cos x + b\sin x$ to the forms $k\cos(x \pm \alpha)$ or $k\sin(x \pm \alpha)$
- how to sketch or recognise graphs of the form $y = a\cos bx + c$ and $y = a\cos(x + d) + c$ or the sine equivalent

Solving trigonometric equations in degrees

Trigonometry is used extensively in science and engineering as well as in mathematics and applied mathematics. While trigonometric equations can be solved **graphically** using, for example, a graphic calculator, all of the equations in this chapter will be solved **algebraically.** It is useful, however, if the solutions can be checked graphically.

Solving linear equations of the form $a\cos x° + b = 0$; $a\sin x° + b = 0$ or $a\tan x° + b = 0$ where a and b are constants.

While this is mainly revision, you will now be required to work in intervals beyond $0 \leq x \leq 360$ and to know:

- the exact value ratios for the sine, cosine and tangent of $30°$, $60°$, $45°$ and their related angles using triangles or otherwise
- the exact values of the sine, cosine and tangent of angles which are multiples of $90°$ from their respective graphs.

Example 8.1

Solve these equations for x.

a $4\tan x° − 3 = 0$, $0 \leq x \leq 360$

b $7\sin t° − 2 = 0$, $0 \leq t \leq 540$

c $\cos x° + 1 = 0$, $0 \leq x \leq 360$

d $2\sin p° + 1 = 0$, $−90 \leq p \leq 270$

a $4\tan x° − 3 = 0$, $0 \leq x \leq 360$

$4\tan x° = 3$

$\tan x° = \frac{3}{4}$

$a = \tan^{-1}\left(\frac{3}{4}\right) = 36.9°$

$x = 36.9$ or $180 + 36.9$

$= 36.9$ or 216.9

> **S** Rearrange to make $\tan x°$ the subject.

> **S** Tan$x°$ is positive, so the solutions are in the 1st and 3rd quadrants.

> **P** Find the related acute angle. This angle will be referred to as a (for acute angle) whether working in degrees or later in radians.

> **S** The solution can be found graphically by tracing the points where the graph $y = 4\tan x° − 3$ cuts the x-axis.

> **C** The question is not to find the angle $x°$, but to find the number x.

b $7\sin t° − 2 = 0$, $0 \leq t \leq 540$

$7\sin t° = 2$

$\sin t° = \frac{2}{7}$

$a = \sin^{-1}\left(\frac{2}{7}\right) = 16.6°$

$t = 16.6$, $180 − 16.6$, $360 + 16.6$ or $540 − 16.6$

$= 16.6$, 163.4, 376.6 or 523.6

> **S** Sin$x°$ is positive, so the solutions are in quadrants 1 and 2.

> **P** As $0 \leq x \leq 540$ you will have to go round the quadrant diagram one and a half times, starting at 0 and finishing at 540.

c $\cos x° + 1 = 0$, $0 \leq x \leq 360$

$\cos x° = −1$

$x = 180$

> **S** 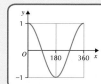 For $\cos x° = −1$ you are dealing with multiples of 90) so read solutions from graph.

d $2\sin p° + 1 = 0$, $−90 \leq p \leq 270$

$2\sin p° = −1$

$\sin p° = −\frac{1}{2}$

$a = \sin^{-1}\left(\frac{1}{2}\right) = 30°$

$p = 0 − 30$ or $180 + 30$

$= −30$ or 210

> **S** Sin$x°$ is negative, so the solutions are in quadrants 3 and 4.

> **S** 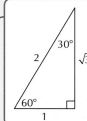 Use the **positive** ratio $\frac{1}{2}$ to find the related acute angle because the sin, cos or tan of an acute angle is always positive.
>
> Use this triangle in a non-calculator situation.

Exercise 8A

Calculators can be used unless stated otherwise.

★ **1** Solve these equations.

    a $3\sin x° - 1 =, 0 \le x \le 360$ **b** $5\cos t° + 2 = 0, 0 \le t \le 180$

    c $2\tan x° = 7, 0 \le x \le 540$ **d** $7\cos x° = 3, 0 \le t \le 720$

> Always look carefully at the interval in which your solutions must lie.

★ **2** Solve these equations **without a calculator**.

    a $2\sin x° - 1 = 0, 0 \le x \le 360$ **b** $\sqrt{2}\cos α° + 1 = 0, 0 \le α \le 180$

    c $\tan x° = \sqrt{3}, 0 \le x \le 360$ **d** $2\sin x° - 1 = 0, 0 \le x \le 540$

    e $2\cos t° + \sqrt{3} = 0, 0 \le t \le 360$ **f** $\tan x° + 1 = 0, 0 \le θ \le 360$

3 The graph is of the form $y = 3\sin x° + 1$, $0 \le x \le 360$

The line $y = 2$ cuts the graph at P and Q. Find the coordinates of P and Q.

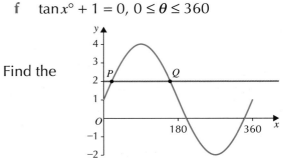

Solving linear equations with compound angles

Consider equations like:

    $3\sin(x - 30)° = 2$ and $5\cos 2x° - 3 = 0$

The angle is no longer simply $x°$, but a function of $x°$, or a **compound angle**.

$3\sin(x - 30)° = 2$ can be solved to find $(x - 30)$ in exactly the same way as $3\sin x° = 2$ is solved to find x.

Linear equations with compound angles can be solved as follows:

- solve to find the compound angle, carefully noting the interval in which the solutions will lie
- use the result for the compound angle to find x.

Example 8.2

Solve these equations

a $5\tan 2x° = 3, 0 \le x \le 360$ **b** $2\sin(t + 60)° - 1 = 0, 0 < t \le 360$

a $5\tan 2x° = 3, 0 \le 2x \le 720$

> Start to solve for $2x$. If $0 \le x \le 360$, then $2 \times 0 \le 2 \times x \le 2 \times 360 \Rightarrow 0 \le 2x \le 720$. **S**

    $\tan 2x° = \frac{3}{5}$

        $a = \tan^{-1}\left(\frac{3}{5}\right) = 31.0°$

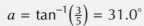

    $2x = 31.0, 180 + 31.0, 360 + 31.0, 540 + 31.0$

> As $0 \le 2x \le 720$ you must go round the quadrant diagram twice starting at 0 and finishing at 720. **S**

$= 31.0, 221.0, 391.0, 571.0$

$x = 15.5, 110.5, 195.5$ or 285.5

> Once you have found $2x$ divide by 2 to find x. **S**

b $2\sin(t + 60)° - 1 = 0, \ 60 \le (t + 60) \le 360 + 60$

$\sin(t + 60)° = \frac{1}{2}$

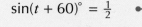

> The solution can be found graphically by tracing the points where the graph $y = 5\tan 2x°$ cuts the line $y = 3$. **C**

$a = \sin^{-1}\left(\frac{1}{2}\right) = 30°$

> It is easier to visualise your solutions on the quadrant diagram if you don't simplify $360 + 60$. **S**

$t + 60 = \cancel{30}, 180 - 30, 360 + 30$

$\qquad = 150$ or 390

$t = 90$ or 330

> Going round the quadrant diagram starting at 0, the first angle 30° is too small so you must discard it. You then collect $180 - 60$ and $360 + 60$ stopping there as the next angle $540 - 60$ would be too big **S**

> Now that you have found $t + 60$ subtract 60 from each side to calculate t. **S**

Example 8.3

The diagram shows the graph of a cosine function.

a State the equation of the function.

b The straight line with equation $y = 2$ cuts the graph at the points A and B.
Find the coordinates of A and B.

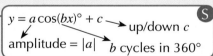

a amplitude $= 2 \Rightarrow a = 2$

1 cycle in $180° \Rightarrow$ 2 cycles in $360° \Rightarrow b = 2$

up $1 \Rightarrow c = 1$

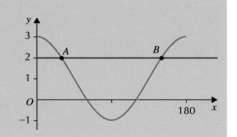

and so the equation is $y = 2\cos 2x° + 1$

> A and B are points of intersection so form a pair of simultaneous equations **S**

b At A and B:

$y = 2\cos 2x° + 1$ the points lie on the curve (1)

$y = 2$ the points lie on the straight line (2)

Solving these equations simultaneously gives:

$2\cos 2x° + 1 = 2, \quad 0 \le 2x \le 360$

> $0 \le x \le 180$ in the diagram so $0 \le 2x \le 360$. **P**

$2\cos 2x° = 1$

$\cos 2x° = \frac{1}{2}$

$\qquad a = \cos^{-1}\left(\frac{1}{2}\right) = 60°$

$2x = 60$ or 300

$x = 30$ or 150

> Remember to list the coordinates in the right order. **C**

A is the point $(30, 2)$ and B is the point $(150, 2)$

Example 8.4

Environmentalists find that the population of some predatory animals varies over time according to the balance between the predators and available prey, or food supply. The number of foxes in a certain woodland area was studied over a period of 5 years and the number of foxes, n, approximated by the formula:

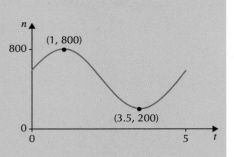

$n = 500 + 300 \cos 72(t - 1)°$

where t denotes the number of years from the start of the study.

Suppose the species is declared vulnerable when the numbers fall below 300. After how long into the 5 year period would the species be declared vulnerable and how long would it remain so?

From the graph you will see that the population will fall below 300 when t lies somewhere between 1 and 5.

To find exactly where, solve the equation:

$500 + 300 \cos 72(t - 1)° = 300$, $2 \leq t \leq 5$ or $0 \leq 72(t - 1) \leq 288$

$300 \cos 72(t - 1)° = -200$

$\cos 72(t - 1)° = -\frac{2}{3}$

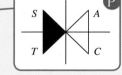

$\quad a = \cos^{-1}\left(\frac{2}{3}\right) = 48.1°$

$72(t - 1) = 180 - 48.1$ or $180 + 48.1$

$\qquad = 131.9$ or 228.1

$t - 1 = 1.83$ or 3.17

$t = 2.83$ or 4.17

To the nearest month the foxes will become vulnerable when t = 2.83, that is after 2 years and ten months. They will recover again when $t = 4.17$ or 4 years 2 months so they remain vulnerable for 1 year and 4 months.

Exercise 8B

A calculator may be used unless stated otherwise.

★ 1 Solve these equations, giving your answer to 1 decimal place where required.

    a $3 \sin 2x° - 1 = 0$, $0 \leq x \leq 360$ **b** $5 \tan 3x° = 2$, $0 \leq x \leq 180$

    c $8 + 3 \cos 4x° = 9$, $0 \leq x \leq 90$ **d** $4 \cos (x + 45)° = 3$, $0 \leq x \leq 360$

    e $3 \tan (x + 60)° + 5 = 0$, $0 \leq x \leq 360$ **f** $3 \sin (x - 20)° = 2$, $0 \leq x \leq 540$

★ **2** Solve these equations **without the use of a calculator**.

a $2\cos(x - 60)° = \sqrt{3},\ 0 \le x \le 360$

b $\sqrt{2}\cos(x - 25)° = 1,\ 0 \le x \le 360$

c $\sqrt{2}\sin(2x - 45)° = -1,\ 0 \le x \le 360$

d $\tan(2x - 50)° + \sqrt{3}, = 0,\ 0 \le x \le 180$

e $2\sin(3x - 70)° + \sqrt{3} = 0,\ 0 \le x \le 180$

f $\cos(2x - 45)° = 1,\ 0 \le x \le 360$

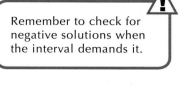

Remember to check for negative solutions when the interval demands it.

3 The graph shows a function of the form $y = p\sin(x - q)°$

a Write down the values of p and q.

b Hence find the coordinates of the points with x coordinates between 0 and 360 where the line with equation $y = -1$ intersects the curve.

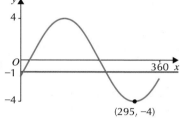

(295, −4)

4 The temperature in a school is controlled by an electronic thermostat The temperature, $T°C$, can be modelled by the equation:

$T = 18 + 6\sin(15t - 105)°$

where t is the time in hours after midnight.

a What is the temperature at 8.30 am, when pupils begin to arrive at school?

b At what time does the temperature first reach 16°C and at what time does it fall below it?

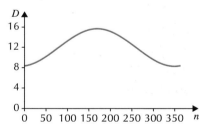

5 At a certain latitude the hours of daylight, D, in each day can be modelled by the function:

$D = 12 + 3.7\sin(n - 80)°$

where n is the number of days after the start of the year with $n = 1$ on the first of January.

a How many hours of daylight are there on the longest day and what is the date?

b Give the dates between which there are more than 14 hours of daylight in each day.

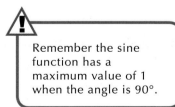

Remember the sine function has a maximum value of 1 when the angle is 90°.

6 The percentage, *P*, of the moon's face visible to a person standing at a certain location on Earth can be approximated by the formula:

$$P = 50\left(\cos\frac{90}{7}t° + 1\right)$$

where *t* is the number of days from the last full moon.

Stargazers prefer there to be less than 10% of the moon visible as the stars are so much clearer in darkness. Between which two days of the lunar cycle should they arrange to study the night sky?

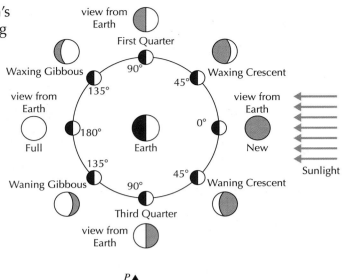

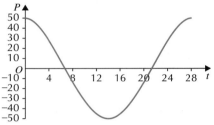

Solving quadratic trigonometric equations in degrees

Quadratic trigonometric equations come in two forms:

- $a\cos^2 x° = b$; $a\sin^2 x° = b$ or $a\tan^2 x° = b$, or their compound angle equivalents where *a* and *b* are constants
- $a\cos^2 x° + b\cos x° + c = 0$ or $a\sin^2 x° + b\sin x° + c = 0$ where *a*, *b* and *c* are constants.

Solving quadratic equations of the form $a\cos^2 x° = b$; $a\sin^2 x° = b$ or $a\tan^2 x° = b$, and their compound angle equivalents where *a* and *b* are constants

Based on your knowledge that if, for example, $\sin^2 x° = c$, then $\sin x° = \pm\sqrt{c}$ these equations are a simple extension of the linear case.

Example 8.5

Solve this equation: $3\tan^2 x° = 1$, $0 \leq x \leq 360$

$\tan^2 x° = \frac{1}{3}$

$\Rightarrow \tan x° = \pm\frac{1}{\sqrt{3}}$

$a = \tan^{-1}\left(\frac{1}{\sqrt{3}}\right) = 30°$

$x = 30, 180 - 30, 180 + 30$ or $360 - 30$

$= 30, 150, 210$ or 330

P Tan$x°$ can be positive or negative so there are 4 solutions, one in each of the four quadrants.

P Use this triangle in a non-calculator situation.

Example 8.6

Solve the equation: $\sin^2(2x - 60)° = 0.25$, $0 \leq x \leq 180$

$\sin^2(2x - 60)° = 0.25$, $-60 \leq (2x - 60) \leq 300$

$\sin(2x - 60)° = \pm 0.5$

$a = \sin^{-1}(0.5) = 30°$

$2x - 60 = -30, 30, 180 - 30, 180 + 30, \cancel{360 - 30}$

$= -30, 30, 150, 210$

$2x = 30, 90, 210, 270$

$x = 15, 45, 105$ or 135

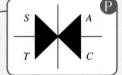

Compound angles are dealt with in the same way as before. Solve to find $2x - 60$ then use your result to find x.

S The interval is $-60 \leq (2x - 60) \leq 300$. Starting at -60, go round the quadrant diagram and finish at 300. The first angle you collect is $-30°$. Sometimes it is better to collect too many angles then discard the ones outside the interval.

Solving quadratic equations of the form $a\cos^2 x° + b\cos x° + c = 0$ or $a\sin^2 x° + b\sin x° + c = 0$ where a, b and c are constants.

The quadratic equation $2S^2 - 3S + 1 = 0$ can be solved to find S: as follows.

$2S^2 - 3S + 1 = 0$

$(2S - 1)(S - 1) = 0$

$S = \frac{1}{2}$ or $S = 1$

The quadratic trigonometric equation $2\sin^2 x° - 3\sin x° + 1 = 0$ can be solved in exactly the same way to find $\sin x°$. Remember $\sin^2 x° = (\sin x°)^2$.

$2\sin^2 x° - 3\sin x° + 1 = 0$

$(2\sin x° - 1)(\sin x° - 1) = 0$

$\sin x° = \frac{1}{2}$ or $\sin x° = 1°$

Example 8.7

Solve these equations.

a $3\sin^2 x° - 2\sin x° - 1 = 0$, $0 \le x \le 360$

b $2\cos^2 x° + 3\cos x° - 2 = 0$, $0 \le x \le 360$

c $3\cos^2 p° - \cos p° - 2 = 0$, $0 \le p < 360$

---

a $3\sin^2 x° - 2\sin x° - 1 = 0$, $0 \le x \le 360$

> Factorise. It may help to use S for $\sin x°$: $3S^2 - 2S - 1 = (3S + 1)(S - 1)$.   Ⓟ

$(3\sin x° + 1)(\sin x° - 1) = 0$

1. $\sin x° = -\frac{1}{3}$

> Form two linear equations and solve separately.   Ⓢ

$a = \sin^{-1}\left(\frac{1}{3}\right) = 19.5°$

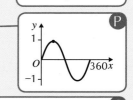

$x = 180 + 19.5$ or $360 - 19.5$

$= 199.5$ or 340.5

or

2. $\sin x° = 1$

$x = 90$

Combining solutions: $x = 90$, 199.5 or 340.5

> Combine the solutions. It is always preferable to arrange the combined solutions in order.   Ⓒ

b $2\cos^2 x° + 3\cos x° - 2 = 0$, $0 \le x \le 360$

> $2C^2 + 3C - 2 = (2C - 1)(C + 2)$ Ⓟ

$(2\cos x° - 1)(\cos x° + 2) = 0$

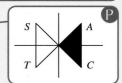

1. $\cos x = \frac{1}{2}$

$$a = \cos^{-1}\left(\tfrac{1}{2}\right) = 60°$$

$x = 60$ or $360 - 60$

$ = 60$ or 300

or

2. $\cos x° = -2$, which has no solution.

> The second equation has no solution as -2 is less than -1 which is the minimum value of the cosine function. **C**

Combining solutions: $x = 60$ or 300

c $\quad 3\cos^2 p° - \cos p° - 2 = 0$, $0 \le p < 360$

$\quad (3\cos p° + 2)(\cos p° - 1) = 0$

1. $\cos p° = -\tfrac{2}{3}$

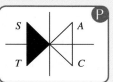

$\quad a = \cos^{-1}\left(\tfrac{2}{3}\right) = 48.2°$

$p = 180 - 48.2$ or $180 + 48.2$

$ = 131.8$ or 228.2

or

2. $\cos p° = 1$

$\Rightarrow p = 0$ or $\cancel{360}$

> The interval for p is $0 \le p < 360$. This strict inequality means that 360 must be eliminated as a solution. If the interval had been $0 \le p \le 360$, 360 would have remained. Always check the interval for x before finalising your answer. A mistake here will lead to the loss of a mark in an exam. **C**

Combining solutions:
$p = 0$, 131.8 or 228.2

Example 8.8

Solve the equation.

a $\quad 6\cos^2 x° - 1 = \cos x°$, $0 \le x \le 360$

a $\quad 6\cos^2 x° - 1 = \cos x°$, $0 \le x \le 360$

> Rearrange in standard quadratic form. **P**

$\quad 6\cos^2 x° - \cos x° - 1 = 0$

$\quad (3\cos x° + 1)(2\cos x° - 1) = 0$

1. $\cos x° = -\tfrac{1}{3}$

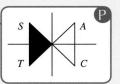

$\quad a = \cos^{-1}\left(\tfrac{1}{3}\right) = 70.5°$

$x = 180 - 70.5$ or $180 + 70.5$

$ = 109.5$ or 250.5

or

2. $\cos° = \tfrac{1}{2}$

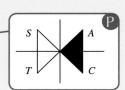

$\quad a = \cos^{-1}\left(\tfrac{1}{2}\right) = 60°$

$x = 60$ or $360 - 60$

$ = 60$ or 300

Combining solutions: $x = 60$, 109.5, 250.5 or 300

Exercise 8C

★ 1 Solve these equations.

 a $3\tan^2 x° = 7, 0 \leq x \leq 270$ b $5\cos^2 t° - 3 = 0, 0 \leq t \leq 360$

 c $\sin^2 x° - \sin x° - 2 = 0, 0 \leq x \leq 360$ d $3\cos^2 x° - 5\cos x° - 2 = 0, 0 \leq x \leq 360$

★ 2 a $5\sin^2 (x - 30)° + 2\sin (x - 30)° - 3 = 0, 0 \leq x \leq 360$

 b $6\cos^2 (2x)° + 5\cos (2x)° - 4 = 0, 0 \leq x \leq 360$

 c $2\sin^2 (x - 60)° + \sin (x - 60)° - 1 = 0, 0 \leq x < 360$

The method for solving quadratic equations extends naturally to cubic and higher forms.

3 Solve these equations. You will need to have completed Chapter 1 before attempting part **c**.

 a $2\sin^3 x° + \sin^2 x° - \sin x° = 0, 0 \leq x \leq 360$

 b $3\cos^3 2x° + \cos^2 2x° - \cos^2 x° = 0, 0 \leq x \leq 360$

 c $6\cos^3 x° - 13\cos^2 x° + 9\cos x° - 2 = 0, 0 \leq x \leq 360$

Solving trigonometric equations in radians

Linear and quadratic trigonometric equations expressed in radians are solved using the same methods as those expressed in degrees. However, it is important to remember that if a question using radians asks for an **exact solution**, then that solution must be presented as a fraction or multiple of π. In general, it is good practice to present radian solutions in this form whenever possible.

$30° = \frac{\pi}{6}$		$90° = \frac{\pi}{2}$	
$60° = \frac{\pi}{6}$		$180° = \pi$	
$45° = \frac{\pi}{4}$		$270° = \frac{3\pi}{2}$	
		$360° = 2\pi$	

Solving linear equations in radians

Solving linear equations in radians uses the same methods as when working in degrees.

Example 8.9

Solve these equations for x.

a $2\sin x + 1 = 0, 0 \leq x \leq 2\pi$ b $3\cos x - 2 = 0, 0 \leq x \leq 2\pi$

a $2\sin x + 1 = 0, 0 \leq x \leq 2\pi$

 $2\sin x = -1$

 $\sin x = -\frac{1}{2}$

> Ⓟ $\sin x$ is negative so solutions are found in quadrants 3 and 4.

 $a = \sin^{-1}\left(\frac{1}{2}\right) = 30° = \frac{\pi}{6}$

 $x = \pi + \frac{\pi}{6}$ or $2\pi - \frac{\pi}{6}$

 $= \frac{7\pi}{6}$ or $\frac{11\pi}{6}$

> Ⓢ If using a calculator, first find the acute angle in degrees. If a calculator is not allowed use your knowledge of exact value ratios. In either case the acute angle = $\sin^{-1}\left(\frac{1}{2}\right) = 30° = \frac{\pi}{6}$

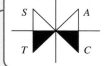

b $3\cos x - 2 = 0$, $0 \le x \le 2\pi$

$3\cos x = 2$

$\cos x = \frac{2}{3}$

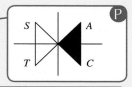

$a = \cos^{-1}\left(\frac{2}{3}\right) = 48.2° = 0.841$

$x = 0.841$ or $2\pi - 0.841$

$ = 0.841$ or 5.442

> $\cos^{-1}\left(\frac{2}{3}\right) = 48.2°$. This cannot be converted to an exact fraction of π. A question like this will only be asked if the use of a calculator is allowed. The simplest approach is to ignore the answer $48.2°$ and start again by resetting your calculator to radian mode to get the solution: $\cos^{-1}\left(\frac{2}{3}\right) = 0.841$. Remember to switch back to degrees when you are finished. **S**

Solving linear equations involving compound angles

Solving linear equations involving compound angles uses two stages:

- solve the equation to find the compound angle, carefully noting the interval in which the solutions will lie
- use your result for the compound angle to find x.

Example 8.10

Solve these equations.

a $\sqrt{3}\tan 2x = 1$, $0 \le x \le 2\pi$ **b** $2\sin\left(\theta + \frac{\pi}{4}\right) - 1 = 0$, $0 < \theta \le 2\pi$

a $\sqrt{3}\tan 2x = 1$, $0 \le 2x \le 4\pi$

> Start to solve for $2x$ noting that if $0 \le x \le 2\pi$, then $(2 \times 0) \le (2 \times x) \le (2 \times 2\pi) \Rightarrow 0 \le 2x \le 4\pi$. **S**

$\tan 2x = \frac{1}{\sqrt{3}}$

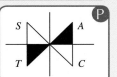

$a = \tan^{-1}\left(\frac{1}{\sqrt{3}}\right) = 30° = \frac{\pi}{6}$

$2x = \frac{\pi}{6},\ \pi + \frac{\pi}{6},\ 2\pi + \frac{\pi}{6},\ 3\pi + \frac{\pi}{6}$

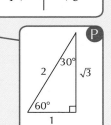

$ = \frac{\pi}{6},\ \frac{7\pi}{6},\ \frac{13\pi}{6},\ \frac{19\pi}{6}$

> Remember that $\pi = \frac{6\pi}{6}$. **C**

$x = \frac{\pi}{12},\ \frac{7\pi}{12},\ \frac{13\pi}{12},\ \frac{19\pi}{12}$

> Once you have found $2x$, divide by 2 to find x **P**

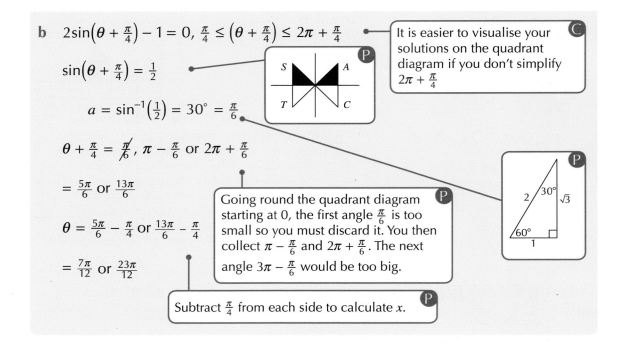

b $2\sin\left(\theta + \frac{\pi}{4}\right) - 1 = 0,\ \frac{\pi}{4} \le \left(\theta + \frac{\pi}{4}\right) \le 2\pi + \frac{\pi}{4}$

It is easier to visualise your solutions on the quadrant diagram if you don't simplify $2\pi + \frac{\pi}{4}$

$\sin\left(\theta + \frac{\pi}{4}\right) = \frac{1}{2}$

$a = \sin^{-1}\left(\frac{1}{2}\right) = 30° = \frac{\pi}{6}$

$\theta + \frac{\pi}{4} = \frac{\pi}{6},\ \pi - \frac{\pi}{6}\ \text{or}\ 2\pi + \frac{\pi}{6}$

$= \frac{5\pi}{6}\ \text{or}\ \frac{13\pi}{6}$

$\theta = \frac{5\pi}{6} - \frac{\pi}{4}\ \text{or}\ \frac{13\pi}{6} - \frac{\pi}{4}$

Going round the quadrant diagram starting at 0, the first angle $\frac{\pi}{6}$ is too small so you must discard it. You then collect $\pi - \frac{\pi}{6}$ and $2\pi + \frac{\pi}{6}$. The next angle $3\pi - \frac{\pi}{6}$ would be too big.

$= \frac{7\pi}{12}\ \text{or}\ \frac{23\pi}{12}$

Subtract $\frac{\pi}{4}$ from each side to calculate x.

Exercise 8D

Calculators can be used unless stated otherwise but exact solutions should be given where possible.

1   Solve the equations **without a calculator**.

  a $2\sin x - 1 = 0,\ 0 \le x \le 2\pi$

  b $\sqrt{2}\cos\alpha + 1 = 0,\ 0 \le \alpha \le \pi$

  c $\tan x - \sqrt{3} = 0,\ 0 \le x \le 2\pi$

  d $\sin x - 1 = 0,\ 0 \le x \le 3\pi$

  e $2\cos\theta + \sqrt{3} = 0,\ 0 \le \theta \le 2\pi$

  f $\tan x + 1 = 0,\ 0 \le \theta \le 2\pi$

★ **2**   Solve the equations for $0 \le x \le 2\pi$. If possible express your answer in terms of π. If that is not possible round your answer to 3 decimal places.

  a $\sqrt{3}\tan x + 1 = 0$

  b $5\tan x - 7 = 0$

  c $2\cos x + \sqrt{2} = 0$

  d $3\tan x - 4 = 0$

  e $5\cos x + 1 = 0$

  f $\sin x - 1 = 0$

  g $2\cos x - \sqrt{3} = 0$

  h $3\tan x - 7 = 0$

3   Solve **without the use of a calculator**

  a $2\sin 2x - 1 = 0,\ 0 \le x \le 2\pi$

  b $\tan 3x = \sqrt{3},\ 0 \le x \le \pi$

  c $2\cos\left(x - \frac{\pi}{6}\right) = -1,\ 0 \le x \le 2\pi$

  d $\sqrt{2}\sin\left(x - \frac{\pi}{4}\right) = 1,\ 0 \le x \le 2\pi$

  e $3\tan(3x) + 3 = 0,\ 0 \le x \le \pi$

  f $2\sin\left(2x + \frac{\pi}{3}\right) + \sqrt{3} = 0,\ 0 \le x \le \pi$

  g $\sqrt{2}\cos\left(x - \frac{\pi}{3}\right) = 1,\ 0 \le x \le 2\pi$

  h $\sqrt{2}\sin\left(2x - \frac{\pi}{4}\right) = -1,\ 0 \le x \le 2\pi$

4 The diagram shows part of the graph of the function $y = 2\cos 2x + 3$

The line $y = 4$ cuts the graph at points A, B, C and D.

Find the coordinates of all four points.

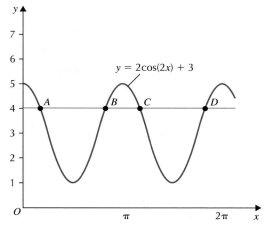

5 The highest tides in the world can be found in the Bay of Fundy off the Atlantic coast of Canada. One hundred billion tonnes of water flow in and out of the bay on an average tide twice a day. The depth of water, d metres, in an inlet on a certain day can be approximated by the formula

$$d = 7.8 - 6.3\cos\left(\tfrac{\pi}{6}t\right),\ 0 \le t \le 24$$

where $t = 0$ at midnight.

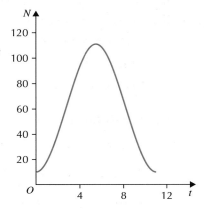

Calculate the four times during that day when the depth of the water is 13.5 metres. Give your answer to the nearest ten minutes.

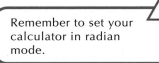

Remember to set your calculator in radian mode.

6 Astronomers keeping track of the number of sun spots that appear on the surface of the sun noticed an eleven year cycle; where the number of sun spots, N, within that cycle could be approximated by the formula

$$N = 60 - 50\cos\left(\tfrac{2\pi}{11}t\right),\ 0 \le t \le 11$$

In which years of the eleven year cycle would they expect to see 100 sunspots? Answer to the nearest year.

7 In Edinburgh, the time of sunrise, t, hours after midnight is approximated by the formula

$$t = \frac{1}{60}\left(394 + 128\cos\left(\frac{2\pi}{365}\right)(d + 10)\right)$$

where d is the day of the year with $d = 1$ on January 1.

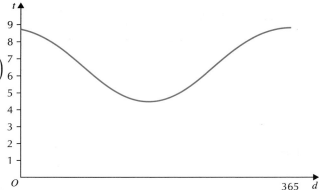

a Giving your answer to the nearest 5 minutes, calculate the time of sunrise on February 6.

b Iain sets out for work each day at 7am. Between which two dates will he set out after sunrise?

Solving quadratic trigonometric equations using radians

Example 8.11

Solve these equations

a $2\cos^2 x - 1 = 0,\ 0 \le x \le \pi$

b $4\sin^2\left(\theta - \frac{\pi}{2}\right) - 3 = 0,\ 0 \le \theta \le 2\pi$

a $2\cos^2\theta - 1 = 0,\ 0 \le x \le \pi$

$\cos^2\theta = \frac{1}{2}$

$\cos\theta = \pm\frac{1}{\sqrt{2}}$

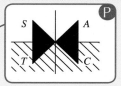

While $\cos\theta$ can be positive or negative the interval is $0 \le \theta \le \pi$ so you can only select angles in the first two quadrants. **S**

$a = \cos^{-1}\left(\frac{1}{\sqrt{2}}\right) = 45° = \frac{\pi}{4}$

$\theta = \frac{\pi}{4}$ or $\pi - \frac{\pi}{4}$

$= \frac{\pi}{4}$ or $\frac{3\pi}{4}$

In a non-calculator question you would use this triangle. $45° = \frac{\pi}{4}$ **S**

b $4\sin^2\left(\theta - \frac{\pi}{2}\right) - 3 = 0,\ -\frac{\pi}{2} \le \left(\theta - \frac{\pi}{2}\right) \le \left(2\pi - \frac{\pi}{2}\right)$

$\sin^2\left(\theta - \frac{\pi}{2}\right) = \frac{3}{4}$

$\sin\left(\theta - \frac{\pi}{2}\right) = \pm\frac{\sqrt{3}}{2}$

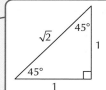

This time there will be a solution for $\theta - \frac{\pi}{2}$ in each of the four quadrants. **C**

$a = \sin^{-1}\left(\frac{\sqrt{3}}{2}\right) = 60° = \frac{\pi}{3}$

$\left(\theta - \frac{\pi}{2}\right) = -\frac{\pi}{3},\ \frac{\pi}{3},\ \pi - \frac{\pi}{3},\ \pi + \frac{\pi}{3},\ 2\pi - \frac{\pi}{3}$

The interval is $-\frac{\pi}{2} \le \left(\theta - \frac{\pi}{2}\right) \le \left(2\pi - \frac{\pi}{2}\right)$. Go round the quadrant diagram, starting at $-\frac{\pi}{2}$ and finishing at $\left(2\pi - \frac{\pi}{2}\right)$. The first angle you collect is $-\frac{\pi}{3}$. $2\pi - \frac{\pi}{3}$ is discarded as too big. **S**

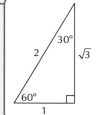

If a calculator is not allowed you can use this triangle. **P**

$$= -\frac{\pi}{3}, \frac{\pi}{3}, \frac{2\pi}{3}, \frac{4\pi}{3}$$

$$\theta = -\frac{\pi}{3} + \frac{\pi}{2}, \frac{\pi}{3} + \frac{\pi}{2}, \frac{2\pi}{3} + \frac{\pi}{2}, \frac{4\pi}{6} + \frac{\pi}{2}$$

$$= \frac{\pi}{6}, \frac{5\pi}{6}, \frac{7\pi}{6}, \frac{3\pi}{2}$$

Example 8.12

Solve these equations.

a $2\cos^2 x + 3\cos x - 2 = 0,\ 0 \le x \le 2\pi$

b $2\sin^2 x = \sin x,\ 0 < x \le 2\pi$

a $2\cos^2 x + 3\cos x - 2 = 0,\ 0 < x \le 2\pi$

> Factorise. It may help to use C for $\cos x$: $2C^2 + 3C - 2 = (2C - 1)(C + 2)$ ⓟ

   $(2\cos x - 1)(\cos x + 2) = 0$

1. $\cos x = \frac{1}{2}$

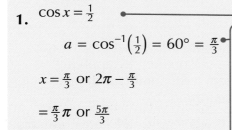

> If a calculator is not allowed you can use this triangle. ⓢ

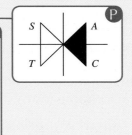

   $a = \cos^{-1}\left(\frac{1}{2}\right) = 60° = \frac{\pi}{3}$

   $x = \frac{\pi}{3}$ or $2\pi - \frac{\pi}{3}$

   $= \frac{\pi}{3}\pi$ or $\frac{5\pi}{3}$

or

2. $\cos x = -2$, which has no solution.

> -2 is less than -1 which is the minimum value of the cosine function. ⓒ

   Combining solutions: $x = \frac{\pi}{3}\pi$ or $\frac{5\pi}{3}$

b $2\sin^2 x = \sin x,\ 0 < x \le 2\pi$

> Rearrange in standard quadratic form. ⓟ

   $2\sin^2 x - \sin x = 0$

   $\sin x(2\sin x - 1) = 0$

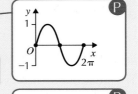

1. $\sin x = 0$

   $x = \cancel{0},\ \pi$ or 2π

> The strict inequality means that 0 must be discarded as a solution ⓒ

or

2. $\sin x = \frac{1}{2}$

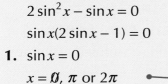

   $a = \sin^{-1}\left(\frac{1}{2}\right) = 30° = \frac{\pi}{6}$

   $x = \frac{\pi}{6}$ or $\pi - \frac{\pi}{6}$

   $= \frac{\pi}{6}$ or $\frac{5\pi}{6}$

   Combining solutions: $x = \frac{\pi}{6}, \frac{5\pi}{6}\ \pi$ or 2π

Exercise 8E

Express your solutions in terms of π where possible, otherwise round to 3 decimal places.

★ **1** Solve:

    a $2\sin^2 x - 1 = 0, \ 0 \le x \le 2\pi$ **b** $4\cos^2\left(\beta - \frac{\pi}{3}\right) = 1, \ 0 \le \beta \le \pi$

    c $\tan^2 x - 3 = 0, \ 0 \le \theta \le 2\pi$ **d** $5\cos^2\left(t + \frac{\pi}{2}\right) = 3, \ 0 \le t \le 2\pi$

★ **2** Solve:

    a $2\cos^2 x + \cos x - 1, \ 0 \le x \le 2\pi$ **b** $2\sin^2 x - \sqrt{3}\sin x = 0, \ 0 \le x \le 2\pi$

    c $6\sin^2 x - \cos x - 1 = 0, \ 0 \le x \le 2\pi$ **d** $2\sin^2 x - 5\sin x + 2 = 0, \ 0 \le x \le 2\pi$

3 Solve $2\sin^2\left(x - \frac{\pi}{6}\right) + \sin\left(x - \frac{\pi}{6}\right) - 1 = 0, \ 0 \le x < \pi$

Solving equations using standard trigonometric identities in degrees

Now that you can solve both linear and quadratic trigonometric equations the next stage is learning how to deal with three standard types of equation by applying trigonometric identities met in chapter 2 to reduce them to quadratic or other factorisable forms. These standard equations are of the forms:

- $a\cos^2 x° + b\sin x° + c = 0$ or $a\sin^2 x° + b\cos x° + c = 0$ where a, b and c are constants
- $a\cos 2x° + b\cos x° + c = 0$ or $a\cos 2x° + b\sin x° + c = 0$ where a, b and c are constants
- $a\sin 2x° + b\cos x° = 0$ or $a\sin 2x° + b\sin x° = 0$, where a, b and c are constants.

Solving equations of the form $a\cos^2 x° + b\sin x° + c = 0$ or $a\sin 2x° + b\cos x° + c = 0$ where a, b and c are constants

Using the identity $\sin^2 x° + \cos^2 x° = 1$ $\cos^2 x°$ and $\sin^2 x°$ can be substituted as follows:

$$\cos^2 x° = 1 - \sin^2 x° \quad \text{and} \quad \sin^2 x° = 1 - \cos^2 x°$$

These substitutions will convert the equations into quadratic forms in either $\cos x°$ or $\sin x°$, which you can already solve.

Example 8.13

Solve these equations.

a $5\sin^2 x° - 2\cos x° - 2 = 0, \ 0 \le x \le 360$ **b** $6\cos^2 x° + \sin x° = 5, \ 0 \le x \le 360$

a $5\sin^2 x° - 2\cos x° - 2 = 0$

    $5(1 - \cos^2 x°) - 2\cos x° - 2 = 0$ •————

> Change $\sin^2 x$ to $1 - \cos^2 x$ to form a quadratic equation in $\cos x$. Ⓢ

> Always remember to use brackets when substituting an expression. Ⓒ

$5 - 5\cos^2 x° - 2\cos x° - 2 = 0$

$-5\cos^2 x° - 2\cos x° + 3 = 0$

$5\cos^2 x° + 2\cos x° - 3 = 0$

$(5\cos x° - 3)(\cos x° + 1) = 0$

> Rearrange in standard form.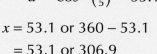

1. $\cos x° = \frac{3}{5}$

$a = \cos^{-1}\left(\frac{3}{5}\right) = 53.1°$

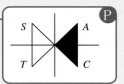

$x = 53.1$ or $360 - 53.1$

$= 53.1$ or 306.9

or

2. $\cos x° = -1$

$\Rightarrow x = 180$

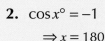

Combining solutions: $x = 53.1$, 180 or 306.9

b $6\cos^2 x° + \sin x° = 5$, $0 \le x \le 360$

> Change $\cos^2 x$ to $1 - \sin^2 x$ to form a quadratic equation in $\sin x$. ⑤

$6(1 - \sin^2 x°) + \sin x° = 5$

$6 - 6\sin^2 x° + \sin x° - 5 = 0$

$-6\sin^2 x° + \sin x° + 1 = 0$

$6\sin^2 x° - \sin x° - 1 = 0$

$(3\sin x° + 1)(2\sin x° - 1) = 0$

1. $\sin x° = -\frac{1}{3}$

$a = \sin^{-1}\left(\frac{1}{3}\right) = 19.5°$

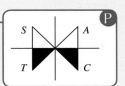

$x = 180 + 19.5 \quad 360 - 19.5$

$= 199.5$ or 340.5

or

2. $\sin x° = \frac{1}{2}$

$a = \sin^{-1}\left(\frac{1}{2}\right) = 30°$

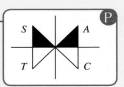

$x = 30$, $180 - 30$

$= 30$ or 150

Combining solutions $x = 30$, 150, 199.5 or 340.5

Solving equations of the form $a\cos 2x° + b\cos x° + c = 0$ or $a\cos 2x° + b\sin x° + c = 0$ where a, b and c are constants

In these equations, there are two different angles $2x°$ and $x°$, where one is double the other. Using the double angle formula (Chapter 2), $\cos 2x°$ can be written in terms of either $\cos^2 x$ or $\sin^2 x$ as follows:

$$\cos 2x° = 2\cos^2 x° - 1 \qquad \text{or} \qquad \cos 2x° = 1 - 2\sin^2 x°$$

The aim is to establish a quadratic equation in either $\cos x°$ or $\sin x°$, so the substitution must be chosen carefully. If the equation already contains $\cos x°$ you will replace $\cos 2x°$ with $2\cos^2 x° - 1$ and if it already contains $\sin x°$ you will use $1 - 2\sin^2 x°$.

Example 8.14

Solve these equations

a $\cos 2x° - 5\sin x° + 2 = 0,\ 0 \leq x \leq 360$

b $3\cos 2x° - 10\sin x° + 1 = 0,\ 0 \leq x \leq 360$

a $\cos 2x° - 5\sin x° + 2 = 0$

$(1 - 2\sin^2 x°) - 5\sin x° + 2 = 0$

$-2\sin^2 x° - 5\sin x° + 3 = 0$

$2\sin^2 x° + 5\sin x° - 3 = 0$

$(2\sin x° - 1)(\sin x° + 3) = 0$

> **S** $\sin x°$ appears in the equation alongside $\cos 2x°$ so use the substitution $\cos 2x° = 1 - 2\sin^2 x°$ to establish a quadratic equation in $\sin x°$.

1. $\sin x° = \frac{1}{2}$

$a = \sin^{-1}\left(\frac{1}{2}\right) = 30°$

$x = 30,\ 180 - 30$

$\quad = 30 \text{ or } 150$

or

2. $\sin x° = -3$, which has no solution

Combining solutions: $x = 30$ or 150

b $3\cos 2x° - 10\sin x° + 1 = 0$

$3(1 - 2\sin^2 x°) - 10\sin x° + 1 = 0$

$3 - 6\sin^2 x° - 10\sin x° + 1 = 0$

$-6\sin^2 x° - 10\sin x° + 4 = 0$

$6\sin^2 x° + 10\sin x° - 4 = 0$

$2(3\sin^2 x° + 5\sin x° - 2)$

$2(3\sin x° - 1)(\sin x° + 2) = 0$

> **S** To establish a quadratic equation in $\sin x°$ change $\cos x°$ to $1 - 2\sin^2 x°$.

1. $\sin x° = \frac{1}{3}$

$a = \sin^{-1}\left(\frac{1}{3}\right) = 19.5°$

$x = 19.5 \text{ or } 180 - 19.5$

$\quad = 19.5 \text{ or } 160.5$

or

2. $\sin x° = -2$, which has no solution.

Combining solutions: $x = 19.5$ or 160.5

Solving trigonometric equations of the form $a \sin 2x° + b \cos x° = 0$ or $a \sin 2x° + b \sin x° = 0$, where a, b and c are constants

This time use the double angle formula for $\sin 2x°$

$$\sin 2x° = 2 \sin x° \cos x°$$

The resulting equation will not be a quadratic, but it will factorise using a common factor, to give two linear equations.

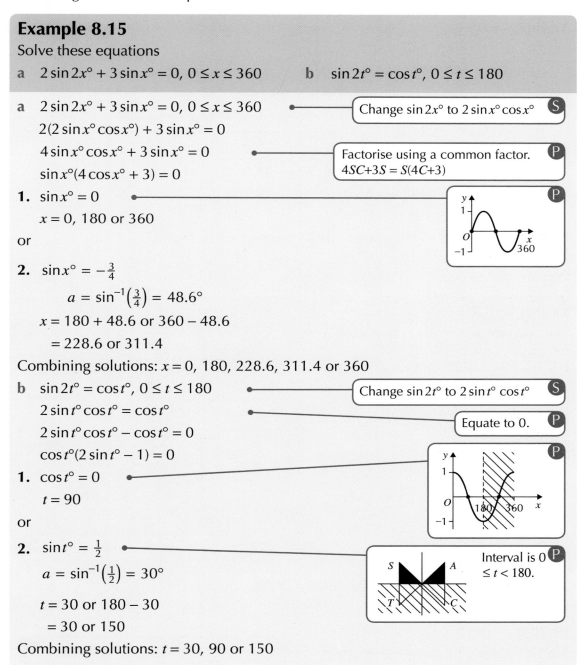

Example 8.15

Solve these equations

a $2 \sin 2x° + 3 \sin x° = 0$, $0 \leq x \leq 360$ **b** $\sin 2t° = \cos t°$, $0 \leq t \leq 180$

a $2 \sin 2x° + 3 \sin x° = 0$, $0 \leq x \leq 360$ — Change $\sin 2x°$ to $2 \sin x° \cos x°$ Ⓢ

$2(2 \sin x° \cos x°) + 3 \sin x° = 0$

$4 \sin x° \cos x° + 3 \sin x° = 0$ — Factorise using a common factor. Ⓟ
$4SC + 3S = S(4C + 3)$

$\sin x°(4 \cos x° + 3) = 0$

1. $\sin x° = 0$ Ⓟ

$x = 0, 180$ or 360

or

2. $\sin x° = -\frac{3}{4}$

$a = \sin^{-1}\left(\frac{3}{4}\right) = 48.6°$

$x = 180 + 48.6$ or $360 - 48.6$

$= 228.6$ or 311.4

Combining solutions: $x = 0, 180, 228.6, 311.4$ or 360

b $\sin 2t° = \cos t°$, $0 \leq t \leq 180$ — Change $\sin 2t°$ to $2 \sin t° \cos t°$ Ⓢ

$2 \sin t° \cos t° = \cos t°$

$2 \sin t° \cos t° - \cos t° = 0$ — Equate to 0. Ⓟ

$\cos t°(2 \sin t° - 1) = 0$ Ⓟ

1. $\cos t° = 0$

$t = 90$

or

2. $\sin t° = \frac{1}{2}$

$a = \sin^{-1}\left(\frac{1}{2}\right) = 30°$

$t = 30$ or $180 - 30$ — Interval is $0 \leq t < 180$. Ⓟ

$= 30$ or 150

Combining solutions: $t = 30, 90$ or 150

Example 8.16

Solve the equation $3\sin 4x° - 5\sin 2x° = 0$, $0 \le x \le 360$ giving solutions correct to the nearest whole number of degrees.

$3\sin 4x° - 5\sin 2x° = 0$ Change $\sin 4x°$ to $2\sin 2x° \cos 2x°$. **S**

$3(2\sin 2x° \cos 2x°) - 5\sin 2x° = 0$, $0 \le 2x \le 720$ Equate to 0. **P**

$6\sin 2x° \cos 2x° - 5\sin 2x° = 0$

$\sin 2x°(6\cos 2x° - 5) = 0$

1. $\sin 2x° = 0$

$\quad 2x = 0, 180, 360, 540, 720$

$\quad x = 0, 90, 180, 270, 360$

or

$\quad \cos 2x° = \frac{5}{6}$

$\quad\quad a = \cos^{-1}\left(\frac{5}{6}\right) = 33.6°$

$\quad 2x = 33.6, 360 - 33.6, 360 + 33.6, 720 - 33.6$

$\quad\quad = 33.6, 326.4, 393.6, 686.4$

$\quad x = 16.8, 163.2, 196.8, 343.2$

As the interval is $0 \le 2x \le 720$, the sine graph must be extended and you will have to go round the quadrant diagram twice. **P**

Combining answers to the nearest whole number: $x = 0, 17, 90, 163, 180, 197, 270, 343$ or 360

Example 8.17

The diagram shows the graphs of $y = \sin 2x°$ and $y = \cos x°$ $0 \le x \le 180$.

Find the coordinates of the point P.

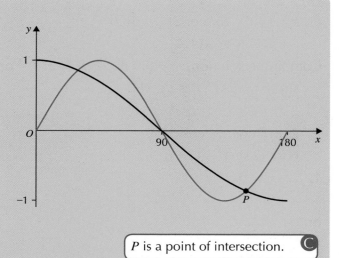

P is a point of intersection. **C**

P lies on both curves, so the y-coordinate of P satisfies both equations:

$y = \sin 2x°$ (1)

$y = \cos x°$ (2)

$\sin 2x° = \cos x°$ Solve the equations simultaneously. **P**

$2\sin x° \cos x° - \cos x° = 0$

$\cos x°(2\sin x° - 1) = 0$

so

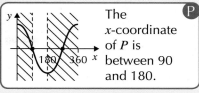

The x-coordinate of P is between 90 and 180.

1. $\cos x° = 0$

This has no solution, as the x-coordinate of P must lie between 90 and 180.

or

2. $\sin x° = \frac{1}{2}$

$\quad a° = \sin^{-1}\left(\frac{1}{2}\right) = 30°$

$x = 180 - 30 = 150$

using equation (2):

$x = 150 \Rightarrow y = \cos 150°$

$\qquad = -\cos 30°$

$\qquad = -\frac{\sqrt{3}}{2}$

150° is in the second quadrant so $\cos 150°$ is negative. $150 = 180 - 30$ so the related acute angle is 30°.

So the coordinates of P are $\left(150, -\frac{\sqrt{3}}{2}\right)$.

Exercise 8F

★ **1** Solve these equations.

    a $\cos^2 x° + 2\sin x° + 2 = 0,\ 0 \leq x \leq 360$ **b** $6\sin^2 x° = 4 - \cos x°,\ 0 \leq x \leq 360$

    c $6\sin^2 t° - \cos t° = 5,\ 0 \leq t \leq 180$ **d** $8\cos^2 x° - 2\sin x° - 5 = 0$

★ **2** Solve these equations for $0 \leq x \leq 360$.

    a $\cos 2x° + 3\cos x° + 2 = 0$ **b** $\cos 2p° - 4\sin p° + 5 = 0$

    c $\cos 2p° - 7\sin p° - 4 = 0$ **d** $\cos 2x° - 3\sin x° - 1 = 0$

    e $5\cos 2x° - \cos x° + 2 = 0$ **f** $3\cos 2t° + \cos t° + 1 = 0$

    g $2\cos 2x° + \cos x° - 1 = 0$ **h** $5\cos 2x° - \cos x° - 2 = 0$

★ **3** Solve these equations.

    a $\sin 2x° + \cos x° = 0,\ 0 \leq x \leq 360$ **b** $\sin 2x° - 3\sin x° = 0,\ 0 \leq x \leq 360$

    c $3\sin 2t° - 2\sin t° = 0,\ 0 \leq t \leq 720$ **d** $5\sin 2t° = 3\cos t°,\ 0 \leq t \leq 360$

4 Solve without the use of a calculator.

    a $2\sin^2(3x)° + \cos(3x)° - 1 = 0,\ 0 \leq x \leq 90$

    b $\cos 4x° + 5\cos 2x° - 2 = 0,\ 0 \leq x < 180$

    c $\sin x° + \sin\left(\frac{x}{2}\right)° = 0,\ 0 \leq x \leq 180$

5 The diagram shows part of the graphs of $y = 2\cos 2x° - 1$ and $y = 3\sin x°$

    Find the x-coordinates of A, B and C.

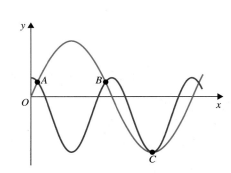

6 The graph shows the functions $y = \sin x$ and $y = p \sin qx$, $0 \le x \le 360$.

a Write down the values of p and q.

b Find the coordinates of all of the points of intersection in the interval $0 \le x \le 360$.

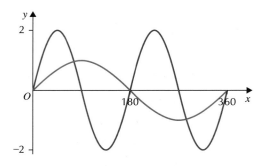

Solving equations in radians using standard trigonometric identities

Methods used for solving equations expressed in radians are the same as used for equations expressed in degrees.

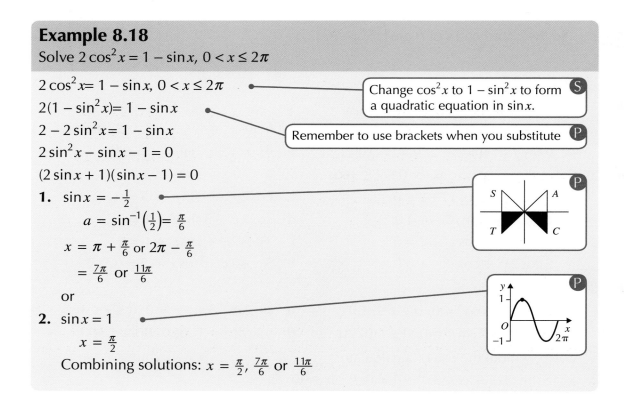

Example 8.18

Solve $2\cos^2 x = 1 - \sin x$, $0 < x \le 2\pi$

$2\cos^2 x = 1 - \sin x$, $0 < x \le 2\pi$

$2(1 - \sin^2 x) = 1 - \sin x$

$2 - 2\sin^2 x = 1 - \sin x$

$2\sin^2 x - \sin x - 1 = 0$

$(2\sin x + 1)(\sin x - 1) = 0$

1. $\sin x = -\frac{1}{2}$

$\quad a = \sin^{-1}\left(\frac{1}{2}\right) = \frac{\pi}{6}$

$\quad x = \pi + \frac{\pi}{6}$ or $2\pi - \frac{\pi}{6}$

$\quad = \frac{7\pi}{6}$ or $\frac{11\pi}{6}$

or

2. $\sin x = 1$

$\quad x = \frac{\pi}{2}$

Combining solutions: $x = \frac{\pi}{2}, \frac{7\pi}{6}$ or $\frac{11\pi}{6}$

> **S** Change $\cos^2 x$ to $1 - \sin^2 x$ to form a quadratic equation in $\sin x$.

> **P** Remember to use brackets when you substitute

Example 8.19

Solve $\cos 2x + 3 = 4\cos x$, $0 < x \leq 2\pi$

$(2\cos^2 x - 1) + 3 = 4\cos x$, $0 < x \leq 2\pi$

$2\cos^2 x - 4\cos x + 2 = 0$

$2(\cos^2 x - 2\cos x + 1) = 0$

$2(\cos x - 1)(\cos x - 1) = 0$

$\cos x = 1$ (twice)

$\qquad x = 0$ or 2π

> ⓢ Cos x appears in the equation alongside $\cos 2x$, so use the substitution $\cos 2x = 2\cos^2 x - 1$ to establish a quadratic equation in $\cos x$.

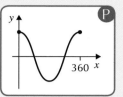

Exercise 8G

★1 Solve these equations.

   a $2\cos^2 x - \sin x + 1 = 0$, $0 \leq x \leq 2\pi$ **b** $1 - 4\sin^2 x = -\cos x$, $0 \leq x \leq 2\pi$

★2 Solve these equations.

   a $\cos 2x + \cos x = 0$, $0 \leq x \leq 2\pi$ **b** $\cos 2x = 3\sin x - 1$, $0 \leq x \leq 2\pi$

   c $\sin 2x = \sin x$, $0 \leq x \leq 2\pi$ **d** $\cos 2x - 1 = \sin x$, $0 \leq x \leq 2\pi$

   e $\sin 2x + \sqrt{3}\cos x = 0$, $0 \leq x \leq 2\pi$ **f** $\cos 2x = 1 - 3\cos x$, $0 \leq x \leq 2\pi$

3 Solve without the use of a calculator.

   a $2\cos^2 2x - \sin 2x - 1 = 0$, $0 \leq x \leq 2\pi$

   b $\cos 4\theta + \cos 2\theta = 0$, $0 \leq x \leq 2\pi$

Solving equations of the form $a\cos x + b\sin x + c = 0$ in both degrees and radians

Equations of the form $a\cos x + b\sin x + c = 0$ contain both the sine and cosine of the same angle. Solutions of these equations make use of the fact that $a\cos x + b\sin x$ can be expressed as $k\cos(x - \alpha)$ for some constants k and α (Chapter 2):

So, the equation

$a\cos x + b\sin x + c = 0$

can be converted to the linear equation

$k\cos(x - \alpha) + c = 0$

The equation could also be converted to $k\cos(x + \alpha) + c = 0$ or $k\sin(x \pm \alpha) + c = 0$.

Example 8.20

a   Rewrite $5\cos x° + 12\sin x°$ in the form $k\cos(x - \alpha)°$ where $k > 0$ and $0 \leq \alpha < 360$

b   Hence solve the equation $5\cos x° + 12\sin x° - 13 = 0$, $0 \leq x \leq 360$.

a   If $5\cos x° + 12\sin x° = k\cos(x - \alpha)°$ it follows that

$5\cos x° + 12\sin x° = k\cos x° \cos\alpha° + k\sin x° \sin\alpha°$

> Expand $k\cos(x - \alpha)°$. **S**

$5\underline{\cos x°} + 12\underline{\underline{\sin x°}} = (k\cos\alpha°)\underline{\cos x°} + (k\sin\alpha°)\underline{\underline{\sin x°}}$

> Rearrange RHS and use single and double underlining to help compare sides. **S**

Comparing sides gives:

$k\cos\alpha° = 5$ and $k\sin\alpha° = 12$

Find k:

$k = \sqrt{(5^2 + 2^2)} = 13$

Find α:

$\tan\alpha° = \frac{12}{5}$ where α is in the first quadrant

> $\tan\alpha = \frac{k\sin\alpha}{k\cos\alpha}$ and α is in the first quadrant as all three of sin, cos and tan are positive **S**

$\alpha° = \tan^{-1}\left(\frac{12}{5}\right) = 67.4°$

$\alpha = 67.4$

and so $5\cos x° + 12\sin x° = 13\cos(x - 67.4)°$

> Note the interval for the compound angle $(x - 67.4)°$. **P**

b $5\cos x° + 12\sin x° - 13 = 0$

$13\cos(x - 67.4)° - 13 = 0$, $-67.4 \leq x - 67.4 \leq 360 - 67.4$

$\cos(x - 67.4)° = 1$

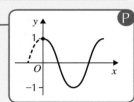

$x - 67.4 = 0$

$\quad x = 67.4$

Example 8.21

a   Rewrite $2\cos x° - 3\sin x°$ in the form $k\sin(x + \alpha)°$ where $k > 0$ and $0 \leq \alpha < 360$.

b   Hence solve the equation $2\cos x° - 3\sin x° = 1$, $0 \leq x \leq 360$.

If $2\cos x° - 3\sin x° = k\sin(x + \alpha)°$ it follows that:

$2\cos x° - 3\sin x° = k\sin x° \cos\alpha° + k\cos x° \sin\alpha°$

$2\underline{\cos x°} - 3\underline{\underline{\sin x°}} = (k\cos\alpha°)\underline{\underline{\sin x°}} + (k\sin\alpha°)\underline{\cos x°}$

> Rearrange RHS and single and double underline as shown. This is especially important here as the $\cos x°$ and $\sin x°$ don't come in the same order on both sides. **S**

Comparing sides:

$k\cos\alpha° = -3$ and $k\sin\alpha° = 2$

Find k:

$k = \sqrt{((-3)^2 + 2^2)} = \sqrt{13}$

Find α:

$\tan\alpha° = -\frac{2}{3}$ where α is in the 2nd quadrant

$\alpha° = 180° - \tan^{-1}\left(\frac{2}{3}\right)$

$\alpha = 146.3$

And so $2\cos x° - 3\sin x° = \sqrt{13}\sin(x + 146.3)°$

> Sin$\alpha°$ is positive but both $\cos\alpha°$ and $\tan\alpha°$ are negative, so α is in the 2nd quadrant. C

b $2\cos x° - 3\sin x° = 1$

$\sqrt{13}\sin(x + 146.3)° = 1$, $146.3 \le (x + 146.3)° \le 360 + 146.3$

$\sin(x + 146.3)° = \frac{1}{\sqrt{13}}$

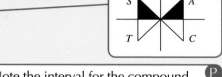

$a = \sin^{-1}\left(\frac{1}{\sqrt{13}}\right) = 16.1°$

$x + 146.3 = 180 - 16.1$ or $360 + 16.1$

$= 163.9$ or 376.1

$x = 17.8$ or 229.8

> Note the interval for the compound angle $(x + 146.3)°$. Going round the quadrant diagram starting at $146.3°$, the first angle collected is $(180 - 16.1)°$, followed by $(360 + 16.1)°$. The next angle would be $(540 - 16.1)°$ but that is eliminated because it is too big. P

Example 8.22

Solve the equation $\sqrt{3}\sin x - \cos x = 2$, $0 \le x \le 2\pi$

This example, unlike examples 8.20 and 8.21, gives you free choice of which of the four addition formulae to use.

To decide which would be the easiest expansion to use, both for comparing the two sides and for ensuring that α is an acute angle, you can compare $\sqrt{3}\sin x - \cos x$ with the four possibilities shown:

$\sqrt{3}\underline{\sin x} - \underline{\cos x}$

$k\cos(x - \alpha) = k\underline{\cos x}\cos\alpha + k\underline{\sin x}\sin\alpha$

$k\cos(x + \alpha) = k\underline{\cos x}\cos\alpha - k\underline{\sin x}\sin\alpha$

$k\sin(x + \alpha) = k\underline{\sin x}\cos\alpha + k\underline{\cos x}\sin\alpha$

$k\sin(x - \alpha) = k\underline{\sin x}\cos\alpha - k\underline{\cos x}\sin\alpha$

$k\sin(x - \alpha)$ gives the most obvious comparison and will ensure that both $\cos\alpha$ and $\sin\alpha$ are positive, making α an acute angle.

Rewrite $\sqrt{3}\sin x - \cos x$ in the form $k\sin(x - \alpha)$:

If $\sqrt{3}\sin x - \cos x = k\sin(x - \alpha)$ it follows that

$\sqrt{3}\sin x - \cos x = k\sin x\cos\alpha - k\cos x\sin\alpha$

$\sqrt{3}\underline{\sin x} - \underline{\cos x} = (k\cos\alpha)\underline{\sin x} - (k\sin\alpha)\underline{\cos x}$

comparing sides gives:

$k\cos\alpha = \sqrt{3}$ and $k\sin\alpha = 1$

Find k:

$k = \sqrt{((\sqrt{3})^2 + 1^2)} = 2$

Find α

$\tan\alpha = \frac{1}{\sqrt{3}}$ with α in the first quadrant

$\alpha = \frac{\pi}{6}$

(continued)

Rewrite the equation:

$\sqrt{3}\sin x - \cos x = 2$

$2\sin\left(x - \frac{\pi}{6}\right) = 2, \; -\frac{\pi}{6} \le x - \frac{\pi}{6} \le 2\pi - \frac{\pi}{6}$

$\sin\left(x - \frac{\pi}{6}\right) = 1$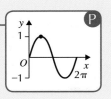

$\left(x - \frac{\pi}{6}\right) = 90° = \frac{\pi}{2}$

$x = \frac{\pi}{2} + \frac{\pi}{6} = \frac{2\pi}{3}$

Example 8.23

Solve the equation $8\cos 2x° - 6\sin 2x° - 5 = 0, \; 0 \le x \le 360$

Rewrite $8\cos 2x° - 6\sin 2x°$ in the form $k\cos(2x + \alpha)°$

If $8\cos 2x° - 6\sin 2x° = k\cos(2x + \alpha)$ it follows that:

$8\cos 2x° - 6\sin 2x° = k\cos 2x° \cos\alpha° - k\sin 2x° \sin\alpha°$

$8\underline{\cos 2x°} - 6\underline{\sin 2x°} = (k\cos\alpha°)\underline{\cos 2x°} - (k\sin\alpha°)\underline{\sin 2x°}$

Comparing sides gives:

$k\cos\alpha° = 8$ and $k\sin\alpha° = 6$

> This equation has a compound angle $2x$ so first find $2x$ and hence x. Although any one of the four expansions could be used, the best match for ease of working is $k\cos$ (sum)

Find k:

$k = \sqrt{(8^2 + 6^2)} = 10$

Find α

$\tan\alpha° = \frac{6}{8}$ with $\alpha°$ in the first quadrant

$\alpha° = \tan^{-1}\left(\frac{6}{8}\right)$

$\alpha = 36.9$

Rewrite the equation:

$8\cos 2x° + 6\sin 2x° - 5 = 0$

$10\cos(2x + 36.9)° = 5, \; 36.9 \le 2x + 36.9 \le 720 + 36.9$

$\cos(2x + 36.9)° = \frac{1}{2}$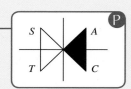

$a = \cos^{-1}\left(\frac{1}{2}\right) = 60°$

$2x + 36.9 = 60, \; 360 - 60, \; 360 + 60, \; 720 - 60, \; \cancel{720 + 60}$

$= 60, \; 300, \; 420, \; 660$

$2x = 23.1, \; 263.1, \; 273.1, \; 623.1$

$x = 11.6, \; 131.6, \; 136.6, \; 311.6$

Example 8.24

The depth of water, d metres, at the entrance to a small harbour varies with the tide. Over a 12 hour period, the depth, d, can be approximated by the formula

$d = 3 + 0.7(\cos 30t° + \sqrt{3}\sin 30t°)$, $0 \le t \le 12$

where t is the number of hours after midnight.

The graph is shown without scales on the axes.

a   Rewrite $\cos 30t° + \sqrt{3}\sin 30t°$ in the form $k\cos(30t - \alpha)°$.

b   State the coordinates of A and explain what they tell you in the context of this problem.

c   Repeat for the point B.

d   A fishing boat needs a depth of at least 2.1 metres to enter the harbour safely without the risk of running aground. Between which times should it avoid entering the harbour?

---

a   If $\cos 30t° + \sqrt{3}\sin 30t° = k\cos(30t - \alpha)°$ it follows that:

$\cos 30t° + \sqrt{3}\sin 30t° = k\cos 30t°\cos\alpha° + k\sin 30t°\sin\alpha°$

$\underline{\cos 30t°} + \sqrt{3}\underline{\sin 30t°} = (k\cos\alpha°)\underline{\cos 30t°} + (k\sin\alpha°)\underline{\sin 30t°}$

Comparing sides gives:

$k\cos\alpha° = 1$ and $k\sin\alpha° = \sqrt{3}$

Find k:

$k = \sqrt{\left(1^2 + (\sqrt{3})^2\right)}$

$\quad = 2$

Find α:

$\tan\alpha° = \sqrt{3}$ and α is in the first quadrant

$\alpha° = \tan^{-1}\sqrt{3}$

$\alpha = 60$

so $\cos 30t° + \sqrt{3}\sin 30t° = 2\cos(30t - 60)°$

b   The formula for the depth is now $d = 3 + 1.4\cos(30t - 60)°$

At A: $30t - 60 = 0 \Rightarrow 30t = 60 \Rightarrow t = 2$ | It is a cosine curve, so it will have its **S** maximum value when the angle is $0°$.

and $d = 3 + 1.4\cos 0° = 4.4$

A is the point $(2, 4.4)$ which means that high tide will be at | Answer in the **C** context of tides.

2 am and the depth of the water at the entrance to the harbour will reach a maximum depth of 4.4 metres at that time.

c   At B: $30t - 60 = 180 \Rightarrow 30t = 240 \Rightarrow t = 8$ | The minimum value will occur **S** when the angle is $180°$.

and $d = 3 + 1.4\cos 180° = 1.6$

B is the point $(8, 1.6)$ and so low tide will be at 8 am when the depth of the water at the entrance to the harbour will fall to 1.6 metres.

(continued)

d The boat cannot enter the harbour between the times shown by C and D.

At C and D:

$d = 3 + 1.4\cos(30t - 60)°$ the points are on the curve (1)

$d = 2.1$ the points are on the straight line (2)

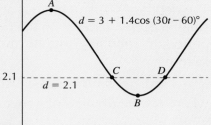

$3 + 1.4\cos(30t - 60)° = 2.1,\ -60 \le (30t - 60) \le 300$

$1.4\cos(30t - 60)° = -0.9$

$\cos(30t - 60)° = -\frac{0.9}{1.4},$

$\quad a = \cos^{-1}\left(\frac{0.9}{1.4}\right) = 50°$

$30t - 60 = 180 - 50 \text{ or } 180 + 50$

$30t = 130 + 60 \text{ or } 230 + 60$

$t = 6.33 \text{ or } 9.67$

> Solve the equations simultaneously. Ⓟ

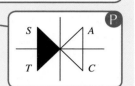

This means that the depth of water is predicted to be 2.1 metres or less between 6.20 am and 9.40 am. However, the question did say that the depth can only be approximated by the curve and so to make sure it would be better to avoid entering the harbour between 6 am and 10 am.

Exercise 8H

1 a Express $7\cos x° + 4\sin x°$ in the form $k\cos(x - \alpha)°$

  b Hence solve the equation $7\cos x° + 4\sin x° = 5,\ 0 \le x < 360$

2 a Express $\sqrt{3}\sin\theta + \cos\theta$ in the form $k\sin(\theta + \alpha)$

  b Hence solve the equation $\sqrt{3}\sin\theta + \cos\theta = 1,\ 0 \le x \le 2\pi$

3 a Express the equation $4\cos x° - 3\sin x°$ in the form $R\cos(x - \beta)°$

  b Hence solve the equation $4\cos x° - 3\sin x° = 2,\ 0 \le x < 360$

★ **4** Solve each of these equations for $0 \le x < 360$.

  a $3\cos x° + \sin x° = 1$ **b** $2\cos x° - \sqrt{5}\sin x° = 2$

  c $2\sin x° + \sin x° = 1$ **d** $3\sin x° - \cos x° + 1 = 0$

  e $2\cos x° + 3\sin x° - 1 = 0$ **f** $3\cos x° + \sin x° = 1$

> ⚠ The equations can be solved using any one of the four addition formulae expansions, but a sensible choice does simplify things.

5 Solve $3\cos 2\theta + 4\sin 2\theta = 5$ for $0 \le \theta \le 2\pi$

6 The displacement, d cm, of a wave after t seconds is given by the formula:

$d = \cos 20t° - \sqrt{3}\sin 20t°$

Find the values of t between 0 and 18 for which the displacement is 1.6 cm

7 The depth, d m, of water at the entrance to a port t hours after midnight can be approximated by the formula:

$d = 2 + 0.4(\sin 30t° + 2\cos 30t°)$.

By first rewriting $\sin 30t° + 2\cos 30t°$ in the form $k\sin(30t° + \alpha)°$ answer these questions:

a What are the times of high and low tide?

b A ship arrives at the harbour at 9 am. If it needs a minimum depth of 1.9 metres to enter the harbour, how long will it have to wait?

Solving further trigonometric equations in degrees or radians

You will be expected to deal with any example which contains a recognisable trigonometric identity from Chapter 2.

Example 8.25

Solve the equation $\cos x° \cos 50° - \sin x° \sin 50° = 0.5$ where $0 \le x \le 360$

$\cos x° \cos 50° - \sin x° \sin 50° = 0.5$ Recognise as cos (sum). **S**

$\cos(x + 50)° = 0.5$ where $50 \le x + 50 \le 410$

$\qquad a = \cos^{-1} 0.5 = 60°$

$x + 50 = 60$ or 300

$x = 10$ or 250

Other equations, while still relying on trigonometric identities to reduce them to linear or quadratic from will have a lead in at the start of the question.

Example 8.26

a Show that $\sin\left(x + \frac{\pi}{3}\right) + \sin\left(x - \frac{\pi}{3}\right) = \sin x$

b Hence solve the equation $\sin\left(x + \frac{\pi}{3}\right) + \sin\left(x - \frac{\pi}{3}\right) = 0.5$ where $0 \le x \le 2\pi$

a LHS $= \sin\left(x + \frac{\pi}{3}\right) + \sin\left(x - \frac{\pi}{3}\right)$ RHS $= \sin x$

Separate LHS and RHS **S** then expand sin (sum) and sin (difference) on the LHS.

$\qquad = \sin x \cos\frac{\pi}{3} + \cos x \sin\frac{\pi}{3} + \sin x \cos\frac{\pi}{3} - \cos x \sin\frac{\pi}{3}$

$\qquad = 2\sin x \cos\frac{\pi}{3}$

$\qquad = 2 \times \sin x \times \frac{1}{2}$

$\qquad = \sin x$

LHS $=$ RHS

b $\sin\left(x + \frac{\pi}{3}\right) + \sin\left(x - \frac{\pi}{3}\right) = 0.5,\ 0 \le x \le 2\pi$ ← Use the identity from part **a**. Ⓢ

$\sin x = 0.5$ ← You now have a simple linear equation. Ⓒ

$a = \sin^{-1} 0.5 = 30° = \frac{\pi}{6}$

$x = \frac{\pi}{6}$ or $\pi - \frac{\pi}{6}$

$= \frac{\pi}{6}$ or $\frac{5\pi}{6}$

Example 8.27

a Show that $(\cos\theta - \sin\theta)^2 = 1 - \sin 2\theta$

b Hence solve the equation:
$(\cos\theta - \sin\theta)^2 = \frac{1}{2},\ 0 \le \theta \le 2\pi$ ← Separate LHS and RHS, multiply out brackets on LHS and use the $\sin$(double angle) formula on the RHS. Ⓢ

a $\text{LHS} = (\cos\theta - \sin\theta)^2$ $\qquad$ $\text{RHS} = 1 - \sin 2\theta$

$\quad = \cos^2\theta - 2\sin\theta\cos\theta + \sin^2\theta$ $\qquad = 1 - 2\sin\theta\cos\theta$

$\quad = (\cos^2\theta + \sin^2\theta) - 2\sin\theta\cos\theta$

$\quad = 1 - 2\sin\theta\cos\theta$ $\qquad$ You now have $-2\sin\theta\cos\theta$ on each side and should know that $(\cos^2\theta + \sin^2\theta) = 1$ Ⓢ

$\text{LHS} = \text{RHS}$

b $(\cos\theta - \sin\theta)^2 = \frac{1}{2},\ 0 \le \theta \le 2\pi$ ← Use part **a**. Ⓢ

$1 - \sin 2\theta = \frac{1}{2},\ 0 \le 2\theta \le 4\pi$ ← You now have a linear equation with a compound angle. Ⓒ

$\sin 2\theta = \frac{1}{2}$

$\quad a = \sin^{-1}\left(\frac{1}{2}\right) = 30° = \frac{\pi}{6}$

$2\theta = \frac{\pi}{6},\ \pi - \frac{\pi}{6},\ 2\pi + \frac{\pi}{6},\ 3\pi - \frac{\pi}{6}$

$\theta = \frac{\pi}{12},\ \frac{5\pi}{12},\ \frac{13\pi}{12},\ \frac{17\pi}{12}$

Exercise 8I

★ **1** Solve:

a $2\cos x \cos\left(\frac{\pi}{3}\right) - 2\sin x \sin\left(\frac{\pi}{3}\right) = 1,\ 0 \le x < 2\pi$

b $3\sin x° \cos 50° - 3\cos x° \sin 50° = 2,\ 0 \le x \le 360$

c $5\cos 2x° \cos 20° + 5\sin 2x° \sin 20° + 3 = 0,\ 0 \le x \le 270$

★ ⚙ **2 a** Show that $(\cos x - \sin x)(\cos x + \sin x) = \cos 2x$

b Hence solve the equation $(\cos x - \sin x)(\cos x + \sin x) + 1 = 0,\ 0 \le x < 3\pi$

⚙ **3 a** Show that $\sin 3x = 3\sin x - 4\sin^3 x$

b Hence solve the equation $3\sin x° - 4\sin^3 x° = \frac{1}{2},\ 0 \le x \le \pi$

4 a Show that $\sin\left(x + \frac{\pi}{6}\right) - \sin\left(x - \frac{\pi}{2}\right) = \sqrt{3}\cos\left(x - \frac{\pi}{6}\right)$

b Hence solve the equation $\sin\left(x + \frac{\pi}{6}\right) - \sin\left(x - \frac{\pi}{2}\right) = \frac{3}{2},\ 0 \le x \le \pi$

⚙ **5 a** Show that $2\sin(x - 60)° + \sin x°$ can be rewritten as $2\sin x° - \sqrt{3}\cos x°$

b Hence solve the equation $2\sin(x - 60)° = 1 - \sin x°,\ 0 \le x \le 360$.

- I can solve linear and quadratic trigonometric equations in degrees, including those involving compound angles. ★ Exercise 8A Q1, 2; Exercise 8B Q1, 2; Exercise 8C Q1, 2
- I can solve linear and quadratic trigonometric equations in radians, including those involving compound angles. ★ Exercise 8D Q2; Exercise 8E Q1, 2
- I can solve equations in degrees using standard trigonometric identities. ★ Exercise 8F QI–3
- I can solve equations in radians using standard trigonometric identities. ★ Exercise 8G Q1, 2
- I can solve equations, in both degrees and radians, involving the wave function $f(x) = a \cos x + b \sin x$. ★ Exercise 8H Q4
- I can solve equations using further trigonometric identities in both degrees and radians. ★ Exercise 8I Q1, 2

For further assessment opportunities, see the Preparation for Assessment for Unit 2 on pages 291–292.

9 Differentiating functions

This chapter will show you how to:

- differentiate an algebraic function which is, or can be simplified to, an expression in powers of x
- differentiate $k\sin x$, $k\cos x$
- differentiate a composite function using the chain rule.

You should already know:

- how to calculate the gradient of a straight line given suitable information
- how to apply the laws of indices to multiplying out brackets and simplifying
- how to solve linear and quadratic equations
- how to perform arithmetic operations on numerical and algebraic fractions
- the key properties of the graphs $y = a\sin(bx + c) + d$ and $y = a\cos(bx + c) + d$
- how to convert between radians and degrees
- how to calculate the exact value of a simple trigonometric expression
- how to determine a composite function $f(g(x))$, given $f(x)$ and $g(x)$.

Introduction to differentiation

Calculus is the branch of mathematics which deals with change. It is probably the most important area of mathematics. There are two types of calculus and this chapter introduces the first of them: differential calculus.

Differential calculus is concerned with **rates of change** and is applied in the real world to model, describe and predict many types of behaviour and phenomena. For example, when a parachutist jumps from an aeroplane, we can use calculus to predict how fast they are falling at a given instant. When there is an outbreak of disease, calculus is used to describe how quickly the infection spreads and can also predict when the outbreak will peak and subsequently decline. Governments use calculus to predict the future population of the country. This is particularly important for long-term planning. To scientists and engineers, calculus is an indispensable tool for analysing and solving all manner of problems. The key process employed in differential calculus is called **differentiation**.

Rate of change and constant speed

Consider an aeroplane flying at a constant speed of 600 miles per hour.

The graph shows how far the plane travels over a period of time.

In each unit of time (1 hour) the plane travels a distance of 600 miles.

The **speed** of the plane is the **rate** at which the **distance** being covered **changes with respect to time**.

In this example, the **rate of change** of distance is 600 miles per hour.

The speed of the plane can be calculated from the graph by finding the **gradient** of the line.

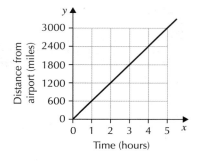

$$\text{Gradient} = \frac{600}{1} = \frac{1200}{2} = \frac{1800}{3} = \frac{2400}{4} = \frac{3000}{5} = 600$$

Gradient = rate of change

In mathematics, the process of determining a **rate of change** is called **differentiation**.

In the above example the equation of the line is $y = 600x$.

The distance y is a function of the time (x).

We could also write $f(x) = 600x$.

When we **differentiate** this function we get the **gradient** of the line. This gives us the **rate of change** of the function. The **rate of change** is known as the **derivative** of the function.

For the function $y = 600x$ the following statements are equivalent:

- the gradient of the graph of the function is 600
- the rate of change of y with respect to x is 600
- the derivative of y with respect to x is 600.

How we write the original function	What we write to show that we have differentiated	What we say
$y = 600x$	$\dfrac{dy}{dx} = 600$	'dee y by dee x' is equal to 600
$f(x) = 600x$	$f'(x) = 600$	f dash of x is equal to 600

In the example above, the speed of the plane at any given time is always 600 miles per hour.

So, when the time x is three hours, we would say:

- when $x = 3$, the speed of the plane is 600 miles per hour.

The following statements are equivalent to the one above:

- when $x = 3$, the gradient of the graph of the function is 600
- when $x = 3$, the rate of change of y (with respect to x) is 600
- when $x = 3$, the derivative of y (with respect to x) is 600
- when $x = 3$, the rate of change of f (with respect to x) is 600
- when $x = 3$, the derivative of f (with respect to x) is 600

We can therefore write:

When $x = 3$, $\frac{dy}{dx} = 600$ or when $x = 3$, $f'(x) = 600$ or simply $f'(3) = 600$.

It is vital that you know the language of differentiation and the notation used to communicate it.

Instantaneous speed and instantaneous rate of change

In this section you should make use of a graphing calculator or computer software to further explore the situation described. Consider the following scenario:

You stand on the edge of a cliff and drop a stone. The height of the cliff is 80 metres.

Assuming the stone falls vertically (and ignoring air resistance) what will its speed be after 2 seconds?

Physics tells us that after x seconds, the distance the stone has fallen, y metres, is given by the equation $y = 5x^2$.

Time in seconds (x)	0	1	2	3	4
Distance fallen in metres (y)	0	5	20	45	80

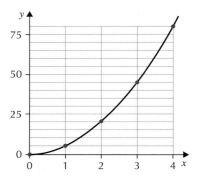

The graph of distance against time is a parabola.

We now know that it takes 4 seconds for the stone to reach the water, but we still don't know what its speed is after 2 seconds.

When $x = 2$ the stone has fallen 20 metres. When $x = 3$ the stone has fallen 45 metres (represented by points A and B in the diagram).

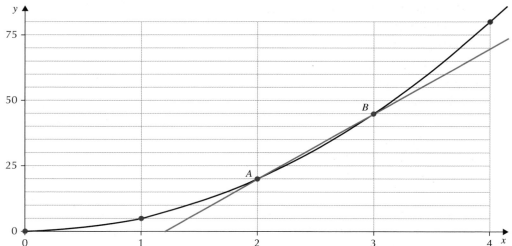

The **average speed** of the stone between $x = 2$ and $x = 3$ can be found by calculating the gradient of AB:

$$m_{AB} = \frac{45 - 20}{3 - 2} = \frac{25}{1} = 25 \text{ ms}^{-1}$$

If the stone travelled at a **constant speed** throughout this one second interval it would fall a further 25 metres as predicted by the equation $y = 5x^2$.

However, the stone is not travelling at constant speed. As it falls, the stone's speed constantly increases because of acceleration due to gravity.

By calculating the average speed of the stone over a shorter time interval we can obtain a better approximation to its **instantaneous speed**. The table shows what happens when we shorten the time interval after $x = 2$.

Time interval	$x = 2$ to $x = 3$	$x = 2$ to $x = 2.5$	$x = 2$ to $x = 2.25$
Duration (seconds)	1	0.5	0.25
Distance travelled (metres)	25	11.25	5.0625
Average speed in metres per second	$\frac{25}{1} = 25$ ms^{-1}	$\frac{11.25}{0.5} = 22.5$ ms^{-1}	$\frac{5.0625}{0.25} = 21.25$ ms^{-1}

The stone travels 11.25 metres in the 0.5 seconds after $x = 2$. If the stone was travelling at constant speed this would be equivalent to travelling 22.5 metres in 1 second, hence an average speed of 22.5 metres per second.

Similarly, we can say that the **average rate of change is 22.5 metres per second**.

Continuing in this way we get the following average speeds: 20.625, 20.3125, 20.15625...

By shortening the time interval further and further the average speed of the stone approaches 20 metres per second (as the time interval **approaches** zero).

We can now say that **at the instant** when $x = 2$ the speed of the stone is 20 metres per second. This is called its **instantaneous speed**.

Shortening the time interval after $x = 2$ (i.e. as B gets closer and closer to A), the gradient of AB changes and gives us a better and better approximation to the instantaneous speed when $x = 2$.

The argument used to calculate the instantaneous speed is called a **limiting argument** because the time interval is tending towards a limit of zero.

Also, as B moves closer and closer to A, the gradient of AB approaches the **gradient of the tangent to the curve** at A.

The instantaneous speed of the stone when $x = 2$ is the same as the gradient of the tangent at A (where $x = 2$) so we can now say that the **gradient of the tangent to the curve** at A is 20.

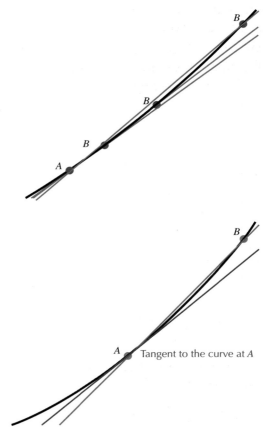

Tangent to the curve at A

We now make a definition:

The gradient of the curve at A is equal to the gradient of the tangent to the curve at A.

Using the notation and language introduced on page *xxx*, the following statements are all equivalent:

- when $x = 2$, the gradient of the tangent to the curve is 20
- when $x = 2$, the gradient of the curve is 20
- when $x = 2$, $\frac{dy}{dx} = 20$
- when $x = 2$, $f'(x) = 20$ or $f'(2) = 20$
- when $x = 2$, the derivative of y (with respect to x) is 20
- when $x = 2$, the derivative of f (with respect to x) is 20
- when $x = 2$, the rate of change of y (with respect to x) is 20.

We can calculate the instantaneous speed of the stone at different times (i.e. for different values of x).

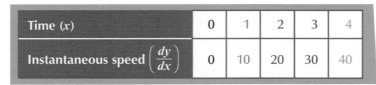

Time (x)	0	1	2	3	4
Instantaneous speed $\left(\dfrac{dy}{dx}\right)$	0	10	20	30	40

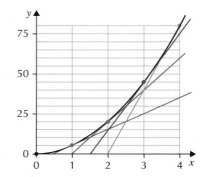

From the previous investigation two points should now be clear:

1. We can calculate the gradient of a curve (at a point) by calculating the gradient of the tangent to the curve (at that point).

2. The gradient of a curve is not constant – the gradient of the tangent changes as we move along the curve.

Gradient function

To find the gradient of the curve $y = 5x^2$ at any point we need only multiply the x coordinate of the point by 10.

For $y = 5x^2$, the **gradient function** is written $\frac{dy}{dx} = 10x$.

Once we choose a value for x (at a **specific** point on the curve), we can calculate the gradient of the curve.

Differentiating an algebraic function which is, or can be simplified to, an expression in powers of x

Differentiation is the process by which we obtain the gradient function of an algebraic function. Before carrying out the differentiation, it may be necessary to carry out some algebraic manipulation, so that the function is expressed correctly in terms of powers of x.

Differentiating a function which is already an expression in powers of x

The limiting argument used previously was a simplified application of the process known as differentiation from first principles. For any function, $f(x)$, we can use differentiation from first principles to find $f'(x)$. Later in this chapter, Exercise 9C explores this in more depth.

Differentiation from first principles is not examinable in Higher, but it is very important for Advanced Higher (AH) mathematics. It is strongly recommended that if you are planning to study AH you attempt Exercise 9C.

In this section, there are 3 rules which will allow you to find the derivative of a function:

Rule 1 $y = x^n \Rightarrow \dfrac{dy}{dx} = nx^{n-1}$

Rule 2 $y = ax^n \Rightarrow \dfrac{dy}{dx} = a \times nx^{n-1} = anx^{n-1}$

Rule 3 $y = f(x) + g(x) \Rightarrow \dfrac{dy}{dx} = f'(x) + g'(x)$

Rule 3 is known as the **sum rule**.

There are 2 other special derivatives you should know. They both follow from basic properties of the straight line.

$y = c$, **where c is a constant** $\Rightarrow \dfrac{dy}{dx} = 0$

$y = cx$, **where c is a constant** $\Rightarrow \dfrac{dy}{dx} = c$

> ⚠ The graph of the function $y = c$, where c is a constant, is a horizontal line. The gradient of a horizontal line is zero.

> ⚠ The straight line $y = cx$, where c is a constant, has a (constant) gradient equal to c.

Example 9.1

Differentiate with respect to x.

a x^5 b x^{-3} c $x^{\frac{5}{3}}$ d $-x^{-\frac{1}{2}}$

a $y = x^5$ — Write down the original function. ⓟ

$\dfrac{dy}{dx} = 5x^4$ — Use Rule 1 where $n = 5$. ⓟ

b $f(x) = x^{-3}$ — Use either $y = \ldots$ or $f(x) = \ldots$ to describe the function. ⓟ

$f'(x) = -3x^{-4}$ — Use Rule 1 where $n = -3$. Be careful – remember that $-3 - 1 = -4$ and not -2. Note the use of $f'(x)$ instead of $\frac{dy}{dx}$. ⓟ

$f'(x) = -3x^{-4} = -\dfrac{3}{x^4}$ — Remember that $x^{-4} = \frac{1}{x^4}$ and so $-3x^{-4} = -3 \times \frac{1}{x^4} = -\frac{3}{x^4}$. ⓟ

$= -\dfrac{3}{x^4}$ — Write the final answer with a positive index. It is good practice to write the final answer with a positive power (index) of x because you may have to go on and calculate the value of $\frac{dy}{dx}$. Ⓒ

(continued)

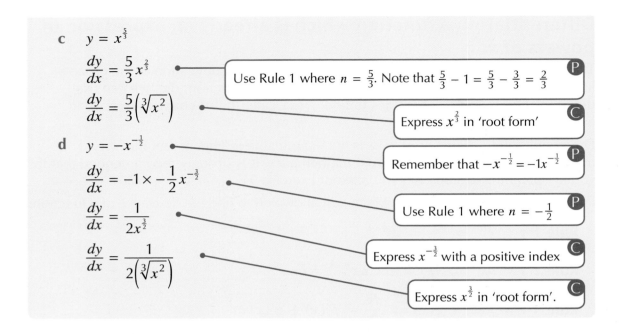

c $y = x^{\frac{5}{3}}$

$\dfrac{dy}{dx} = \dfrac{5}{3}x^{\frac{2}{3}}$ — Use Rule 1 where $n = \frac{5}{3}$. Note that $\frac{5}{3} - 1 = \frac{5}{3} - \frac{3}{3} = \frac{2}{3}$ **P**

$\dfrac{dy}{dx} = \dfrac{5}{3}\left(\sqrt[3]{x^2}\right)$ — Express $x^{\frac{2}{3}}$ in 'root form' **C**

d $y = -x^{-\frac{1}{2}}$ — Remember that $-x^{-\frac{1}{2}} = -1x^{-\frac{1}{2}}$ **P**

$\dfrac{dy}{dx} = -1 \times -\dfrac{1}{2}x^{-\frac{3}{2}}$ — Use Rule 1 where $n = -\frac{1}{2}$ **P**

$\dfrac{dy}{dx} = \dfrac{1}{2x^{\frac{3}{2}}}$ — Express $x^{-\frac{3}{2}}$ with a positive index **C**

$\dfrac{dy}{dx} = \dfrac{1}{2\left(\sqrt[3]{x^2}\right)}$ — Express $x^{\frac{3}{2}}$ in 'root form'. **C**

Example 9.2

Differentiate with respect to x.

a $3x^4$ **b** $\dfrac{3}{4}x^{-2}$ **c** $6x^{\frac{3}{2}}$ **d** $\dfrac{1}{4}x^{-\frac{1}{2}}$

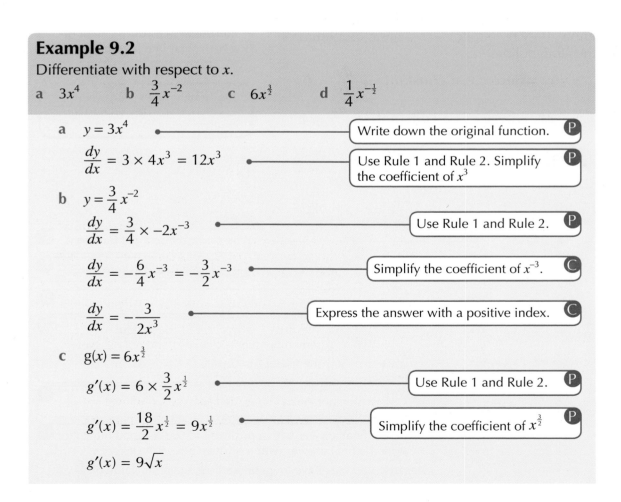

a $y = 3x^4$ — Write down the original function. **P**

$\dfrac{dy}{dx} = 3 \times 4x^3 = 12x^3$ — Use Rule 1 and Rule 2. Simplify the coefficient of x^3 **P**

b $y = \dfrac{3}{4}x^{-2}$

$\dfrac{dy}{dx} = \dfrac{3}{4} \times -2x^{-3}$ — Use Rule 1 and Rule 2. **P**

$\dfrac{dy}{dx} = -\dfrac{6}{4}x^{-3} = -\dfrac{3}{2}x^{-3}$ — Simplify the coefficient of x^{-3}. **C**

$\dfrac{dy}{dx} = -\dfrac{3}{2x^3}$ — Express the answer with a positive index. **C**

c $g(x) = 6x^{\frac{3}{2}}$

$g'(x) = 6 \times \dfrac{3}{2}x^{\frac{1}{2}}$ — Use Rule 1 and Rule 2. **P**

$g'(x) = \dfrac{18}{2}x^{\frac{1}{2}} = 9x^{\frac{1}{2}}$ — Simplify the coefficient of $x^{\frac{3}{2}}$ **P**

$g'(x) = 9\sqrt{x}$

d $y = \frac{1}{4}x^{-\frac{1}{2}}$

$\frac{dy}{dx} = \frac{1}{4} \times -\frac{1}{2}x^{-\frac{3}{2}}$ Use Rule 1 and Rule 2. **P**

$\frac{dy}{dx} = -\frac{1}{8}x^{-\frac{3}{2}}$

$\frac{dy}{dx} = -\frac{1}{8}x^{-\frac{3}{2}} = -\frac{1}{8x^{\frac{3}{2}}} = -\frac{1}{8\left(\sqrt{x^3}\right)}$ Express the answer with a positive index, writing $x^{\frac{3}{2}}$ in root form. **C**

Differentiating an expression with several terms

When you are differentiating an expression which contains several terms, you should proceed in this order:

1. Use the sum rule to differentiate term by term, applying Rules 1 and 2.
2. Simplify the resulting coefficients.
3. Ensure that all terms contain positive indices.
4. Express any fractional powers using 'root form'.

Example 9.3

Differentiate with respect to x.

a $4x^2 - 6x + 3$ **b** $\frac{1}{5}x^{-5} + 8 - 4x^{-\frac{3}{4}}$

a $y = 4x^2 - 6x + 3$

$\frac{dy}{dx} = 2 \times 4x^1 - 6 + 0$ Use the sum rule to differentiate term by term, applying Rules 1 and 2 first. The terms $6x$ and 3 should be treated as special derivatives, you do not need to apply Rule 1 or 2 to them **P**

$\frac{dy}{dx} = 8x - 6$

b $h(x) = \frac{1}{5}x^{-5} + 8 - 4x^{-\frac{3}{4}}$ Simplify the coefficient of x. **C**

$h'(x) = \frac{1}{5} \times -5x^{-6} + 0 - 4 \times -\frac{3}{4}x^{-\frac{7}{4}}$ Use the sum rule to differentiate term by term, applying rules 1 and 2 first. **P**

$h'(x) = -x^{-6} + 0 + 3x^{-\frac{7}{4}}$ Be careful with signs. Note that the final term in the answer is $+3x^{-\frac{7}{4}}$ and not $-3x^{-\frac{7}{4}}$ **P**

$h'(x) = -\frac{1}{x^6} + 0 + \frac{3}{x^{\frac{7}{4}}}$

$h'(x) = -\frac{1}{x^6} + \frac{3}{\sqrt[4]{x^7}}$

209

Example 9.4

Differentiate with respect to the given variable.

a $5p^7$ **b** $3t^{-2} - 6 - \dfrac{5}{2}t^6$

a $y = 5p^7$

$\dfrac{dy}{dp} = 5 \times 7p^6 = 35p^6$

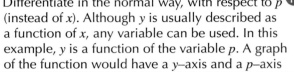

Differentiate in the normal way, with respect to p (instead of x). Although y is usually described as a function of x, any variable can be used. In this example, y is a function of the variable p. A graph of the function would have a y–axis and a p–axis

b $f(t) = 3t^{-2} - 6 - \dfrac{5}{2}t^6$

$f'(t) = 3 \times -2t^{-3} - 0 - \dfrac{5}{2} \times 6t^5$

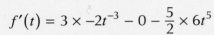

We write $f'(t)$ because we are differentiating with respect to t

$f'(t) = -6t^{-3} - 15t^5$

$f'(t) = -\dfrac{6}{t^3} - 15t^5$

Exercise 9A

1 Differentiate with respect to x.

   a x^8 **b** $-x^4$ **c** x **d** x^{-1} **e** x^{-6} **f** $-x^{-9}$

   g 4 **h** $4x^3$ **i** $\dfrac{3}{4}x^8$ **j** $9x^{-2}$ **k** -5 **l** $-\dfrac{1}{6}x^{-3}$

   m $x^{\frac{3}{2}}$ **n** $x^{\frac{5}{3}}$ **o** $x^{\frac{7}{5}}$ **p** $x^{\frac{1}{2}}$ **q** $x^{\frac{1}{4}}$ **r** $x^{\frac{2}{3}}$

   s $x^{-\frac{1}{2}}$ **t** $-x^{-\frac{2}{3}}$ **u** $8x^{\frac{1}{2}}$ **v** $6x^{-\frac{1}{3}}$ **w** $-\dfrac{5}{3}x^{\frac{3}{2}}$

   x $-\dfrac{1}{4}$ **y** $\dfrac{3}{7}x^{-\frac{7}{4}}$

2 Differentiate with respect to x.

   a $x^2 + 2x - 1$ **b** $-2x^2 - 8x + 3$ **c** $3x - 4$

   d $x^3 - 4x^2 + 8x - 10$ **e** $\dfrac{1}{3}x^3 + 2x^2 - 12x + 4$ **f** $x^4 - 2x^3 + 3x^2 - x - 6$

   g $-5x^3 - 2x^{-6}$ **h** $4 - 2x - x^{-6}$ **i** $\dfrac{1}{2}x^4 + 2x - 3x^{-2}$

   j $\dfrac{3}{10}x^5 - 7 - x^{-1}$

3 Differentiate with respect to x.

   a $x^{\frac{1}{2}} + x^{-\frac{1}{2}}$ **b** $12x^{\frac{3}{2}} - x^{\frac{5}{2}}$ **c** $6x^3 - 2x^{\frac{1}{2}}$

   d $\dfrac{2}{5}x^{\frac{5}{4}} - 3x^{-\frac{1}{3}}$ **e** $4x^{-\frac{1}{2}} + \dfrac{4}{3}x^{-6} - 8x$ **f** $\dfrac{2}{3}x^{-6} + 8x^{\frac{1}{4}}$

   g $8x^{\frac{3}{4}} - \dfrac{2}{5}x^{-\frac{1}{2}}$ **h** $-4x^{-\frac{1}{4}} + \dfrac{4}{3}x^{-6} - 8x$ **i** $\dfrac{4}{5}x^{\frac{1}{4}} - \dfrac{5}{4}x^{-\frac{1}{5}}$

4 **a** Given $y = x^3 - 4x^2 + x - 8$, find $\frac{dy}{dx}$

  b Given $f(x) = x^{\frac{4}{3}}$, find $f'(x)$.

  c $y = 6x^{\frac{1}{2}}$. Find the derivative of y with respect to x.

  d $f(x) = 2x^2 - 3x - 4$. Find the rate of change of f.

  e Differentiate $x^{\frac{3}{4}} - 8x$.

5 Differentiate with respect to the given variable.

  a $p^4 - 6p^2 + 9$ **b** $3p^5 - p^{-2}$ **c** $w^3 - 3w + 6$ **d** $5t - 10t^3$

  e $8t^{\frac{3}{2}} - 4t^{-\frac{1}{2}}$ **f** $4u^{\frac{2}{3}} - u$ **g** $6t^{-4} - 8t + 7$

★ 6 The table below contains eight functions and their derivatives. Five of the eight derivatives in the second column contain mistakes. Find all the mistakes and correct them.

	$f(x)$	$f'(x)$
a	$4x^{-2} - 5x - 3$	$-\dfrac{8}{x^3} - 8$
b	$8x^{\frac{5}{2}} - 2x^{-1}$	$20\left(\sqrt{x^3}\right) + \dfrac{2}{x^2}$
c	$4 - 5x - 3x^{-4}$	$-5x - \dfrac{3}{x^5}$
d	$\dfrac{3}{4}x^{\frac{4}{3}}$	$\sqrt[2]{x^3}$
e	$9 + 7x - \dfrac{1}{2}x^{\frac{1}{2}}$	$7 - \dfrac{1}{4\sqrt{x}}$
f	$3x^{\frac{2}{3}} - 6x^{-\frac{1}{3}}$	$\dfrac{2}{\sqrt[3]{x}} + \dfrac{2}{\sqrt[4]{x^3}}$
g	$\dfrac{3}{5}x^{\frac{5}{3}}$	$\sqrt[3]{x^2}$
h	$1 - 2x^{-1} - 3x^{-2}$	$\dfrac{2}{x} + \dfrac{3}{2x^3}$

⚙ 7 Make up five functions to differentiate. Include a mixture of positive, negative and fractional powers.

  a Work out the correct derivatives for your five functions (keep these secret).

  b Create a 'wrong' answer for the derivative of each function.

  c Present your partner with the five original functions and the corresponding wrong answers. Have them correct your answers.

Differentiating a function which is not already an expression in powers of x

A function which is already an expression in powers of x is said to be in 'differentiable form', because it can be differentiated without any manipulation. Often, however, functions have to be rewritten to make them in differentiable form.

An expression is in differentiable form if – ignoring constant terms and terms in x – all other terms are of the form ax^n, where a and n are constants and which may be integers or fractions, positive or negative.

Example 9.5

Given $y = \frac{5}{x^2}$, find $\frac{dy}{dx}$.

$$y = \frac{5}{x^2}$$

$$y = 5x^{-2}$$

> Remember that $\frac{5}{x^2} = 5 \times \frac{1}{x^2} = 5 \times x^{-2}$. $5x^{-2}$ is in the form ax^n and so is in differentiable form. **P**

$$\frac{dy}{dx} = -10x^{-3} = -\frac{10}{x^3}$$

> Differentiate, remembering to write the answer with a positive power of x. **P**

Example 9.6

Given $f(x) = \frac{1}{4x^8}$, find $f'(x)$.

$$f(x) = \frac{1}{4x^8}$$

$$f(x) = \frac{1}{4}x^{-8}$$

$$\frac{dy}{dx} = \frac{1}{4} \times -8x^{-9} = -2x^{-9} = -\frac{2}{x^9}$$

Example 9.7

Given $y = 6\sqrt{x}$, find $\frac{dy}{dx}$.

$$y = 6\sqrt{x}$$

$$y = 6x^{\frac{1}{2}}$$

$$\frac{dy}{dx} = 6 \times \frac{1}{2}x^{-\frac{1}{2}} = 3x^{-\frac{1}{2}} = \frac{3}{x^{\frac{1}{2}}} = \frac{3}{\sqrt{x}}$$

> Remember to express the denominator in root form. **P**

Example 9.8

Differentiate $9\left(\sqrt[3]{x^2}\right)$.

$$y = 9\left(\sqrt[3]{x^2}\right)$$

$$y = 9x^{\frac{2}{3}}$$

$$\frac{dy}{dx} = 9 \times \frac{2}{3}x^{-\frac{1}{3}} = 6x^{-\frac{1}{3}} = \frac{6}{x^{\frac{1}{3}}} = \frac{3}{\sqrt[3]{x}}$$

Example 9.9

$y = (3x - 2)(x + 5)$ Find $\frac{dy}{dx}$.

$$y = 3x^2 + 15x - 2x - 10 \quad \bullet\!\!-\!\!-\!\!-\!\!\boxed{\text{Multiply out the brackets} \quad \text{P}}$$

$$y = 3x^2 + 13x - 10 \quad \bullet\!\!-\!\!-\!\!\boxed{\text{Simplify, noting that } y \text{ is now in differentiable form.} \quad \text{P}}$$

$$\frac{dy}{dx} = 6x + 13$$

Example 9.10

$f(x) = \frac{x^3 - 4x + 5}{x^2}$. Find $f'(x)$.

Method 1

$$f(x) = \frac{x^3 - 4x + 5}{x^2}$$

$$f(x) = \frac{x^3}{x^2} - \frac{4x}{x^2} + \frac{5}{x^2} \quad \bullet\!\!-\!\!\boxed{\text{Divide each term of the numerator by the denominator} \quad \text{P}}$$

$$f(x) = x - \frac{4}{x} + \frac{5}{x^2} \quad \bullet\!\!-\!\!-\!\!\boxed{\text{Simplify} \quad \text{P}}$$

$$f(x) = x - 4x^{-1} + 5x^{-2} \quad \bullet\!\!-\!\!\boxed{\text{Express } f(x) \text{ in differentiable form} \quad \text{P}}$$

$$f'(x) = 1 + 4x^{-2} - 10x^{-3} = 1 + \frac{4}{x^2} - \frac{10}{x^3} \quad \bullet\!\!-\!\!\boxed{\begin{array}{l}\text{Differentiate, writing the answer} \quad \text{P}\\ \text{using positive indices.}\end{array}}$$

Method 2

$$f(x) = \frac{x^3 - 4x + 5}{x^2}$$

$$f(x) = \frac{1}{x^2} \times \frac{x^3 - 4x + 5}{1} \quad \bullet\!\!-\!\!-\!\!\boxed{\text{Remember that } \frac{1}{a} \times \frac{b}{1} = \frac{b}{a} \quad \text{P}}$$

(Continued)

$$f(x) = x^{-2}\left(x^3 - 4x + 5\right)$$

$$f(x) = x^{-2} \times x^3 - 4x^{-2} \times x + 5 \times x^{-2}$$ — Multiply out. (P)

$$f(x) = x^{-2+3} - 4x^{-2+1} + 5x^{-2}$$

$$f(x) = x - 4x^{-1} + 5x^{-2}$$ — Express $f(x)$ in differentiable form. Now differentiate exactly as shown in Method 1. (P)

Example 9.11

Differentiate $\dfrac{24}{\sqrt{t}}$ with respect to t.

$$f(t) = \frac{24}{\sqrt{t}}$$

$$f(t) = \frac{24}{t^{\frac{1}{2}}} = 24t^{-\frac{1}{2}}$$ — Express $f(t)$ in differentiable form (P)

$$f'(t) = -\frac{1}{2} \times 24t^{-\frac{3}{2}} = -12t^{-\frac{3}{2}} = -\frac{12}{t^{\frac{3}{2}}} = -\frac{12}{\sqrt{t^3}}$$

Example 9.12

$g(x) = \dfrac{1 - 5x^2}{x\sqrt{x}}$. Find $g'(x)$.

$$g(x) = \frac{1 - 5x^2}{x^1 \times x^{\frac{1}{2}}} = \frac{1 - 5x^2}{x^{\frac{3}{2}}}$$ — Express the denominator as a power of x. (P)

Method 1

$$g(x) = \frac{1}{x^{\frac{3}{2}}} - \frac{5x^2}{x^{\frac{3}{2}}}$$ — Divide each term of the numerator by the denominator. (P)

$$g(x) = \frac{1}{x^{\frac{3}{2}}} - 5x^{\frac{1}{2}}$$

$$g(x) = x^{-\frac{3}{2}} - 5x^{\frac{1}{2}}$$

$$g'(x) = -\frac{3}{2}x^{-\frac{5}{2}} - 5 \times \frac{1}{2}x^{-\frac{1}{2}}$$

$$g'(x) = -\frac{3}{2x^{\frac{5}{2}}} - \frac{5}{2x^{\frac{1}{2}}} = -\frac{3}{2\sqrt{x^5}} - \frac{5}{2\sqrt{x}}$$

Method 2

$$g(x) = \frac{1 - 5x^2}{x^1 \times x^{\frac{1}{2}}} = \frac{1 - 5x^2}{x^{\frac{3}{2}}}$$ — Express the denominator as a power of x. (P)

$$g(x) = \frac{1}{x^{\frac{3}{2}}} \times \frac{1 - 5x^2}{1}$$

$$g(x) = x^{-\frac{3}{2}}\left(1 - 5x^2\right)$$

$$g(x) = x^{-\frac{3}{2}} \times 1 - x^{-\frac{3}{2}} \times 5x^2$$

$$g(x) = x^{-\frac{3}{2}} - 5x^{-\frac{3}{2}+2}$$

$$g(x) = x^{-\frac{3}{2}} - 5x^{\frac{1}{2}} \qquad \text{Now differentiate in the usual way.}$$

Exercise 9B

★ **1** Express these functions in differentiable form by multiplying out and simplifying. Do not differentiate.

a $\quad y = (2x - 3)(x + 4)$
b $\quad f(x) = (3 - 4x)(1 - x)$

c $\quad g(x) = (x - 3)(2x^2 - 5x + 2)$
d $\quad h(x) = (x^2 - 3x - 4)(x - 2)$

e $\quad y = (x + 2)(x - 1)(x - 1)$
f $\quad y = x(2x - 1)^2$

g $\quad y = (x + 2)(x - 1)^2$

★ **2** Express in differentiable form.

a $\quad \dfrac{3}{x}$
b $\quad \dfrac{8}{x^3}$
c $\quad \dfrac{5}{x^6}$
d $\quad \dfrac{1}{3x^2}$
e $\quad \dfrac{1}{6x^4}$

f $\quad \dfrac{2}{5x}$
g $\quad \dfrac{4}{3x^9}$
h $\quad \dfrac{1}{2x^7}$
i $\quad \dfrac{3}{x^{\frac{1}{2}}}$
j $\quad \dfrac{1}{x^{\frac{3}{4}}}$

k $\quad \dfrac{5}{x^{\frac{4}{3}}}$
l $\quad \dfrac{4}{5x^{\frac{3}{2}}}$
m $\quad \dfrac{2}{x^{\frac{5}{8}}}$
n $\quad 2\sqrt{x}$
o $\quad \dfrac{5}{\sqrt{x}}$

p $\quad \dfrac{1}{2\sqrt{x}}$
q $\quad \dfrac{3}{4}\left(\sqrt[5]{x}\right)$
r $\quad \dfrac{1}{\sqrt{x^8}}$
s $\quad \dfrac{1}{\sqrt[6]{x}}$
t $\quad \dfrac{5}{3\left(\sqrt[4]{x}\right)}$

u $\quad \dfrac{6x^4}{x^2}$
v $\quad \dfrac{2x^2}{x^5}$
w $\quad \dfrac{3x}{8x^6}$
x $\quad \dfrac{4x^{\frac{1}{2}}}{x}$
y $\quad \dfrac{x^2}{5x^{\frac{2}{3}}}$
z $\quad \dfrac{7x^{-\frac{1}{2}}}{4x^{\frac{3}{2}}}$

3 For each of these functions:

i express the function in differentiable form

ii differentiate the function with respect to x.

a $\quad y = (x - 3)(x + 5)$
b $\quad f(x) = (4x + 1)(2x - 3)$

c $\quad y = x(x + 3)(x - 2)$
d $\quad y = (x + 2)(x^2 + 3x - 4)$

e $\quad g(x) = 2x^2(x - 1)^2$
f $\quad y = (x + 1)(x - 3)^2$

g $\quad y = (x + 4)(x + 1)(x - 2)$

★ **4** For each of these functions:

   i express the function in differentiable form

   ii differentiate the function with respect to x, expressing the answer with positive indices.

a $y = \dfrac{5}{x^2}$ **b** $y = \dfrac{7}{x^4}$ **c** $y = \dfrac{1}{2x^3}$

d $f(x) = \dfrac{1}{6x^2}$ **e** $g(x) = 4x^3 - \dfrac{2}{x^5}$ **f** $y = \dfrac{4}{3x}$

g $y = 8x + 5 - \dfrac{1}{x^2}$ **h** $y = \dfrac{4}{x^3} - \dfrac{3}{x}$ **i** $f(x) = \dfrac{3}{2x^4} - 5x - 6$

★ **5** For each of these functions:

   i express the function in differentiable form

   ii differentiate the function with respect to x, expressing the answer in root form.

a $y = 8\sqrt{x}$ **b** $y = \sqrt[3]{x^2}$ **c** $f(x) = 12\left(\sqrt[4]{x^3}\right)$

d $g(x) = 4\left(\sqrt{x^5}\right)$ **e** $y = \sqrt{x^7}$ **f** $y = \dfrac{1}{\sqrt{x}}$

g $y = \dfrac{1}{x\sqrt{x}}$ **h** $y = \dfrac{4}{\sqrt[4]{x}}$ **i** $\dfrac{10}{\sqrt{x^3}}$

j $y = \dfrac{9}{\sqrt[6]{x^5}}$ **k** $y = \dfrac{3}{2}\left(\sqrt[9]{x^4}\right)$ **l** $y = \dfrac{1}{8\left(\sqrt[5]{x^6}\right)}$

m $g(x) = \dfrac{5}{2\left(\sqrt[3]{x^2}\right)}$

6 For each of these functions:

   i express the function in differentiable form

   ii differentiate the function with respect to x, expressing the answer with positive indices

a $y = \dfrac{x^2 - 4}{x}$ **b** $y = \dfrac{3x^2 - 5x}{x^2}$ **c** $f(x) = \dfrac{4 - x^3}{x}$

d $g(x) = \dfrac{x^2 + 5x - 3}{x^2}$ **e** $y = \dfrac{1 - x^4}{2x^3}$ **f** $y = \dfrac{x^2 - 3x - 2}{6x^3}$

g $y = \dfrac{(x + 1)(x + 4)}{x^2}$ **h** $y = \dfrac{(3 - x)(1 + 2x)}{x^2}$ **i** $y = \dfrac{(x - 2)(2x + 1)^2}{x^3}$

7 Differentiate these functions.

a $y = \sqrt{x}(x - 3)$ **b** $f(x) = \dfrac{3}{x^2}\left(x^2 - \dfrac{1}{x}\right)$

c $g(x) = \left(1 - \sqrt{x}\right)\left(2 - \dfrac{1}{\sqrt{x}}\right)$ **d** $h(x) = \dfrac{2}{\sqrt{x}}\left(1 - \dfrac{1}{\sqrt{x}}\right)^2$

e $y = \left(\dfrac{1}{2x} - 2\sqrt{x} \right)\left(\dfrac{2}{\sqrt{x}} - x \right)$ **f** $y = \left(\dfrac{3}{x} - \dfrac{x}{3} \right)^2$

8 Differentiate these functions.

a $f(x) = \dfrac{2 - x^3}{\sqrt{x}}$

b $y = \dfrac{(x - 2)^2}{4\sqrt{x}}$

c $g(x) = \dfrac{\left(1 - \dfrac{1}{x}\right)\left(1 - \dfrac{2}{x}\right)}{2x^2}$

d $h(x) = \dfrac{(x - 2)(\sqrt{x} + 5)}{x\sqrt{x}}$

e $f(x) = \dfrac{\left(\dfrac{4}{x} - \dfrac{3}{2x}\right)^2}{4\left(\sqrt[3]{x^2}\right)}$

⚙ **9** Make up five functions to differentiate. Choose from the types of functions covered in Questions 3 – 8.

a Express your functions in differentiable form but make sure that each contains a mistake. Do not differentiate your functions.

b Present your partner with the five original functions and the corresponding wrong differentiable form. Have them correct your answers.

⚙ **10** Make up five functions to differentiate. Choose from the types of functions covered in Questions 3 – 8.

a Work out the correct derivatives for your five functions (keep these secret).

b Create a 'wrong' answer for the derivative of each function.

c Present your partner with the five original functions and the corresponding wrong answers. Have them correct your answers.

Differentiation from first principles

Earlier in the chapter, a limiting argument was used to find the instantaneous speed of a falling stone. The limiting idea can be explored more mathematically to develop the process of differentiation from first principles. This shows clearly where the main rules for differentiation actually come from.

Note that differentiation from first principles is not examinable in Higher maths, but it will be extremely important in Advanced Higher maths.

Consider the function $f(x) = x^2$.

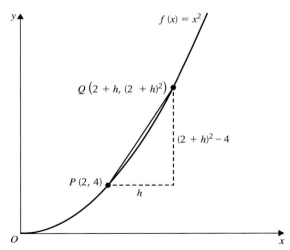

P has coordinates (2, 4) and the point Q has an x coordinate which is just a little bigger than 2, i.e. $x_Q = 2 + h$. The coordinates of Q are therefore $(2 + h, (2 + h)^2)$.

The gradient of PQ is given by $m_{PQ} = \dfrac{(2+h)^2 - 4}{(2+h) - 2} = \dfrac{(2+h)^2 - 4}{h}$.

As the value of h gets smaller and smaller, i.e. as $h \to 0$ (on the graph this means that Q **tends towards** P) the gradient of PQ becomes closer and closer to the gradient of the curve at P. This is the same argument used earlier in the chapter to find the instantaneous speed of a falling stone.

This is used to give a definition of the gradient when $x = 2$.

$$f'(2) = \lim_{h \to 0} \frac{(2+h)^2 - 4}{h}$$

> When $f(x) = x^2$ this is how we define the gradient at the point where $x = 2$ i.e. $f'(2)$

$$f'(2) = \lim_{h \to 0} \frac{4 + 4h + h^2 - 4}{h}$$

$$f'(2) = \lim_{h \to 0} \frac{4h + h^2}{h}$$

$$f'(2) = \lim_{h \to 0} \frac{h(4 + h)}{h}$$

$$f'(2) = \lim_{h \to 0} (4 + h)$$

As $h \to 0$, $f'(2) \to 4$

This shows that at the point P (where $x = 2$), the gradient of the curve is equal to 4.

Example 9.13

Show that when $f(x) = x^2$ and R and S have coordinates (3, 9) and $(3 + h, (3 + h)^2)$ respectively then the gradient of the curve at R is equal to 6.

$$f'(3) = \lim_{h \to 0} \frac{(3+h)^2 - 9}{h}$$

> Write down the definition of $f'(3)$ i.e. the gradient of the curve at R. **P**

$$f'(3) = \lim_{h \to 0} \frac{9 + 6h + h^2 - 9}{h}$$

> Multiply out the brackets in the numerator to begin the process of finding $f'(3)$. **P**

$$f'(3) = \lim_{h \to 0} \frac{6h + h^2}{h}$$

> Simplify numerator. **P**

$$f'(3) = \lim_{h \to 0} \frac{h(6 + h)}{h}$$

> Factorise numerator. **P**

$$f'(3) = \lim_{h \to 0} (6 + h)$$

> Divide numerator and denominator by h. **P**

As $h \to 0$, $f'(3) \to 6$

This shows that at the point R (where $x = 3$), the gradient of the curve is equal to 6.

Example 9.14

Show that when $A(x, x^2)$ is any point on the graph with equation $f(x) = x^2$, the gradient of the curve at A is $2x$.

$$f'(x) = \lim_{h \to 0} \frac{(x + h)^2 - x^2}{h}$$

$$f'(x) = \lim_{h \to 0} \frac{x^2 + 2xh + h^2 - x^2}{h}$$

$$f'(x) = \lim_{h \to 0} \frac{2xh + h^2}{h}$$

$$f'(x) = \lim_{h \to 0} \frac{h(2x + h)}{h}$$

$$f'(x) = \lim_{h \to 0} (2x + h)$$

As $h \to 0$, $f'(x) \to 2x$

This shows that at the point A the gradient of the curve is equal to $2x$.

The process of obtaining $f'(x)$ by working from the definition through the steps shown above is called *differentiation from first principles*.

More generally, for any function f where $y = f(x)$ we define the derived function as:

$$\frac{dy}{dx} = f'(x) = \lim_{h \to 0} \frac{f(x + h) - f(x)}{h}$$

Exercise 9C

1   Using differentiation from first principles, show that if $f(x) = x$ then $f'(x) = 1$.

2   Adapt the solution from Question 1 to show that if $f(x) = cx$, where c is a constant, then $f'(x) = c$.

3   Using differentiation from first principles, show that:
    a   when $f(x) = x^3$, $f'(x) = 3x^2$ b   when $f(x) = x^4$, $f'(x) = 4x^3$

4   Using differentiation from first principles try to show that if $f(x) = x^n$, where n is a positive integer, then: $f'(x) = nx^{n-1}$.

     Find out about Pascal's Triangle.

6   a   Show that $\dfrac{1}{x + h} - \dfrac{1}{x} = -\dfrac{h}{x(x + h)}$.

    b   Using the answer to part **a** show that if $f(x) = \dfrac{1}{x}$ then $f'(x) = \lim_{h \to 0} -\left(\dfrac{1}{x(x + h)} \right)$.

    c   Deduce that $f'(x) = x^{-\frac{1}{2}}$.

    d   When $f(x) = \dfrac{k}{x}$, where k is a constant, show that $f'(x) = -\dfrac{k}{x^2}$.

    e   Find – using differentiation from first principles – the derivatives of $\frac{1}{x^2}$ and $\frac{1}{x^3}$.

7 a Show that $\sqrt{x+h} - \sqrt{x} = \dfrac{h}{\sqrt{x+h}+\sqrt{x}}$.

> ⚠ Multiply $\sqrt{x+h} - \sqrt{x}$
> by $\dfrac{\sqrt{x+h}+\sqrt{x}}{\sqrt{x+h}+\sqrt{x}}$

b Hence show that if $f(x) = \sqrt{x}$ then:

$$f'(x) = \lim_{h \to 0} \frac{1}{\sqrt{x+h}+\sqrt{x}}.$$

c Deduce that $f'(x) = \dfrac{1}{2\sqrt{x}}$.

8 Using differentiation from first principles show that if $f(x) = \frac{1}{\sqrt{x}}$ then $f'(x) = -\dfrac{1}{2\sqrt{x^3}}$.

9 Using differentiation from first principles show that when $y = f(x) + g(x)$,
$\dfrac{dy}{dx} = f'(x) + g'(x)$.

10 One alternative definition of $f'(x)$ is given by:

$$f'(x) = \lim_{h \to 0} \frac{f(x+h) - f(x-h)}{2h}$$

a Can you interpret this alternative definition geometrically?

b Using the alternative definition for $f'(x)$ obtain derivatives for x^2 and x^3.

Evaluating derivatives – rates of change and gradients of tangents and curves

The process of differentiation gives rise to the **derivative** or **gradient function**. This function can then be used to find the gradient of a curve at any point. Recall that the phrases **find the gradient of the tangent to the curve, find the gradient of the curve** and **find the rate of change** are all instructing you to differentiate.

Example 9.15

A parabola has equation $y = x^2 - 4x + 6$.

A tangent to the curve is drawn at the point $P(4, 6)$.

Find the gradient of this tangent.

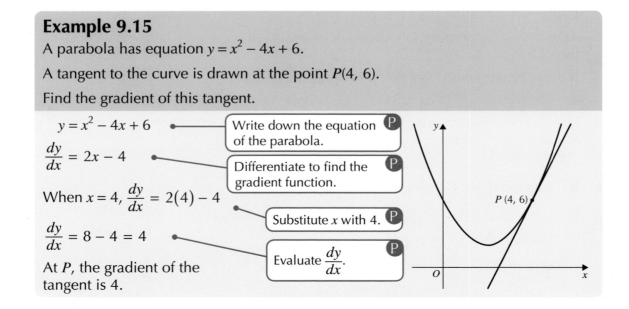

$y = x^2 - 4x + 6$ — Write down the equation of the parabola.

$\dfrac{dy}{dx} = 2x - 4$ — Differentiate to find the gradient function.

When $x = 4$, $\dfrac{dy}{dx} = 2(4) - 4$ — Substitute x with 4.

$\dfrac{dy}{dx} = 8 - 4 = 4$ — Evaluate $\dfrac{dy}{dx}$.

At P, the gradient of the tangent is 4.

Example 9.16

A curve has equation $y = 6\sqrt{x}$. Find the gradient of the tangent to the curve at the point where $x = 36$.

$y = 6\sqrt{x} = 6x^{\frac{1}{2}}$ — Express y in differentiable form. **P**

$\dfrac{dy}{dx} = 6 \times \dfrac{1}{2}x^{-\frac{1}{2}} = 3x^{-\frac{1}{2}} = \dfrac{3}{x^{\frac{1}{2}}} = \dfrac{3}{\sqrt{36}}$ — Differentiate, remembering to express the answer in root form. **P**

When $x = 36$, $\dfrac{dy}{dx} = \dfrac{3}{\sqrt{x}}$ — Substitute x with 36. **P**

$\dfrac{dy}{dx} = \dfrac{3}{6} = \dfrac{1}{2}$

Example 9.17

Find the gradient of the tangent to the curve $y = \frac{8}{\sqrt{x}}$ at the point where $x = 4$.

$y = \dfrac{8}{\sqrt{x}}$

$y = \dfrac{8}{x^{\frac{1}{2}}} = 8x^{-\frac{1}{2}}$ — Express y in differentiable form. **P**

$\dfrac{dy}{dx} = 8 \times -\dfrac{1}{2}x^{-\frac{3}{2}} = -4x^{-\frac{3}{2}} = -\dfrac{4}{x^{\frac{3}{2}}} = -\dfrac{4}{\sqrt{x^3}}$ — Differentiate, remembering to express the answer in root form. **P**

When $x = 4$, $\dfrac{dy}{dx} = \dfrac{4}{\sqrt{4^3}}$ — Substitute x with 4. **P**

$\dfrac{dy}{dx} = \dfrac{4}{\sqrt{64}} = -\dfrac{4}{8} = -\dfrac{1}{2}$

In non–calculator work it important to note that $\sqrt[n]{x^m}$ is the same as $\left(\sqrt[n]{x}\right)^m$. In Example 9.17 $\sqrt{4^3}$ could have been found by writing it as $\left(\sqrt{4}\right)^3$ and then doing $\left(\sqrt{4}\right)^3 = 2^3 = 8$. To evaluate e.g. $\sqrt[5]{32^3}$ it is much easier to write $\left(\sqrt[5]{32}\right)^3$ and then do $\left(\sqrt[5]{32}\right)^3 = 2^3 = 8$, rather than do $\sqrt[5]{32768}$.

Example 9.18

A curve has equation $y = 3x^2 - 12x + 6$. Find the x–coordinate of the point at which the tangent to the curve has gradient 6.

$y = 3x^2 - 12x + 6$

$\dfrac{dy}{dx} = 6x - 12$

(Continued)

$6 = 6x - 12$ •────────

$6x - 12 = 6$

> The gradient of the tangent is 6 so replace $\frac{dy}{dx}$ with 6.
> (Do not replace x with 6.)

$6x = 18$

$x = \dfrac{18}{6} = 3$ The gradient of the curve is equal to 6 at the point where $x = 3$.

Exercise 9D

 1 For each of the following, find the gradient of the curve at the given point.

a

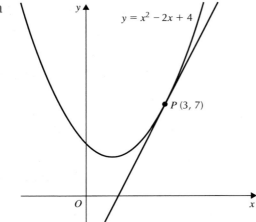

$y = x^2 - 2x + 4$

$P\,(3, 7)$

b

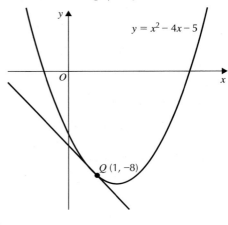

$y = x^2 - 4x - 5$

$Q\,(1, -8)$

c

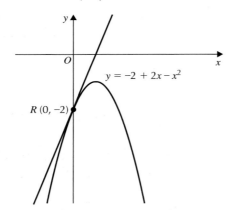

$y = -2 + 2x - x^2$

$R\,(0, -2)$

d

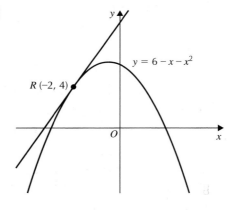

$y = 6 - x - x^2$

$R\,(-2, 4)$

★ **2 a** Find the gradient of the tangent to the curve $y = x^2 + 4x + 2$ at the point where $x = 3$.

   b A curve has equation $y = 5x^2 - 15x$. Find the gradient of the curve at the point where $x = 2$.

   c Given $f(x) = x^3 - 4x^2 + 5x + 3$, find the rate of change of f when $x = 1$.

   d Find the gradient of the curve $y = (x + 2)(x + 5)$ at the point where $x = -3$

   e Given $g(x) = 6x - x^3$, find the value of $g'(-2)$.

   f A curve has equation $y = 4x(x^2 - 2)$. Find $\frac{dy}{dx}$ when $x = -1$.

3 A curve has equation $y = \frac{2}{x}$ where $x \neq 0$. Find the gradient of the curve when

    a $x = 1$ **b** $x = -3$ **c** $x = \frac{1}{2}$

4 On a suitable domain, the function f is defined by $f(x) = 3\sqrt{x}$

    a   Find the gradient of the tangent to the curve $y = f(x)$ at the point where $x = 4$.

    b   Find the rate of change of f when $x = 9$.

    c   Evaluate $f'\left(\frac{1}{16}\right)$.

5 The diagram shows part of the graph of the cubic function with equation $f(x) = x(x^2 - 4)$. A tangent to the graph is drawn at P.

Find the gradient of this tangent.

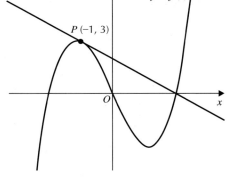

6 A curve has equation $y = \frac{5}{4x^2}$ where $x \neq 0$.

Find the gradient of the curve at the point

where $x = -10$.

7 **a**   Find the x–coordinate of the point where the tangent to the curve $y = x^2 + 8x - 3$ has gradient 2.

    b   The function f is defined by $f(x) = 5 - 4x - x^2$. Determine the value of p, given that $f'(p) = 2$.

8 Find the coordinates of the point where the tangent to the curve $y = 3x^2 - 4x + 1$ has gradient -10.

9 Find the x–coordinate of the point where the tangent to the curve $y = x^4 + 20x$ has gradient -12.

10 **a**   Determine the x–coordinates of the points where the tangent to the curve $y = \frac{1}{3}x^3 - 3x^2 + 12x + 2$ has gradient 4.

    b   Determine the x–coordinates of the points where the tangent to the curve $y = x^3 + 2x^2 - 7x + 1$ has gradient -3.

11 A cubic function has equation $f(x) = x^3 + kx^2 + 2x + 3$, where k is a constant. Given that $(4, 11)$ is on the graph of $y = f(x)$, find the value of k and hence find the gradient of the curve when $x = 2$.

12 A curve has equation $y = \frac{8}{x\sqrt{x}}$ where $x > 0$. Find algebraically the value of x for which $\frac{dy}{dx} = -\frac{3}{8}$.

13 The function f is defined by $f(x) = \frac{x-3}{\sqrt{x}}$ where $x > 0$. Given that $f'\left(\frac{9}{16}\right) = \frac{a}{b}$, where a and b are integers, determine possible values for a and b.

14 Find the coordinates of the point where the tangent to the curve $y = x^2 - 2x + 6$ is parallel to the line with equation $x - \frac{1}{4}y + 2 = 0$

15 P is the point of intersection of the lines with equations $y = -\frac{2}{3}x + 9$ and $4x + y + 1 = 0$

    a   Find the gradient of the tangent to the curve $y = 8 - 4x - x^2$ at P.

    b   Show that the tangent at P is perpendicular to the line with equation $x + 2y = 5$

16 A parabola has equation $f(x) = (x - 4)^2 + 1$.

At Q the gradient of the parabola is 4.

a Determine the coordinates of Q.

b Hence, write down the coordinates of the point where the gradient of the parabola is -4.

The function h is defined by $h(x) = f(x - 2) - 5$.

c Write down the coordinates of the point where the gradient of the tangent to the curve $y = h(x)$ has gradient -4.

17 a Express $3x^2 - 6x + 4$ in the form $a(x - b)^2 + c$.

A cubic has equation $f(x) = x^3 - 3x^2 + 2x - 5$.

b Using your answer to part **a**, or otherwise, show that there are no points on the graph of $y = f(x)$ where the gradient of the curve is equal to -2.

18 Find the range of values of x for which the gradient of the curve $y = x^3 + x^2 - 5x + 2$ is greater than 3.

19 A tangent is drawn to the curve $y = x^2 - 8x + 18$ at P.

This tangent makes an angle of $a°$ with the positive direction of the x–axis.

a Given that $\cos 2a° = -\frac{4}{5}$, find the value of $\cos a°$.

b Hence find the value of $\tan a°$.

c Determine the coordinates of P.

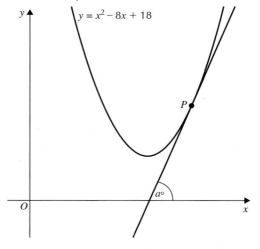

Differentiate $k\sin x$, $k\cos x$

Differentiating trigonometric functions is vitally important in physics and other sciences.

In physics, a simple pendulum exhibits what is known as **simple harmonic motion**. The distance (displacement), x, of the pendulum from its **equilibrium** position can be described by the function $x = A\cos pt + B\sin pt$, where A, B and p are constants and t represents time.

In order to determine the velocity and acceleration of the pendulum we need to be able to differentiate this function.

To start with we will look at the function $f(x) = \sin x$.

The diagram shows the graph of $f(x) = \sin x$, $0 \le x \le \pi$.

Tangents have been drawn at a selection of points.

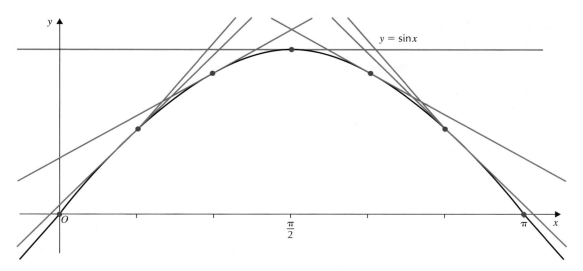

If we measure the gradient of the tangents we get the following information:

x–coordinate	0	$\frac{\pi}{6}$	$\frac{\pi}{3}$	$\frac{\pi}{2}$	$\frac{2\pi}{3}$	$\frac{5\pi}{6}$	π
Gradient of tangent	1	0.9	0.5	0	−0.5	−0.9	−1

If we continue the above analysis of the sine graph right up to $x = 2\pi$ we can plot a series of points which will be on the graph of $y = f'(x)$.

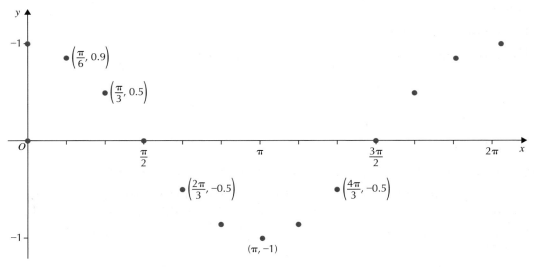

The graph above suggests that if $y = \sin x$ then $\frac{dy}{dx} = \cos x$ and this is indeed true, provided x is measured in radians.

It can also be shown that if $y = \cos x$ then $\frac{dy}{dx} = -\sin x$.

When k is any constant we have the following two key facts:

$f(x)$	$f'(x)$
$k\sin x$	$k\cos x$
$k\cos x$	$-k\sin x$

Example 9.19

Differentiate

a $8\sin x$ b $-2\sin x$ c $\dfrac{3}{4}\sin x$

  a $y = 8\sin x$

     $\dfrac{dy}{dx} = 8\cos x$

  b $y = -2\sin x$

     $\dfrac{dy}{dx} = -2\cos x$

  c $y = \dfrac{3}{4}\sin x$

     $\dfrac{dy}{dx} = \dfrac{3}{4}\cos x$

Example 9.20

Differentiate

a $4\cos x$ b $-3\cos x$ c $-\dfrac{1}{5}\cos x$

  a $y = 4\cos x$

     $\dfrac{dy}{dx} = -4\sin x$

  b $f(x) = -3\cos x$

     $f'(x) = 3\sin x$ •————— Note the answer is $3\sin x$ and not $-3\sin x$. Ⓟ

  c $g(x) = -\dfrac{1}{5}\cos x$

     $g'(x) = \dfrac{1}{5}\sin x$

Example 9.21

Find the gradient of the tangent to the curve $y = 3\sin x$ at the point where $x = \frac{\pi}{4}$

$y = 3\sin x$

$\dfrac{dy}{dx} = 3\cos x$

When $x = \dfrac{\pi}{4}$, $\dfrac{dy}{dx} = 3\cos\dfrac{\pi}{4}$ •————— Substitute x with $\dfrac{\pi}{4}$ Ⓟ

$\dfrac{dy}{dx} = 3\left(\dfrac{1}{\sqrt{2}}\right) = \dfrac{3}{\sqrt{2}}$ •————— Find the gradient of the tangent. Ⓟ

Example 9.22

Find the gradient of the tangent to the curve with equation $f(x) = 4\cos x$ at the point where $x = \frac{4\pi}{3}$.

$f(x) = 4\cos x$

$f'(x) = -4\sin x$ •————————— Note the answer is $-4\sin x$ and not $4\sin x$. Ⓟ

When $x = \frac{4\pi}{3}$, $f'(x) = -4\sin\frac{4\pi}{3}$

$f'(x) = -4 \times -\frac{\sqrt{3}}{2} = \frac{4\sqrt{3}}{2} = 2\sqrt{3}$

It is vital that you are able to quickly calculate trigonometric exact values, where the angles may be in degrees or radians.

Example 9.23

A curve has equation $y = 2\sin x$, $0 \le x \le \pi$. Find the x coordinates of the point where the gradient of the curve is $-\sqrt{3}$.

$y = 2\sin x$

$\frac{dy}{dx} = 2\cos x$

$2\cos x = -\sqrt{3}$ •————————— Substitute $\frac{dy}{dx}$ with $-\sqrt{3}$ Ⓢ

$\cos x = -\frac{\sqrt{3}}{2}$

$x = \pi - \frac{\pi}{6} = \frac{5\pi}{6}$ ⟵ Since the value of cos is negative, you need to find an angle in the 2nd quadrant or 3rd quadrant. However, since x cannot exceed π radians (180 degrees) you need only look for a solution in the 2nd quadrant. Ⓟ

Exercise 9E

1   Differentiate:

a $8\sin x$

b $3\cos x$

c $-\sin x$

d $\frac{1}{2}\sin x$

e $\frac{2}{3}\cos x$

f $-\frac{5}{8}\cos x$

g $6x^2 + 7\sin x$

h $3\sin x + 7\cos x$

i $\cos x + 6\sin x$

j $\sin x - \cos x$

k $\frac{3}{x^2} - \cos x$

l $\frac{4}{5}\sin x - 6\sqrt{x}$

m $-5\cos x + \frac{3}{4x}$

n $5x^3 - \frac{1}{\sqrt[3]{x^5}} + 9\sin x$

o $\frac{1 - 9x^2\cos x}{3x^2}$

p $\frac{4 - x^2}{x^3} - \frac{1}{5}\cos x$

q $\frac{6}{\sqrt{x}} - 4\sin x$

r $-\frac{5}{6}\cos x - \frac{5 - \sqrt{x}}{x^2}$

s $\frac{\sin x - 3\cos x}{5}$

★ **2 a** Given $f(x) = 6\sin x$, find the value of $f'\left(\frac{\pi}{3}\right)$.

b $y = 2\cos x$ Find $\frac{dy}{dx}$ when $x = \frac{\pi}{6}$.

c Find the gradient of the tangent to the curve with equation $y = \frac{1}{2}\sin x$ at the point where $x = \frac{\pi}{4}$.

d On a suitable domain, the function f is defined by $f(x) = -4\cos x$. Find the rate of change of f when $x = \frac{\pi}{3}$.

 3 a A curve has equation $y = 4\sin x$. Find the gradient of the curve at the point where $x = \frac{5\pi}{3}$.

b On a suitable domain, the function g is defined by $g(x) = 5\cos x$. Find $g'\left(\frac{7\pi}{6}\right)$.

c Find the gradient of the tangent to the curve $y = \frac{3}{4}\sin x$ at the point where $x = \frac{2\pi}{3}$.

d $y = -6\sin x$ Find $\frac{dy}{dx}$ when $x = \frac{5\pi}{4}$.

 4 A function is defined by $y = 4\sin x - \cos x$. Find $\frac{dy}{dx}$ when $x = \frac{3\pi}{2}$.

★ **5** **You may use a calculator in this question.** Give your answers to 2 decimal places.

a Find the gradient of the tangent to the curve with equation $y = 3\sin x - 4$ when $x = 2$.

b On a suitable domain, the function f is defined by $f(x) = -4\cos x + 1$. Evaluate $f'(4)$.

c Find the gradient of the curve with equation $y = 7\sin x - 5\cos x$ at the point where $x = 2.2$.

d On a suitable domain the function g is defined by $4x - \frac{1}{2}\sin x$. Find the rate of change of g when $x = -1.6$.

 6 The function f is defined by $f(x) = 6\sin x$ where $0 \leq x < 2\pi$. Find the values of x for which the rate of change of f is equal to 3.

 7 A curve has equation $y = 8\cos x$. Find the values of x, $0 \leq x < 2\pi$, for which
$\frac{dy}{dx} = -4$

 8 The function g is defined by $g(x) = -2\cos x$ where $\frac{\pi}{2} \leq x < \pi$. Find the value of x for which $g'(x) = \sqrt{3}$.

 9 A curve has equation $y = \frac{1}{2}\sin x$ where $0 \leq x \leq \frac{\pi}{2}$. Find the coordinates of the point on the curve at which the tangent to the curve has gradient $\frac{\sqrt{2}}{4}$.

 10 A curve has equation $y = 4\sin x$ where $0 \leq x \leq 2\pi$. Determine algebraically the coordinates of the points where the gradient of the curve is -2.

11 On a suitable domain, the function h is defined by $h(x) = \frac{1}{x} - 2\cos x$. Show that
$$h'\left(\frac{5\pi}{6}\right) = \frac{(5\pi - 6)(5\pi + 6)}{25\pi^2}.$$

12 Functions f and g are defined, on suitable domains, by $f(x) = 3x - 2$ and $g(x) = \cos x$.

    a Write down an expression for $f(g(x))$

       The function h is defined by $h(x) = f(g(x))$

    b Solve the equation $2h'(x) = 3$ where $-\pi \le x \le \pi$

13 The function f is defined, on a suitable domain, by $f(x) = x + 8\cos x$

    Functions p and q are defined by $p(x) = 3 + f'(x)$ and $q(x) = \frac{1}{x}$.

    Determine algebraically the value of x, $\frac{\pi}{2} < x < \pi$ for which $q(p(x))$ is undefined.

14 A curve has equation $y = 5\sin x + 2x$, $\pi < x < \frac{3\pi}{2}$. At the point P, the tangent to this curve makes an angle of $135°$ with the positive direction of the x–axis.

    Determine algebraically the coordinates of P.

15 A curve has equation $y = 3 - 2\cos x$ where $0 < x < 2\pi$. Determine the coordinates of the points on the curve where the tangent is perpendicular to the straight line with equation $\sqrt{3}y - x - 2 = 0$.

16 The graph of a trigonometric function has equation $y = f(x)$.

    Another function, h, is defined by $h(x) = 3f\left(x - \frac{\pi}{4}\right) - 4$.

    $P\left(\frac{2\pi}{3}, 11\right)$ is on the graph of $y = h(x)$. At P, the tangent to the graph of $y = h(x)$ has gradient -15.

Write down the coordinates of any point on the graph of $y = f(x)$ where the tangent to the curve has gradient -5.

Exercise 9F

Only attempt this exercise if you have already done Exercise 9C.

Recall that $f'(x) = \lim\limits_{h \to 0} \dfrac{f(x + h) - f(x)}{h}$.

When $f(x) = \sin x$ we have $f'(x) = \lim\limits_{h \to 0} \dfrac{\sin(x + h) - \sin(x)}{h}$.

1 By expanding the numerator, show that $f'(x)$ can be written as

$$f'(x) = \lim_{h \to 0} \frac{\sin x(\cos h - 1) + \sin x \cos h}{h}$$

We can re–arrange $f'(x)$ further to obtain

$$f'(x) = \lim_{h \to 0} \frac{(\cos h - 1)}{h}\sin x + \lim_{h \to 0} \frac{\sin h}{h}\cos x$$

2 Using a graphing calculator or computer software, investigate the value of $\frac{\cos h - 1}{h}$ as h approaches the value zero. Do you think it would be OK to start with a negative value for h?

3 Repeat 2 for $\frac{\sin h}{h}$.

4 The diagram shows two triangles *OAB* and *OCD*. *O* is the centre of a circle, radius 1 unit and angle $AOB = h°$. *AD* is an arc of the circle.

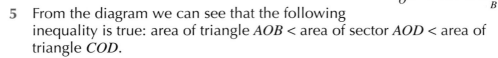

   a Show that the area of sector *AOD* is $\frac{\pi h}{360}$ square units.

   b Hence show that, if *h* is measured in radians, the area of sector *AOD* is $\frac{h}{2}$ square units.

   c Write down expressions for *OA* and *OB* and hence find the area of triangle *OAB*.

   d Find the area of triangle *COD*.

5 From the diagram we can see that the following inequality is true: area of triangle *AOB* < area of sector *AOD* < area of triangle *COD*.

   a Show that $\frac{1}{2}\sin h \cos h < \frac{h}{2} < \frac{1}{2}\tan h$.

   b Deduce that $\frac{1}{\cos h} > \frac{\sin h}{h} > \cos h$.

   c As $h \to 0$, what value does $\cos h$ approach?

6 a By multiplying numerator and denominator by $\cos h + 1$ show that $\frac{\cos h - 1}{h}$ can be written as $\frac{-\sin^2 h}{h(\cos h+1)}$.

     It now follows that: $\lim\limits_{h\to 0} \dfrac{-\sin^2 h}{h(\cos h + 1)} = -\lim\limits_{h\to 0}\left(\dfrac{\sin h}{h} \times \sin h \times \dfrac{1}{\cos h + 1}\right)$
     Can you see why?

   b Deduce that $\lim\limits_{h\to 0} \dfrac{\cos h - 1}{h} = 0$.

7 Go back to Question 1 and show now that when $f(x) = \sin x$, $f'(x) = \cos x$.

8 Try to prove, using differentiation from first principles, that when $f(x) = \cos x$, $f'(x) = -\sin x$.

9 Suppose that *h* was measured in degrees and not radians. Explain why the derivative of $\sin x°$ would not be $\cos x°$. Write down the derivative of $\sin x°$.

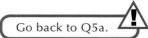

Go back to Q5a. ⚠

Differentiate a composite function using the chain rule

The function $y = (3x + 1)^5$ is of the form $y = f(g(x))$, and is known as a **composite function**. Composite functions of the form $f(g(x))$ can be differentiated using the **chain rule**.

$$y = f(g(x)) \Rightarrow \frac{dy}{dx} = f'(g(x)) \times g'(x)$$

The diagram shows the chain of functions used to make $(3x + 1)^5$, together with the required derivatives.

g is the function 'treble and add 1' i.e. $g(x) = 3x + 1$.

f is the function 'raise to the power of 5' i.e. $y = f(3x + 1) = (3x + 1)^5$

$$x \xrightarrow{\ g\ } (3x + 1) \xrightarrow{\ f\ } (3x + 1)^5$$

$$\underset{g'(x)}{3} \qquad \underset{f'(g(x))}{5(3x + 1)^4}$$

$$\frac{dy}{dx} = f'(g(x)) \times g'(x)$$

$$= 5(3x + 1)^4 \times 3$$

$$= 15(3x + 1)^4$$

Example 9.24

Differentiate $(5x - 2)^7$.

$y = (5x - 2)^7$

$y = f(g(x))$, where $g(x) = 5x - 2$ and $f(g(x)) = (5x - 2)^7$

$\dfrac{dy}{dx} = f'(g(x)) \times g'(x)$ •————— Write down the chain rule. **C**

$\dfrac{dy}{dx} = 7(5x - 2)^6 \times 5$ •————— Differentiate f and g. **P**

$\dfrac{dy}{dx} = 35(5x - 2)^6$ •————— Simplify your answer. **C**

Example 9.25

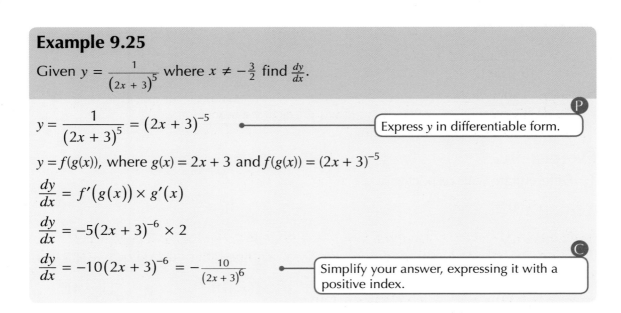

Given $y = \dfrac{1}{(2x + 3)^5}$ where $x \neq -\frac{3}{2}$ find $\frac{dy}{dx}$.

P

$y = \dfrac{1}{(2x + 3)^5} = (2x + 3)^{-5}$ •————— Express y in differentiable form.

$y = f(g(x))$, where $g(x) = 2x + 3$ and $f(g(x)) = (2x + 3)^{-5}$

$\dfrac{dy}{dx} = f'(g(x)) \times g'(x)$

$\dfrac{dy}{dx} = -5(2x + 3)^{-6} \times 2$

C

$\dfrac{dy}{dx} = -10(2x + 3)^{-6} = -\dfrac{10}{(2x + 3)^6}$ •————— Simplify your answer, expressing it with a positive index.

Example 9.26

Differentiate $(x^3 - 4x + 1)^8$.

$y = (x^3 - 4x + 1)^8$

$y = f(g(x))$, where $g(x) = x^3 - 4x + 1$ and $f(g(x)) = (x^3 - 4x + 1)^8$

$\frac{dy}{dx} = f'(g(x)) \times g'(x)$

$\frac{dy}{dx} = 8(x^3 - 4x + 1)^7 \times (3x^2 - 4)$ ⟵ Differentiate f and g. Note that brackets are needed around $(3x^2 - 4)$. **P**

$\frac{dy}{dx} = 8(x^3 - 4x + 1)^7 (3x^2 - 4)$ ⟵ $\frac{dy}{dx}$ cannot be simplified further. **P**

Example 9.27

Given $y = \sqrt[4]{(5 - 8x)^3}$, find $\frac{dy}{dx}$

$y = \sqrt[4]{(5 - 8x)^3} = (5 - 8x)^{\frac{3}{4}}$

$y = f(g(x))$, where $g(x) = 5 - 8x$ and $f(g(x)) = (5 - 8x)^{\frac{3}{4}}$

$\frac{dy}{dx} = f'(g(x)) \times g'(x)$

$\frac{dy}{dx} = f'(g(x)) \times g'(x) = \frac{3}{4}(5 - 8x)^{-\frac{1}{4}} \times -8$ ⟵ Differentiate f and g. Be careful. $g'(x) = -8$ and not 5. **P**

$\frac{dy}{dx} = \frac{3}{4} \times (-8) \times (5 - 8x)^{-\frac{1}{4}}$

$= -6(5 - 8x)^{-\frac{1}{4}}$

$= -\frac{6}{(5 - 8x)^{\frac{1}{4}}}$

$= -\frac{6}{\sqrt[4]{5 - 8x}}$ ⟵ Simplify the answer, expressing it in root form. **C**

Exercise 9G

1 Differentiate with respect to x

a $(x + 4)^3$ b $(x - 2)^6$ c $(x + 3)^9$ d $(x - 1)^5$

e $5(x + 1)^4$ f $8(x - 3)^6$ g $\frac{1}{2}(x + 5)^8$ h $\frac{2}{7}(x - 5)^7$

i $(x + 2)^{-1}$ j $(x - 5)^{-4}$ k $(x + 6)^{-7}$ l $(x - 3)^{-4}$

m $4(x - 2)^{-3}$ n $9(x - 7)^{-2}$ o $\frac{3}{4}(x + 1)^{-8}$ p $\frac{5}{6}(x - 4)^{-9}$

q $6 + (x - 1)^4$ r $3 - (x + 4)^5$ s $2(x + 2)^9 - 4x^3$

t $6\sqrt{x} + 5(x - 1)^4$ u $3(x + 4)^{-1} - \frac{5}{x^2}$ v $\frac{4}{7x^3} + 8(x - 4)^{-2}$

2 Differentiate with respect to x

a $(3x + 1)^6$ b $(5x - 2)^4$ c $(2x - 7)^5$ d $(4x + 1)^9$

e $2(3x - 4)^7$ f $10(6x + 2)^3$ g $8(5x - 4)^8$ h $6(7x - 1)^4$

i $(4x - 1)^{-2}$ j $(2x + 5)^{-8}$ k $(9x - 2)^{-1}$ l $(5x + 4)^{-6}$

m $3(2x - 1)^{-4}$ n $5(7x + 1)^{-2}$ o $(2x + 5)^{-3} - 8x$

p $\dfrac{1}{\sqrt[4]{x^5}} - 4(3x - 1)^{-1}$ q $(x - 4)^{-6} + \dfrac{4x^2 - 1}{x}$ r $3(8x - 1)^{-2} - \dfrac{4}{\sqrt{x}}$

s $\sqrt{x}\left(2x - \dfrac{1}{x^2}\right) + 6(4x - 1)^{-2}$ t $-2(3x + 4)^{-5} - \dfrac{2}{5x^{10}}$

3 Differentiate with respect to x

a $(1 - x)^5$ b $(5 + x)^{-3}$ c $(3 + 7x)^4$ d $\left(\frac{2}{3}x - 4\right)^6$

e $(2 - 5x)^6$ f $\left(\frac{3x}{5} + 2\right)^{10}$ g $\dfrac{(2 - 3x)^4}{5}$ h $(6 - x)^{-1}$

i $5(1 - 2x)^{-3} - 7x^2$ j $6x\sqrt{x} - \dfrac{(1 - 2x)^{-4}}{8}$

4 Differentiate with respect to x

a $\dfrac{1}{(x + 1)^4}$ b $\dfrac{1}{(x - 5)^2}$ c $\dfrac{1}{(4x + 1)^5}$ d $\dfrac{1}{3x - 4}$

e $\dfrac{5}{x - 3}$ f $\dfrac{2}{(x + 1)^3}$ g $\dfrac{6}{(2x - 5)^2}$ h $\dfrac{8}{(5x - 1)^6}$

i $\dfrac{1}{3(x + 2)}$ j $\dfrac{1}{4(x - 1)^2}$ k $\dfrac{3}{2(x + 1)^4}$ l $\dfrac{2}{9(3x - 2)^5}$

m $\dfrac{1}{2 - x}$ n $\dfrac{8}{5 - 3x}$ o $\dfrac{1}{(5 - 8x)^2}$ p $\dfrac{9}{4(1 - 2x)^2}$

5 Differentiate with respect to x

a $(x + 5)^{\frac{4}{3}}$ b $(x - 1)^{\frac{3}{2}}$ c $(4x + 3)^{\frac{5}{2}}$ d $(7x - 2)^{\frac{5}{4}}$

e $(x - 2)^{\frac{1}{3}}$ f $(2x - 5)^{\frac{1}{4}}$ g $(x + 4)^{-\frac{1}{3}}$ h $(5x + 6)^{-\frac{3}{2}}$

i $\dfrac{1}{(x - 1)^{\frac{2}{3}}}$ j $\dfrac{1}{4(x + 2)^{\frac{3}{2}}}$ k $\sqrt[3]{(x - 4)^2}$ l $\sqrt{(6x + 1)^5}$

m $\dfrac{1}{\sqrt{x + 2}}$ n $\dfrac{8}{\sqrt[4]{(x - 3)^3}}$ o $\dfrac{1}{\sqrt[5]{4 - x}}$ p $6\left(\sqrt{(2 - 5x)^3}\right)$

q $\dfrac{4}{\sqrt[4]{(5 - 2x)^3}}$

6 Differentiate with respect to x

a $(x^2 - 3)^4$

b $(x^3 - 2x^2 + 1)^5$

c $(x^4 - 5x - 2)^3$

d $(4 - 3x^2)^6$

e $(2x^2 + 5x - 3)^{-1}$

f $(3 - 2x - x^3)^{-4}$

g $\dfrac{1}{x^2 - 5}$

h $\sqrt{2x^2 - x + 5}$

i $\dfrac{1}{\sqrt{2 - 3x^3}}$

j $\sqrt{(x^4 - 1)^3}$

k $\sqrt[3]{(x^3 + x^2 + x + 1)}$

l $(\sqrt{x} - 2)^6$

★ **7** Find the gradient of the tangent to the curve $y = (4x + 5)^6$ at the point where $x = -1$.

★ **8** The function f is defined by $f(x) = \frac{1}{x-3}$ where $x \neq 3$. Find the rate of change of f when $x = 0$.

★ **9** Given $y = 3(2x + 1)^4$, find $\frac{dy}{dx}$ when $x = -\frac{3}{2}$.

10 The function g is defined by $g(x) = \sqrt{(1 - 4x)}$.

    a Suggest a suitable domain for g.

    b Find the gradient of the tangent to the curve $y = g(x)$ at the point where $x = -2$.

    c Determine algebraically the x–coordinate of the point at which the gradient of the curve $y = g(x)$ is -4.

11 Given $y = (x^2 - 2x - 8)^4$, find the values of x for which $\frac{dy}{dx} = 0$.

12 On a suitable domain, the function h is defined by $h(x) = \sqrt{x^2 - 5}$. Determine algebraically the coordinates of the points where the tangent to the curve $y = h(x)$ has gradient $-\frac{3}{2}$.

13 The function g is defined by $g(x) = \sqrt{x^2 - x + 3}$.

    a Show that g is defined for all real values of x.

    b Find algebraically the values of x for which $g'(x) = \frac{5}{6}$.

14 The function f is defined by $f(x) = \left(px^2 - 4x + p\right)^3$, where p is a constant and $p \neq 0$.

    a Find $f'(x)$.

    b Find algebraically the values of p for which the equation $f'(x) = 0$ has exactly 2 real roots and hence determine the value of these roots.

15 The function g is defined on a suitable domain by $g(x) = \frac{1}{x^3+(p-6)x^2-9p+2}$

    a Find $g'(x)$.

    b Determine algebraically the value of p for which the tangent to the curve $y = g(x)$ is parallel to the x–axis at only one point.

16 a Show that $(x + 1)$ is a factor of $x^3 + x^2 - 4x - 4$ and hence factorise $x^3 + x^2 - 4x - 4$ completely.

    On a suitable domain the function f is defined by $f(x) = \frac{1}{3x^4+4x^3-24x^2-48x}$

    b Find $f'(x)$.

    c Determine algebraically the values of x for which the tangent to the curve $y = f(x)$ is perpendicular to the line with equation $x = -2$.

Example 9.28

Given $y = 5\cos 2x$, find $\frac{dy}{dx}$

$y = 5\cos(2x)$ ——————————————— Put brackets around the $2x$. **S**

$y = f(g(x))$, where $g(x) = 2x$ and $f(g(x)) = 5\cos(2x)$

$\frac{dy}{dx} = f'(g(x)) \times g'(x)$

$\frac{dy}{dx} = 5 \times -\sin(2x) \times 2$ ——————— Differentiate f and g. Remember to include the 5. **P**

$\frac{dy}{dx} = -10\sin 2x$ ——————————————— Simplify your answer. **C**

Example 9.29

Differentiate $\sin^4 x$.

$y = (\sin x)^4$ ——————————————— Put brackets around $\sin x$. Be careful not to confuse $\sin^4 x$ with $\sin 4x$. **S**

$y = f(g(x))$, where $g(x) = \sin x$ and $f(g(x)) = (\sin x)^4$

$\frac{dy}{dx} = f'(g(x)) \times g'(x)$

$\frac{dy}{dx} = 4 \times (\sin x)^3 \times \cos x$ ——————— Differentiate f and g. **P**

$\frac{dy}{dx} = 4\sin^3 x \cos x$ ——————————————— Simplify your answer. **C**

Exercise 9H

1 Differentiate with respect to x

   a $\sin 2x$ **b** $\cos 5x$ **c** $3\sin 4x$

   d $6\cos 3x$ **e** $\cos\left(2x + \frac{\pi}{6}\right)$ **f** $\sin(6x - \pi)$

   g $\frac{3}{2}\sin 4x$ **h** $\frac{3}{5}\cos 10x$ **i** $\frac{1}{3}\sin(6x + 2)$

   j $-8\cos\left(\frac{1}{4}x\right)$ **k** $\sin(2 - x)$ **l** $-\frac{1}{3}\sin(1 - 9x)$

   m $\cos 5x - 2\sin 3x$ **n** $\sin x + 8\cos 3x$ **o** $\sin 3x - 2\cos x$

   p $\frac{5}{4}\sin(2x - \pi) - \cos\left(\frac{3}{2}x\right)$

★ 2 **a** Find the gradient of the tangent to the curve $y = \sin 2x$ at the point where $x = \frac{\pi}{6}$

b The function f is defined by $f(x) = 2\cos\left(x - \frac{\pi}{2}\right)$ Find $f'\left(\frac{2\pi}{3}\right)$

c A curve has equation $y = \cos 3x$. Find the gradient of the curve at the point where $x = \frac{2\pi}{9}$

d The function g is defined by $g(x) = 4\sin\left(\frac{1}{2}x\right)$. Find the rate of change of g when $x = \frac{\pi}{2}$

e Given $y = 6\cos\left(2x + \frac{\pi}{6}\right)$, find $\frac{dy}{dx}$ when $x = \frac{3\pi}{2}$

 ★ 3 **You will need a calculator for this question.** Give your answers to 2 decimal places.

a Given $y = \sin 5x$, find $\frac{dy}{dx}$ when $x = 1$.

b The function h is defined by $h(x) = 5\cos(3x - 1)$. Evaluate $h'(-3)$.

c Find the gradient of the tangent to the curve with equation $y = 2\sin(5 - 3x)$ at the point where $x = \frac{\pi}{2}$

d The function f is defined by $f(x) = 2\cos(\pi - 3x)$. Find the rate of change of f when $x = -2$.

 4 Differentiate with respect to x.

a $\sin^2 x$ **b** $\cos^3 x$ **c** $2\sin^3 x$

d $5\cos^6 x$ **e** $-\frac{1}{2}\sin^4 x$ **f** $\frac{6}{5}\cos^5 x$

g $(\sin x + \cos x)^3$ **h** $(\sin(2x - 1))^5$

 5 Find the gradient of the tangent to the curve with equation $y = \sin^2 x$ at the point where $x = \frac{\pi}{4}$

 6 Differentiate with respect to x.

a $\dfrac{1}{\sin x}$ **b** $\dfrac{3}{\cos x}$ **c** $\sqrt{\sin x}$ **d** $\dfrac{1}{\cos^2 x}$

e $\dfrac{1}{\cos(3x - \pi)}$ **f** $\sin\sqrt{x}$ **g** $\sin(\sin x)$ **h** $\cos(\sin x)$

 7 Show that when $y = \sin x°$, $\frac{dy}{dx} = \frac{\pi}{180}\cos x°$ ⟶ ⟨ Convert $x°$ into radians. ⚠

8 Repeat question 7 for $y = \cos x°$

9 Find the x–coordinates of the point on the curve $y = 2\sin 3x$, $0 \le x \le \pi$, at which the gradient of the tangent is -3.

10 A curve has equation $y = 4\cos 2x$ where $0 \le x \le \frac{\pi}{2}$. Find the coordinates of the points where the gradient of the curve is -4.

11 A curve has equation $\cos(3x - \pi)$ where $0 \le x \le \pi$. Find the x–coordinates of the points on the curve where the tangent is parallel to the straight line with equation $3x + 2y = 0$

12 Functions f and g are defined by $f(x) = 3\cos 2x$ and $g(x) = 6\sin x$. Solve the equation $f'(x) = g'(x)$ where $0 \le x \le 2\pi$

13 Functions f and g are defined by $f(x) = \cos x$ and $g(x) = 3x^2 - 1$

The function h is defined by $h(x) = f(g(x))$.

  a Write down an expression for $h(x)$.

  b Find $g'(x)$

  c Show that $g(f(x)) = 3\cos 2x + 2$

  d At the point where $x = \frac{2\pi}{3}$, the tangent to the curve $y = g(f(x))$ is parallel to the straight line $ax - \sqrt{3}y = 0$. Determine algebraically the value of a.

14 A curve has equation $y = \frac{2}{9}\sin\left(3x + \frac{\pi}{2}\right)$. At the point where $x = \frac{4\pi}{9}$, the tangent to the curve makes an angle of $a°$ with the positive direction of the x–axis. Find the value of a.

🚲 STEP BACK IN TIME

The development of differential calculus is attributed to two people: the English mathematician Sir Isaac Newton (1642 – 1727) and his German contemporary, Gottfried Leibniz (1646 – 1716).

Newton's calculus was borne out of an attempt to calculate the instantaneous velocities of objects, whilst Leibnitz approached differentiation from a purely mathematical perspective.

For a long time there was debate about who should receive credit for being the first to discover calculus. Nowadays, both men are rightly lauded and afforded equal status for their efforts. Universally acknowledged, however, is that Leibniz produced the better 'notation' for calculus – the use of $\frac{dy}{dx}$ is entirely down to him – and this notation is favoured by mathematicians, physicists and other scientists.

- I can differentiate an expression which is already in differentiable form. ★ Exercise 9A Q6
- I can express a function in differentiable form. ★ Exercise 9B Q1, 2
- I can differentiate an expression. ★ Exercise 9B Q4, 5
- I can evaluate derivatives to find the gradient of a curve ★ Exercise 9D Q2
- I can differentiate expressions of the form $k\sin x$ and $k\cos x$. ★ Exercise 9E Q2, 5
- I can differentiate a composite function. ★ Exercise 9G Q 7–9 ★ Exercise 9H Q2, 3

For further assessment opportunities, see the Preparation for Assessment for Unit 2 on pages 291–292.

10 Using differentiation to investigate the nature and properties of functions

In this chapter students will learn how to:

- determine the equation of a tangent to a curve at a given point by differentiation
- determine where a function is strictly increasing, decreasing
- sketch the graph of an algebraic function.

Students should already know:

- how to differentiate an algebraic function which is reducible to sums of powers of x
- how to differentiate $k\sin x$ and $k\cos x$
- how to differentiate simple composite functions.

Determine the equation of a tangent to a curve by differentiation

In Chapter 9 you learned about the concept of instantaneous change and how this relates to gradient. Using the process of differentiation you can determine the gradient of a tangent to a curve at any specified point.

With a known gradient at a given point, this can be extended to determine the equation of a tangent to a curve. Any line which isn't horizontal or vertical has an equation of the form $y - b = m(x - a)$. For the equation of a tangent to a curve, the point with coordinates (a, b) is the point of contact and m is the gradient. The gradient of the tangent to the curve can be found using differentiation.

Note that throughout this chapter you should make use of a graphing calculator or computer software to promote discussion and enhance understanding.

> a **x-coordinate of the point of contact**
> b **y-coordinate of the point of contact**
> m **gradient of the tangent when $x = a$**

Example 10.1

Find the equation of the tangent to the curve $y = x^2 - 4x + 7$ at the point $(3, 4)$.

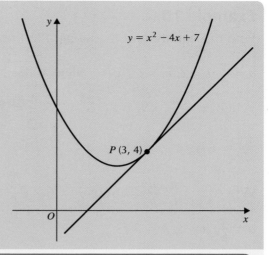

$y = x^2 - 4x + 7$

$P(3, 4)$

| $a = 3, b = 4, m = ?$ | Write down what you already know about a, b and m. **S** |

$\dfrac{dy}{dx} = 2x - 4$ — Differentiate to find the gradient function. **P**

$\dfrac{dy}{dx} = 2(3) - 4$ — Substitute x with 3. **P**

$\dfrac{dy}{dx} = 6 - 4 = 2$ — Find the gradient m of the tangent at $(3, 4)$. **P**

$y - 4 = 2(x - 3)$ — Write down the equation of the tangent. **S**

$y = 2x - 2$ — Express the equation in the form $y = mx + c$ **C**

Example 10.2

At P, the tangent to the curve $y = x^2 - 3x + 4$ has gradient $= 5$. Find the equation of the tangent at P.

$a = ?, b = ?, m = 5$

$\dfrac{dy}{dx} = 2x - 3$ — Differentiate to find the gradient function. **P**

$2x - 3 = 5$ — $m = 5$ so $\dfrac{dy}{dx} = 5$. So substitute $\dfrac{dy}{dx}$ with 5. **P**

$2x = 8, x = 4$ — Solve the equation to find the x–coordinate of the point of contact, to give the value of a. **P**

When $x = 4$, $y = 4^2 - 3(4) + 4$ — Substitute x with 4 to begin finding the y-coordinate of the point of contact, to give the value of b. **P**

$y = 16 - 12 + 4 = 8$

$a = 4, b = 8, m = 5$ — Obtain the value of b. **P**

$y - 8 = 5(x - 4)$

$y = 5x - 12$ — Write down the equation of the tangent. **C**

Example 10.3

Find the equation of the tangent to the curve $y = 6\sin x$ at the point where $x = \frac{\pi}{6}$.

$a = \frac{\pi}{6}, b = ?, m = ?$ — Write down what you already know about a, b and m. **S**

When $x = \frac{\pi}{6}$, $y = 6\sin\frac{\pi}{6} = 6 \times \frac{1}{2} = 3$ — Substitute x with $\frac{\pi}{6}$ to find the y-coordinate of the point of contact to give b. **P**

$\dfrac{dy}{dx} = 6\cos x$

When $x = \frac{\pi}{6}$, $\dfrac{dy}{dx} = 6\cos\frac{\pi}{6} = 6 \times \frac{\sqrt{3}}{2} = \frac{6\sqrt{3}}{2} = 3\sqrt{3}$ — To find m, differentiate and substitute x with $\frac{\pi}{6}$. At the point $\left(\frac{\pi}{6}, 3\right)$ the gradient of the tangent is $3\sqrt{3}$ so $m = 3\sqrt{3}$. **P**

$a = \frac{\pi}{6}, b = 3, m = 3\sqrt{3}$

$y - 3 = 3\sqrt{3}\left(x - \frac{\pi}{6}\right).$ — Write down the equation of the tangent. **C**

Example 10.4

Find the equation of the tangent to the curve $y = 3\cos\left(2x - \frac{\pi}{2}\right)$ at the point where $x = \frac{2\pi}{3}$

$a = \frac{2\pi}{3}, b = ?, m = ?$ — Write down what you already know about a, b and m. **S**

When $x = \frac{2\pi}{3}$, $y = 3\cos\left(\frac{4\pi}{3} - \frac{\pi}{2}\right) = 3\cos\left(\frac{5\pi}{6}\right) = 3 \times -\frac{\sqrt{3}}{2} = -\frac{3\sqrt{3}}{2}$

$\dfrac{dy}{dx} = -6\sin\left(2x - \frac{\pi}{2}\right)$ — Substitute x with $\frac{2\pi}{3}$ to find the y-coordinate of the point of contact to give b. **P**

When $x = \frac{2\pi}{3}$, $\dfrac{dy}{dx} = 6\sin\left(\frac{5\pi}{6}\right) = 6 \times \frac{1}{2} = 3$ — To find m, differentiate and substitute x with $\frac{2\pi}{3}$. At the point of contact the gradient of the tangent is 3, so $m = 3$. **P**

$a = \frac{2\pi}{3}, b = -\frac{3\sqrt{3}}{2}, m = 3$

$y + \frac{3\sqrt{3}}{2} = 3\left(x - \frac{2\pi}{3}\right)$ — Write down the equation of the tangent. **C**

Exercise 10A

1 For each of these functions, find the equation of the tangent at the given point.

   a $y = x^2 + 3x + 6$; $(1, 10)$ **b** $y = x^2 - 6x + 4$; $(3, -5)$

   c $y = 2x^2 + 3x - 5$; $(-1, -6)$ **d** $y = x^3 + 3x^2 - 2x + 4$; $(2, 20)$

   e $y = x^3 - 2x$; $(-3, -21)$ **f** $y = 4 - 3x - x^2$ $(2, -6)$

2 For each of these functions, find the equation of the tangent at the given point.

a $y = \cos x;\ \left(\dfrac{\pi}{6},\ \dfrac{\sqrt{3}}{2}\right)$

b $y = \sin 3x;\ \left(\dfrac{\pi}{9},\ \dfrac{\sqrt{3}}{2}\right)$

c $y = 4\sin x;\ \left(\dfrac{\pi}{4},\ 2\sqrt{2}\right)$

d $y = 2\cos\left(x - \dfrac{\pi}{4}\right);\ \left(\dfrac{7\pi}{12},\ 1\right)$

e $y = 5\sin\left(x + \dfrac{\pi}{3}\right);\ \left(0,\ \dfrac{5\sqrt{3}}{2}\right)$

f $y = -\dfrac{1}{2}\cos\left(x - \dfrac{\pi}{6}\right);\ \left(\dfrac{\pi}{2},\ -\dfrac{1}{4}\right)$

 ★ **3** For each of these functions, find the equation of the tangent at the given point

a $y = x^2 + 5x + 1;\ x = 1$

b $y = x^2 - x + 6;\ x = 3$

c $y = 3x^2 + 8x - 5;\ x = -2$

d $y = x^3 - 3x^2 - 4x + 8;\ x = 4$

e $y = 8 - x^3;\ x = -3$

f $y = 6 + x - 2x^2;\ x = -1$

 ★ **4** For each of these functions, find the equation of the tangent at the given point.

a $y = \cos x;\ x = \dfrac{\pi}{3}$

b $y = \sin 2x;\ x = \dfrac{\pi}{6}$

c $y = 8\cos x;\ x = \dfrac{2\pi}{3}$

d $y = 6\sin\left(x - \dfrac{\pi}{4}\right);\ x = \dfrac{\pi}{4}$

e $y = -4\cos\left(x - \dfrac{\pi}{6}\right);\ x = \dfrac{\pi}{3}$

f $y = 2\sin\left(x + \dfrac{\pi}{8}\right);\ x = \dfrac{5\pi}{8}$

 5 a A curve has equation $y = x^2 + 4x - 6$. At P, the gradient of the curve is -6. Find the equation of the tangent to the curve at P.

b A curve has equation $y = 3x^2 - 7x + 9$. At Q the tangent to the curve has gradient $= 5$. Find the equation of the tangent at Q.

c A curve has equation $y = 5x^2 - x + 3$. At R, the tangent to the curve has gradient $= -6$. Find the equation of the curve at R.

d A curve has equation $y = 8\sin x$, $0 \le x \le \dfrac{\pi}{2}$ At A, the gradient of the curve is 4. Find the equation of the tangent to the curve at A.

e A curve has equation $y = \cos 4x$, $0 \le x \le \dfrac{\pi}{6}$ At B, the tangent to the curve has gradient $= -2$. Find the equation of the tangent to the curve at B.

f A curve has equation $y = 2\sin\left(x - \dfrac{\pi}{6}\right)$, $0 \le x \le \dfrac{\pi}{2}$ At C, the tangent to the curve has gradient $= \sqrt{3}$. Find the equation of the tangent to the curve at C.

★ **6** For each of these functions, find the equation of the tangent at the given point. **You will need to express each function in differentiable form first.**

a $y = (x - 2)(x + 5);\ x = 4$

b $y = 2x^2(x - 3);\ x = -1$

c $y = (x - 2)(x^2 + x - 1);\ x = 1$

d $y = x(x + 2)^2;\ x = -2$

e $y = \dfrac{4x - 1}{x};\ x = 2$

f $y = 4\sqrt{x};\ x = 36$

7 The function f is defined by $f(x) = 8 - 2(x - 4)^2$.
The graph of $y = f(x)$ intersects the x-axis
at A and B.

Determine the equation of the tangent at B.

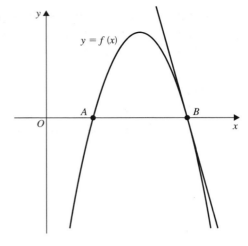

8 The function g is defined by $g(x) = x^3 + ax^2 + bx + 4$.

$P(2, -6)$ and $Q(-1, 6)$ both lie on the graph of $y = g(x)$.

a Find the values of a and b.

b Find the equation of the tangent to the curve at the point where $x = 3$.

c Find the x-coordinates of the points at which the tangent to the graph of
$y = g(x)$ has gradient 4.

9 The diagram shows part of graph of the cubic
function $f(x) = x^3 + 3x^2 - 21x + 6$. At P and Q,
the tangent to the graph of $y = f(x)$ has
gradient $= 3$.

a Find the equation of the tangent at Q.

b Determine the coordinates of P.

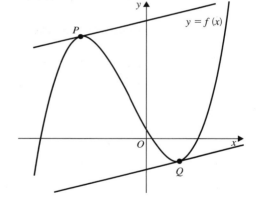

10 Find the equation of the tangent to the curve $y = 4\sin(3x - \pi)$ at the point
where $x = \frac{\pi}{4}$

11 The function f is defined by $f(x) = 2\cos\left(2x + \frac{\pi}{2}\right)$, $0 \le x \le \pi$. At P, the
tangent to the graph of $y = f(x)$ has gradient 2. Find the equation of the
tangent at P.

12 An old road can be represented by the curve with equation $x^3 - 2x^2 + 6x + 7$ as shown in the diagram.

A new road is represented by the dashed line. The line is a tangent to the curve at the point P, where $x = 2$.

Find the equation of the dashed line.

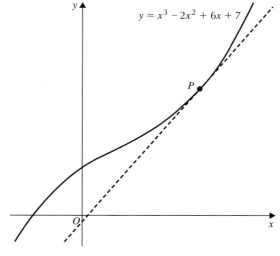

$y = x^3 - 2x^2 + 6x + 7$

 13 In a chemical reaction, the curve with equation $y = \frac{1}{2x+5}$ can be used to describe the concentration, y, of a certain chemical present x seconds after the start of the reaction.

    a Find the rate of change of y after 1 second.

    b Find the equation of the tangent to the curve at the point where $x = 3$.

 14 Find the equation of the tangent to the curve $y = (x - 3)^5$ at the point where $x = 2$.

 15 Find the equation of the tangent to the curve $y = (3x + 2)^5$ at the point where $x = -1$.

 16 At P the tangent to the curve with equation $y = \frac{1}{6}(2x - 3)^6$ has gradient $= -2$. Find the equation of the tangent at P.

 17 Find the equation of the tangent to the curve with equation $y = (x^2 - 3x + 2)^4$ at the point where $x = 3$.

 18 A curve has equation $y = 4x^2 - 5x + 3$. At the point A, the tangent to the curve makes an angle of $\frac{\pi}{4}$ radians with the positive direction of the x-axis.

    a Determine algebraically the x–coordinate of A.

    b Hence, find the equation of the tangent to the curve at A.

 19 Find the equation of the tangent to the curve with equation $y = \sin^3 x$ at the point where $x = \frac{7\pi}{6}$.

 20 A curve has equation $y = \frac{12}{\sqrt{x}}$, $x \neq 0$. At P the gradient of the tangent to the curve is $-\frac{3}{4}$. Determine the equation of the tangent to the curve at P.

 21 The function h is defined by $h(x) = f(4x - 9) + 5$. $P\,(5, 7)$ is on the graph of $y = h(x)$. At P the tangent to the curve $y = h(x)$ has gradient $= -1$.

Find the equation of the tangent to the curve with equation $y = f(x)$ at the point where $x = 11$.

22 Functions f and g are defined on suitable domains by $f(x) = \sqrt{x}$ and $g(x) = 5 + 8\sin x$.

   a Write down an expression for $f(g(x))$.

      The function h is defined by $h(x) = f(g(x))$, $0 \le x \le \frac{\pi}{2}$ At T, the tangent to the curve with equation $y = h(x)$ has gradient $\frac{2\sqrt{3}}{3}$

   b **i** Find $h'(x)$.

      ii Hence, show that $12\cos^2 x - 8\sin x - 5 = 0$.

   c Determine the equation of the tangent to the curve $y = h(x)$ at T.

23 Functions f and g are defined, on suitable domains, by $f(x) = \frac{4}{x}$ and $g(x) = x^2 - 2x - 15$.

   a Determine the largest possible domain for $f(g(x))$.

   b The function $k(x)$ is defined by $k(x) = f(g(x))$. Determine the equation of the tangent to the curve $y = k(x)$ at the point where $x = 6$.

24 The diagram shows the graph of the quadratic function $y = (x - 2)^2 + 5$. The graph intersects the y-axis at A.

   a Find the equation of the tangent at A.

      Another tangent passes through the origin and meets the curve at B.

   b Determine algebraically the equation of this tangent and hence find the coordinates of B.

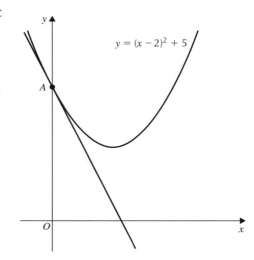

   c The point C lies on the tangent OB and $AB = AC$. Find the coordinates of C.

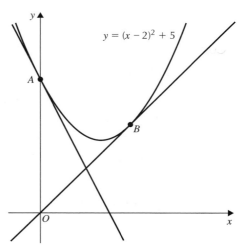

25 The diagram shows part of the parabola with equation $y = x^2 - 6x + 14$. At P, the tangent to the parabola makes an angle of $a°$ with the positive direction of the x-axis.

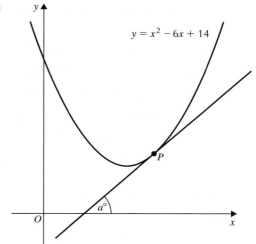

  a Given that $\sin a° = \frac{2}{\sqrt{5}}$ determine the equation of the tangent at P.

The tangent at P intersects the x-axis at R. Q is the turning point of the parabola. The straight line passing through R and Q meets the parabola again at S.

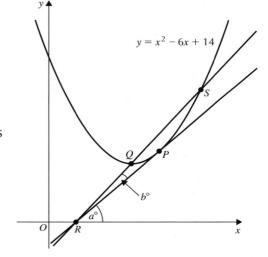

  b Given that angle $PRQ = b°$, find the exact value of $\cos b°$.

  c Determine algebraically the coordinates of S.

▲ Challenge

Consider the function $f(x) = x^3$. The coordinates of any point on the graph of $y = f(x)$ are (k, k^3).

1 When $x = -k$ what is the value of f?

2 Using your answer to Question 1 explain why the graph of $y = x^3$ possesses half-turn symmetry. Where is the centre of rotation?

3 Show that the graph of $y = ax^3$ has half–turn symmetry about the origin.

The question now is whether or not the graph of **any** cubic has half-turn symmetry. Consider first the function $g(x) = ax^3 + cx$.

(continued)

4 a Show that the graph of $y = g(x)$ has half-turn symmetry about the origin.

b Would the graph of $y = g(x) + d$ still have half-turn symmetry? If so, where would the centre of rotation be?

The only cubic we haven't yet dealt with is the function
$h(x) = ax^3 + bx^2 + cx + d$.

5 Show that $h\left(x - \frac{b}{3a}\right)$ can be written in the form $ax^3 + px + q$, expressing p and q in terms of a, b, c and d.

6 Explain why the graph of $y = h\left(x - \frac{b}{3a}\right) - q$ has half–turn symmetry about the origin.

7 Hence, explain why the graph of $h(x) = ax^3 + bx^2 + cx + d$ has half–turn symmetry and state the coordinates of the centre of rotation.

Determining where a function is strictly increasing or decreasing

When we want to get an idea of the shape of a graph of a function it is useful to know where the function is increasing or decreasing. Knowing where a function is increasing or decreasing helps us to determine the shape of the graph and also gives information about how quickly one quantity is changing with respect to another.

We make the following definitions:

When $f'(a) > 0$ the function $f(x)$ is increasing **(at the point where $x = a$).**

When $f'(a) < 0$ the function $f(x)$ is decreasing **(at the point where $x = a$).**

When $f'(a) = 0$ the function $f(x)$ is stationary **(at the point where $x = a$).**

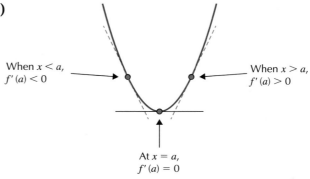

When $x < a$, $f'(a) < 0$

When $x > a$, $f'(a) > 0$

At $x = a$, $f'(a) = 0$

Example 10.5

The function f is defined by $f(x) = x^3 - 5x + 6$. Show that $f(x)$ is increasing when $x = 2$.

$f(x) = x^3 - 5x + 6$

$f'(x) = 3x^2 - 5$ ————————————— Differentiate. **S**

$f'(2) = 3(2)^2 - 5 = 12 - 5 = 7$ ————— Evaluate $f'(x)$ when $x = 2$. **P**

$f'(2) > 0$ therefore $f(x)$ is increasing when $x = 2$ ——— Communicate clearly. **C**
$f'(2) > 0$ must be stated.

Example 10.6

The function g is defined by $g(x) = x + \frac{20}{\sqrt{x}}$ Show that $g(x)$ is decreasing when $x = 4$.

$$g(x) = x + \frac{20}{\sqrt{x}}$$

$$g(x) = x + 20x^{-\frac{1}{2}}$$

Express $g(x)$ in differentiable form. P

$$g'(x) = 1 - 10x^{-\frac{3}{2}} = 1 - \frac{10}{x^{\frac{3}{2}}} = 1 - \frac{10}{\sqrt{x^3}}$$

Differentiate. S

$$g'(4) = 1 - \frac{10}{\sqrt{4^3}} = 1 - \frac{10}{\sqrt{64}} = 1 - \frac{10}{8} = -\frac{1}{4}$$

Evaluate $g(x)$ when $x = 4$. P

$g'(4) < 0$ therefore $g(x)$ is decreasing when $x = 4$

Communicate clearly. C
$g'(4) < 0$ must be stated.

Example 10.7

The function h is defined by $h(x) = x(x^3 - 32)$. Find the range of values of x for which $h(x)$ is increasing.

$$h(x) = x(x^3 - 32)$$

$$h(x) = x^4 - 32x$$

Express $h(x)$ in differentiable form. P

$$h'(x) = 4x^3 - 32$$

$$4x^3 - 32 > 0$$

$h'(x) > 0$ since we want $h(x)$ to be increasing. S

$$4x^3 > 32$$

$$x^3 > 8$$

$$x > 2$$

Solve the inequality to find the range of values for x. P

Example 10.8

The function p is defined by $p(x) = x^3 - 3x^2 - 9x + 4$. Find the range of values of x for which $p(x)$ is decreasing.

$$p(x) = x^3 - 3x^2 - 9x + 4$$

$$p'(x) = 3x^2 - 6x - 9$$

$$3x^2 - 6x - 9 < 0$$

$p'(x) < 0$ since we want $p(x)$ to be decreasing. S

(*continued*)

RELATIONSHIPS AND CALCULUS

Method 1 Factorise and use a table of signs

$3(x^2 - 2x - 3) = 3(x - 3)(x + 1)$ •————(Factorise $3x^2 - 6x - 9$. Ⓟ

	$x < -1$	$x = -1$	$-1 < x < 3$	$x = 3$	$x > 3$
$(x - 3)$	–	–	–	0	+
$(x + 1)$	–	0	+	+	+
$3(x - 3)(x + 1)$	+	0	–	0	+

Use a table of Ⓟ signs to find when $p'(x) < 0$.

$p(x)$ is decreasing when $-1 < x < 3$ •———(Communicate your final answer clearly. Ⓒ

Method 2 Graphical

$3(x^2 - 2x - 3) = 0$

$3(x + 1)(x - 3) = 0$

$x = -1$ or $x = 3$ •————————————(Solve $p'(x) = 0$. Ⓟ

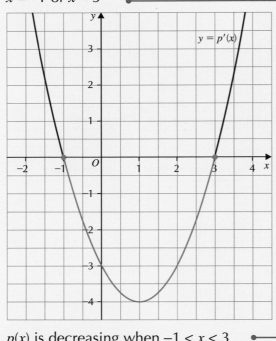

$y = p'(x)$

Sketch the graph of $y = p'(x)$. Ⓢ We want $p'(x) < 0$ so look for where the graph of $y = p'(x)$ lies below the x-axis.

$p(x)$ is decreasing when $-1 < x < 3$ •———(Communicate your final answer clearly. Ⓒ

Example 10.9

The function f is given by $f(x) = x^3 + 3x^2 + 5x - 8$. Show that $f(x)$ is always increasing.

$f(x) = x^3 + 3x^2 + 5x - 8$

$f'(x) = 3x^2 + 6x + 5$ •—————————————————————(Differentiate. Ⓟ

To show that $f(x)$ is always increasing we have to show that $f'(x) > 0$ for all values of x. Since $f'(x)$ is a quadratic function there are three methods we can use.

NaN248

Method 1 Completing the square

$$3x^2 + 6x + 5 = 3(x^2 + 2x) + 5$$

$$3(x^2 + 2x) + 5 = 3[(x + 1)^2 - 1] + 5$$

$$3[(x + 1)^2 - 1] + 5 = 3(x + 1)^2 - 3 + 5 = 3(x + 1)^2 + 2$$ — Complete the square. **P**

The minimum value of $3(x + 1)^2 + 2$ is 2 and so $f'(x) > 0$ for all values of x. — Communicate your final answer clearly. **C**

Method 2 Graphical

For a quadratic function $y = ax^2 + bx + c$, the turning point occurs when $x = -\frac{b}{2a}$

$$-\frac{b}{2a} = -\frac{6}{2(3)} = -\frac{6}{6} = -1$$ — Find x-coordinate of turning point of $y = f'(x)$. **P**

When $x = -1$, $f'(x) = 3(-1)^2 + 6(-1) + 5 = 3 - 6 + 5 = 2$. — Find y-coordinate of turning point of $y = f'(x)$. **P**

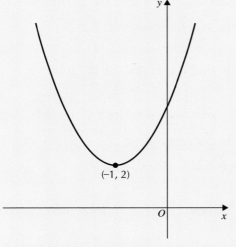

(−1, 2)

The two important features are the shape of the graph and the turning point. **S**

Since the graph of $y = f'(x)$ lies entirely above the x-axis $f'(x) > 0$ for all values of x. Hence, $f(x)$ is always increasing. — Communicate clearly. **C**

Method 3 Using the discriminant

$$b^2 - 4ac = 6^2 - 4(3)(5) = 36 - 60 = -24$$ — Find the discriminant of $f'(x)$ **S** **P**

$b^2 - 4ac < 0$ and so the equation $f'(x) = 0$ has no real roots. — Communicate clearly. **C**

The graph of $y = f'(x)$ never intersects the x-axis and has a minimum turning point (since $a > 0$). — Communicate clearly the significance of $b^2 - 4ac < 0$ and why the graph has a minimum turning point. **C**

It therefore lies entirely above the x-axis. — Either make a statement about the graph of $y = f'(x)$ or else make a sketch. **C**

Exercise 10B

★ **1 a** Show that the function $f(x) = x^3 - 2x^2 + 6x + 3$ is increasing when $x = 1$.

   b Show that the function $g(x) = 2x^3 + x^2 - 8x - 1$ is decreasing when $x = -1$.

   c Show that the function $h(x) = x^4 - 2x^3 - 2x^2 - 3$ is decreasing when $x = -1$.

   d Show that the function $p(x) = 6x - x^4$ is increasing when $x = -2$.

   e Show that the function $q(x) = 2x + \frac{3}{x}$ is decreasing when $x = -1$.

★ **2 a** Find the range of values of x for which the function $f(x) = \frac{1}{3}x^3 - x^2 - 3x + 2$ is increasing.

   b Find the range of values of x for which the function $g(x) = \frac{1}{3}x^3 + 2x^2 + 3x + 1$ is decreasing.

   c Find the range of values of x for which the function $h(x) = x^3 - 6x + 5$ is increasing.

   d Find the range of values of x for which the function $k(x) = x^4 + 32x$ is decreasing.

   e Find the range of values of x for which the function $f(x) = x^3 - 6x^2$ is decreasing.

3 The function f is defined by $f(x) = x^2 - \frac{3}{x}$ where $x \neq 0$. Show that $f(x)$ is increasing when $x = -1$.

4 The function f is defined by $f(x) = x(2x - 1)^2$. Show that $f(x)$ is decreasing when $x = \frac{1}{3}$.

5 A function is given by $h(x) = (3x - 1)^4$.

   a Show that $h(x)$ is increasing when $x = \frac{1}{2}$

   b Find the range of values of x for which $h(x)$ is decreasing.

6 The function g is defined by $g(x) = \frac{1}{(3-2x)^2}$ where $x \neq \frac{3}{2}$. Show that $g(x)$ is decreasing when $x = 2$.

7 The function k is defined by $k(x) = (x^3 - 4x + 1)^3$. Show that $k(x)$ is decreasing when $x = -1$.

8 The function h is defined by $h(x) = 4\sin x + 3\cos\left(x + \frac{\pi}{6}\right)$
Show that $h(x)$ is increasing when $x = \frac{\pi}{6}$.

9 Given $g(x) = \cos 4x + 4\sin 2x$, show that $g(x)$ is increasing when $x = \frac{\pi}{6}$

10 The function p is defined by $p(x) = \sin 3x + \cos\left(2x + \frac{\pi}{2}\right)$. Show that $p(x)$ is increasing when $x = \frac{\pi}{12}$

11 Given $f(x) = \frac{1}{3}x^3 - 3x^2 + 10x - 5$, show that $f(x)$ is always increasing.

12 The function h is defined by $h(x) = \frac{2}{3}x^3 + 2x^2 + 2x - 3$. Show that $h(x)$ is never decreasing.

13 On a suitable domain the function f is defined by $f(x) = \frac{1}{\sqrt{4x-1}}$

   a Suggest a suitable domain for $f(x)$.

   b Show that when $x = \frac{5}{16}$ the gradient of the tangent to the curve $y = f(x)$ is -16.

   c Explain why, on a suitable domain, $f(x)$ is always decreasing.

14 The function f is defined by $f(x) = x^3 + kx^2 - 5x + 4$. When $x = -2$ determine the least positive integer value of k for which $f(x)$ is decreasing.

 15 The function f is defined by $f(x) = 2\sin^3 x$. Show that $f(x)$ is decreasing when $x = \frac{2\pi}{3}$

 16 The function g is defined by $g(x) = \frac{x^2-6}{x}$ where $x \neq 0$. Show that when $x \neq 0$, $g(x)$ is always increasing.

 17 Functions f and g are defined on suitable domains by $f(x) = x^2 + 4x + 9$ and $g(x) = \frac{3}{x}$.

    a **i** Write down an expression for $g(f(x))$.

        ii Show that $g(f(x))$ is defined for all values of x.

        The function h is defined by $h(x) = g(f(x))$.

    b Find $h'(x)$ and hence show that $h(x)$ is increasing when $x = -3$

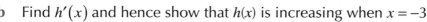 **18** The following information is known about the cubic function $f(x)$:

- $f'(-2) > 0$
- At the point where $x = 1$, the tangent to the graph of $y = f(x)$ is horizontal.
- When $x = 4$, $\frac{dy}{dx} = 0$

Below are five statements about $f(x)$. Only three are correct.
Identify the three correct statements.

    A $f(x)$ is increasing when $x = -3$

    B $f(x)$ is decreasing when $x = 5$

    C The range of values of x for which $f(x)$ is increasing is $1 < x < 4$

    D $f'(6) < f'(5)$

    E $f'(0) > 0$

 19 The function g is defined by $g(x) = 9x + \frac{1}{x-1}$, $x \neq 1$

    a Find $g'(x)$.

    b Show that when $g(x)$ is increasing, $9x^2 - 18x + 8 > 0$.

    c Hence, or otherwise, find the range of values of x for which $g(x)$ is increasing.

 20 The cubic function f is defined by $f(x) = x^3 + px^2 + 15x - 9$. Determine algebraically the largest integer value of p such that $f(x)$ is always increasing.

21 The function f is defined by $f(x) = p\sin^2(ax + b) + p\cos^2(ax + b)$.

    a Show that $f(x)$ is never increasing or decreasing.

    b It is possible to do part a without differentiating. Can you explain how?

Sketching the graph of an algebraic function

Knowledge of differentiation can be used to sketch an algebraic function. This section will focus on polynomials of degree 3 and 4, known as **cubic** and **quartic** functions.

Before sketching the graph of a cubic or quartic function, it is essential to determine any stationary points. It is strongly recommended that you make use of a graphing calculator or computer software.

Determining the stationary points of a function

The function $f(x)$ is said to be stationary whenever $f'(x) = 0$. This means that the tangent to the graph of $y = f(x)$ is horizontal, so the gradient is zero.

Any point on the graph where the tangent is horizontal is called a **stationary point**, (sometimes shortened to **SP**). Stationary points are often referred to as **turning points**.

There are three types of stationary point:

- minimum
- maximum
- horizontal inflection.

A **nature table** is used to help determine the nature of a stationary point. A nature table shows the overall shape of a curve, near to the turning points and also more generally, by making use of the gradient function.

Minimum stationary point (or minimum turning point)

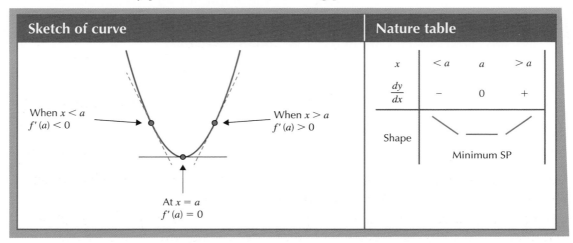

The first two rows of the nature table provide information about the gradient of the curve at key points. When $x < a$, the gradient of the curve is negative; when $x = a$ the gradient of the curve is 0 and when $x > a$ the gradient of the curve is positive.

The information from the first two rows is then used to complete the third row – this confirms the nature of the turning point as well as telling us about the shape of the curve.

Maximum stationary point (or maximum turning point)

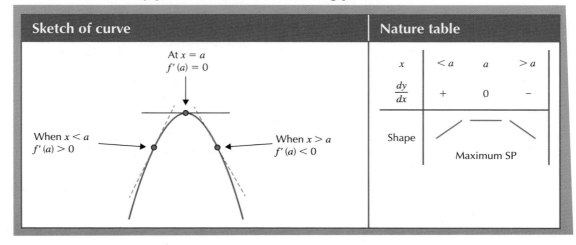

The gradient **changes sign** moving left to right (or right to left) along the curve through the turning point.

Point of horizontal inflection

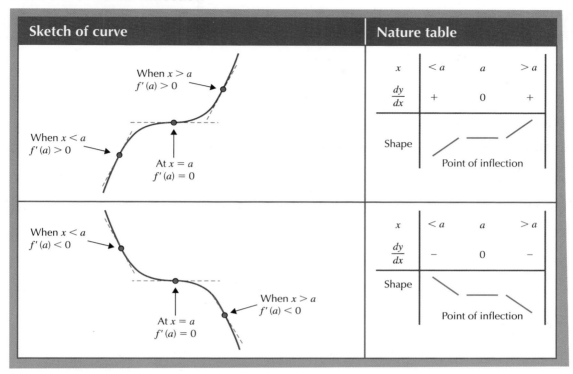

Sketch of curve	Nature table

At a point of **horizontal inflection** $f'(a) = 0$ and the gradient **does not** change sign moving left to right (or right to left) through this stationary point.

A **point of horizontal inflection** is a special example of the more general **point of inflection**. If you study Advanced Higher mathematics you will study points of inflection in more depth.

Example 10.10

Find the coordinates of the stationary points of the curve with equation $y = \frac{1}{3}x^3 - 2x^2 - 12x + 5$ and determine their nature.

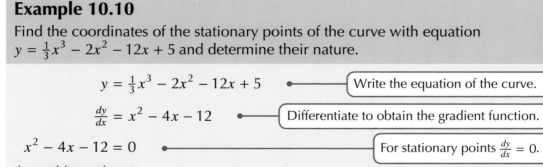

$$y = \tfrac{1}{3}x^3 - 2x^2 - 12x + 5$$ — Write the equation of the curve. **P**

$$\tfrac{dy}{dx} = x^2 - 4x - 12$$ — Differentiate to obtain the gradient function. **P**

$$x^2 - 4x - 12 = 0$$ — For stationary points $\frac{dy}{dx} = 0$. **C**

$$(x + 2)(x - 6) = 0$$

$$x = -2 \text{ or } x = 6$$ — Solve $\frac{dy}{dx} = 0$ to find the x-coordinates of the stationary points. **P**

(continued)

When $x = -2$, $y = \frac{1}{3}(-2)^3 - 2(-2)^2 - 12(-2) + 5 = 18\frac{1}{3}$

When $x = 6$, $y = \frac{1}{3}(6)^3 - 2(6)^2 - 12(6) + 5 = -67$

> Find the y-coordinates of the **P** stationary points. Use the equation of the curve, not $\frac{dy}{dx}$

The stationary points are $\left(-2, 18\frac{1}{3}\right)$ and $(6, -67)$

> Give the x- and y-coordinates of the stationary points.

Work out key information to construct a nature table.

When $x = -3$, $\frac{dy}{dx} = (-3)^2 - 4(-3) - 12 = 9$

> Choose a value of x less than **P** -2, such as $x = -3$. Determine the gradient of the curve when $x = -3$. Note that $\frac{dy}{dx} > 0$.

When $x = 1$, $\frac{dy}{dx} = 1^2 - 4(1) - 12 = -15$

When $x = 7$, $\frac{dy}{dx} = 7^2 - 4(7) - 12 = 9$

> Choose a value of x greater than 6, **P** such as $x = 7$. Determine the gradient of the curve when $x = 7$. Note that $\frac{dy}{dx} > 0$.

> Choose a value of x between -2 and 6, **P** such as $x = 1$. Determine the gradient of the curve when $x = 1$. Note that $\frac{dy}{dx} < 0$.

Nature table

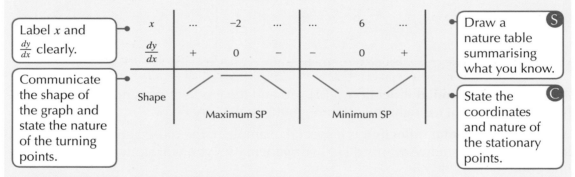

> Label x and $\frac{dy}{dx}$ clearly.

> Communicate the shape of the graph and state the nature of the turning points.

> Draw a **S** nature table summarising what you know.

> State the **C** coordinates and nature of the stationary points.

Maximum SP at $\left(-2, 18\frac{1}{3}\right)$ and minimum SP at $(6, -67)$.

The y-coordinates of stationary points are referred to as the stationary **values** of the function.

Example 10.11

Find the coordinates of the stationary points of the curve with equation $y = x^2(18 - x^2)$ and determine their nature.

$$y = x^2(18 - x^2)$$
$$y = 18x^2 - x^4$$

> Express y in differentiable form. **P**

$$\frac{dy}{dx} = 36x - 4x^3$$

> Differentiate to obtain the gradient function. **P**

$$36x - 4x^3 = 0$$

> For stationary points $\frac{dy}{dx} = 0$. **P**

$$4x\left(9 - x^2\right) = 0$$

$$4x(3 - x)(3 + x) = 0$$

$x = 0,\ x = 3,\ x = -3$

> Solve $\frac{dy}{dx} = 0$ to find the x-coordinates of the stationary points. **P**

When $x = 0,\ y = 18(0)^2 - (0)^4 = 0$

When $x = 3,\ y = 18(3)^2 - 3^4 = 81$

When $x = -3,\ y = 18(-3)^2 - (-3)^4 = 81$

> Find the y-coordinates of the stationary points. **P**

The stationary points are $(0, 0),\ (3, 81)$ and $(-3, 81)$.

Use key information to construct a nature table.

When $x = -4,\ \dfrac{dy}{dx} = 36(-4) - 4(-4)^3 = 112$

> Choose a value of x less than -3, such as $x = -4$. Determine the gradient of the curve when $x = -4$. Note that $\frac{dy}{dx} > 0$ **P**

When $x = -1,\ \dfrac{dy}{dx} = 36(-1) - 4(-1)^3 = -32$

> Choose a value of x between -3 and 0, such as $x = -1$. Determine the gradient of the curve when $x = -1$. Note that $\frac{dy}{dx} < 0$ **P**

When $x = 1,\ \dfrac{dy}{dx} = 36(1) - 4(1)^3 = 32$

When $x = 4,\ \dfrac{dy}{dx} = 36(4) - 4(4)^3 = -112$

> Choose a value of x between 0 and 3, such as $x = 1$. Determine the gradient of the curve when $x = 1$. Note that $\frac{dy}{dx} > 0$ **P**

> Choose a value of x greater than 3, such as $x = 4$. Determine the gradient of the curve when $x = 4$. Note that $\frac{dy}{dx} < 0$ **P**

x	...	-3	...	...	0	...	...	3	...
$\dfrac{dy}{dx}$	$+$	0	$-$	$-$	0	$+$	$+$	0	$-$
	/	—	\	\	—	/	/	—	\
		Maximum SP			Minimum SP			Maximum SP	

Maximum SP at $(-3, 81)$, minimum SP at $(0, 0)$ and maximum SP at $(3, 81)$.

Example 10.12

Find the coordinates of the stationary points of the curve with equation $y = \frac{1}{4}x^4 - \frac{4}{3}x^2 + 2x^2 + 3$ and determine their nature.

$$y = \tfrac{1}{4}x^4 - \tfrac{4}{3}x^3 + 2x^2 + 3$$

$$\dfrac{dy}{dx} = x^3 - 4x^2 + 4x$$

(*continued*)

$$x^3 - 4x^2 + 4x = 0$$

$$x\left(x^2 - 4x + 4\right) = 0$$

$$x(x - 2)(x - 2) = 0$$

$$x = 0, \ x = 2$$

When $x = 0$, $y = \frac{1}{4}(0)^4 - \frac{4}{3}(0)^2 + 2(0)^2 + 3 = 3$.

When $x = 2$, $y = \frac{1}{4}(2)^4 - \frac{4}{3}(2)^2 + 2(2)^2 + 3 = 9\frac{2}{3}$

The stationary points are $(0, 3)$ and $\left(2, 9\frac{2}{3}\right)$.

Use key information to construct a nature table.

> Choose a value of x less than ⓟ 0, such as $x = -1$. Determine the gradient of the curve when $x = -1$. Note that $\frac{dy}{dx} < 0$

When $x = -1$, $\frac{dy}{dx} = (-1)^3 - 4(-1)^2 + 4(-1) = -9$

> Choose a value of x between 0 and 2, ⓟ such as $x = 1$. Determine the gradient of the curve when $x = 1$. Note that $\frac{dy}{dx} > 0$

When $x = 1$, $\frac{dy}{dx} = 1^3 - 4(1)^2 + 4(1) = 1$

> Choose a value of x greater than 2, ⓟ such as $x = 3$. Determine the gradient of the curve when $x = 3$. Note that $\frac{dy}{dx} > 0$

When $x = 3$, $\frac{dy}{dx} = 3^3 - 4(3)^2 + 4(3) = 3$

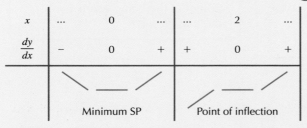

x	...	0	...	...	2	...
$\frac{dy}{dx}$	−	0	+	+	0	+

Minimum SP   Point of inflection

Minimum SP at $(0, 3)$. Point of inflection at $\left(2, 9\frac{2}{3}\right)$.

Exercise 10C

 ★ **1** For each of these curves:

     i determine the coordinates of the stationary points

     ii determine the nature of the stationary points.

     a $y = \frac{1}{3}x^3 - x + 2$ **b** $y = \frac{1}{3}x^3 - 3x^2 + 9x - 4$

     c $y = \frac{1}{3}x^3 - 3x^2 + 9x - 4$ **d** $y = x^3 - 2x^2 - 4x + 1$

     e $y = x^3 - \frac{1}{2}x^2 - 4x + 5$ **f** $y = 2x^3 + \frac{7}{2}x^2 - 3x + 1$

2 Determine the coordinates and nature of the stationary points for each of these curves.

  a $y = 2 + 5x - x^2 - x^3$ **b** $y = 2 - 8x + 7x^2 - x^3$ **c** $y = 4x(x^2 - 12)$

  d $y = x^2(2 - x)$ **e** $y = \frac{1}{3}x(27 - x^2)$ **f** $y = \frac{1}{6}x^2(15 - 4x)$

3 Determine the coordinates and nature of the stationary points for each of these curves.

  a $y = x^3(x - 2)$ **b** $y = x^3(6 - x)$ **c** $y = x^4 - 32x^2$

  d $y = \frac{1}{2}x^2(2 - x^2)$ **e** $y = 3x^4 + 8x^3 + 6x^2 - 6$

4 A function is defined by $f(x) = (4x-1)^5$.

  a Find the coordinates of the stationary point of the curve $y = f(x)$.

  b Determine the nature of this stationary point.

5 A curve has equation $y = (x^2 - 2x - 8)^3$.

  a Find $\frac{dy}{dx}$

  b **i** Find the coordinates of the stationary points of the curve.

   ii Determine their nature.

6 Functions f and g are defined by $f(x) = \frac{1}{x}$, $x > 0$ and $g(x) = x^2 - 6x$.

  a Write down an expression for $f(g(x))$.

  b Find the values of x for which $f(g(x))$ is undefined.

  c **i** Given $h(x) = f(g(x))$, find $h'(x)$.

   ii Hence, determine the coordinates and nature of the stationary point of the curve $y = h(x)$.

7 A curve has equation $y = 1 + 3x - \frac{12}{x^3}$ where $x \neq 0$.

  Determine the coordinates and nature of the stationary points of the curve.

8 **You will need a calculator for this question**. The function f is defined by $f(x) = 2\sin^2 x - x$ where $0 \leq x \leq \pi$

  a Find $\frac{dy}{dx}$

  b Determine the coordinates and nature of the turning points of the curve $y = f(x)$.

9 **a** Express $3x^2 - 12x + 7$ in the form $a(x + b)^2 + c$.

  b A curve has equation $y = x^3 - 6x^2 + 7x - 3$. Show that the curve has no stationary points.

10 A curve has equation $y = (x^2 - 5x + 8)^3$.

    a   Show that the curve has only one stationary point.

    b   Determine the coordinates and nature of this stationary point.

Only attempt the remaining questions if you have already completed Chapters 1 and 7.

11 a   **i**   Show that $(x - 2)$ is a factor of $x^3 - 4x^2 + x + 6$.

       ii   Hence factorise fully $x^3 - 4x^2 + x + 6$.

    b   Determine the x–coordinates of the stationary points of the curve

$$y = x^4 - \tfrac{16}{3}x^3 + 2x^2 + 24x - 10.$$

    c   Determine the nature of the stationary points.

12 The function f is defined by $f(x) = x^3 - 4x^2 + x + 6$.

    a   **i**   Show that $x = -1$ is a root of the equation $f(x) = 0$.

       ii   Determine algebraically the remaining roots of the equation $f(x) = 0$.

    b   Determine the coordinates and nature of the stationary points of the curve
$y = x^4 - 16x^3 + 6x^2 + 72$.

13 Functions f and g are defined by $f(x) = x^3 - 7x^2$ and $g(x) = x + 2$.

    a   Show that $f(g(x)) = x^3 - x^2 - 16x - 20$

    b   The function h is defined by $h(x) = f(g(x))$.

       i   Show that $x = -2$ is a root of the equation $h(x) = 0$.

       ii   Fully factorise $h(x)$.

    c   Find the stationary points of the curve with equation $y = h(x)$ and determine their nature.

Sketching the graph of an algebraic function

There are three main features to consider when sketching the graph of an algebraic function:

- any stationary points and their nature
- points of intersection with the x- and y-axes
- the behaviour of the function for large positive and negative values of x.

Example 10.13

A curve has equation $y = x^3 - 6x^2$.

a Find where the curve meets the x- and y-axes.

b Find the coordinates of the stationary points of the curve and determine their nature.

c Sketch the graph of the curve.

a When $x = 0$, $y = 0^3 - 0(0)^2 = 0$

    y-intercept is at $(0, 0)$

> On the y-axis, $x = 0$. Substitute $x = 0$ to find the y-intercept. **P**

    $x^3 - 6x^2 = 0$

> Write down the coordinates of the y-intercept. **C**

    $x^2(x - 6) = 0$

       $x^2 = 0$ or $x - 6 = 0$

> On the x-axis, $y = 0$. **C**

       $x = 0$ or $x = 6$

    $(0, 0)$, $(6, 0)$

> Write down where the curve meets the x-axis. **C**

b $y = x^3 - 6x^2$

    $\dfrac{dy}{dx} = 3x^2 - 12x$

    $3x^2 - 12x = 0$

    $3x(x - 4) = 0$

    $x = 0$, $x = 4$

    When $x = 0$, $y = 0$

    When $x = 4$, $y = 4^3 - 6(4)^2 = -32$

> Find the coordinates of the stationary points **P**

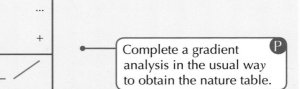

x	...	0	...	...	4	...
$\dfrac{dy}{dx}$	+	0	–	–	0	+
	/	—	\	\	—	/
		Maximum SP			Minimum SP	

> Complete a gradient analysis in the usual way to obtain the nature table. **P**

Maximum SP at $(0, 0)$, minimum SP at $(4, -32)$.

c As $x \to +\infty$, $y \to +\infty$

    As $x \to -\infty$, $y \to -\infty$

> As x becomes larger and larger so does the value of y. This is because the value of x^3 will dominate over the values of the other terms. **C**

> As x becomes large negative x^3 becomes large negative and so the value of y becomes large negative. **C**

(continued)

Use these steps to sketch the graph in a non–calculator paper:

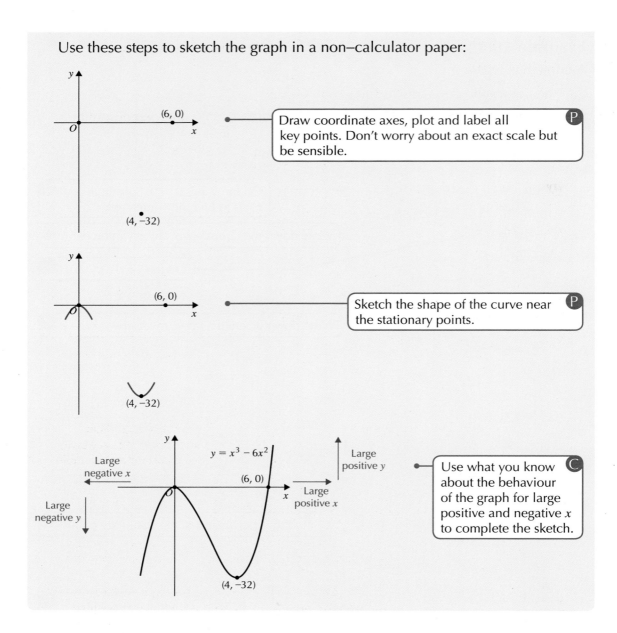

Draw coordinate axes, plot and label all key points. Don't worry about an exact scale but be sensible. **P**

Sketch the shape of the curve near the stationary points. **P**

Use what you know about the behaviour of the graph for large positive and negative x to complete the sketch. **C**

Example 10.14

A curve has equation $y = 2x^4 - 8x^3$.

a Find where the curve meets the x- and y-axes.

b Find the coordinates of the stationary points of the curve and determine their nature.

c Sketch the graph of the curve.

a When $x = 0$, $y = 2(0)^4 - 8(0)^3 = 0$

$(0, 0)$

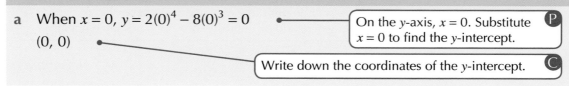

On the y-axis, $x = 0$. Substitute $x = 0$ to find the y-intercept. **P**

Write down the coordinates of the y-intercept. **C**

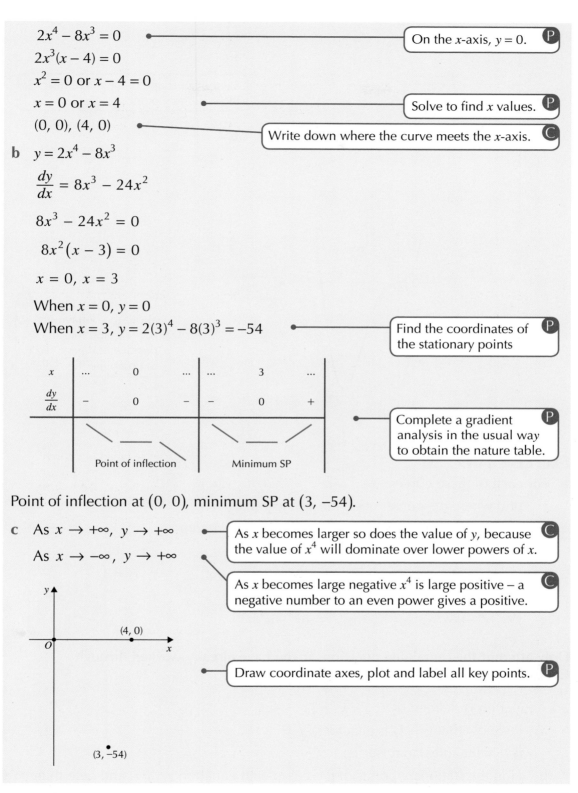

$2x^4 - 8x^3 = 0$ — On the x-axis, $y = 0$. **P**

$2x^3(x - 4) = 0$

$x^2 = 0$ or $x - 4 = 0$

$x = 0$ or $x = 4$ — Solve to find x values. **P**

$(0, 0), (4, 0)$ — Write down where the curve meets the x-axis. **C**

b $y = 2x^4 - 8x^3$

$\dfrac{dy}{dx} = 8x^3 - 24x^2$

$8x^3 - 24x^2 = 0$

$8x^2(x - 3) = 0$

$x = 0, x = 3$

When $x = 0, y = 0$

When $x = 3, y = 2(3)^4 - 8(3)^3 = -54$ — Find the coordinates of the stationary points **P**

x	...	0	...	...	3	...
$\dfrac{dy}{dx}$	–	0	–	–	0	+

Point of inflection   Minimum SP

Complete a gradient analysis in the usual way to obtain the nature table. **P**

Point of inflection at $(0, 0)$, minimum SP at $(3, -54)$.

c As $x \to +\infty, y \to +\infty$ — As x becomes larger so does the value of y, because the value of x^4 will dominate over lower powers of x. **C**

As $x \to -\infty, y \to +\infty$ — As x becomes large negative x^4 is large positive – a negative number to an even power gives a positive. **C**

Draw coordinate axes, plot and label all key points. **P**

(continued)

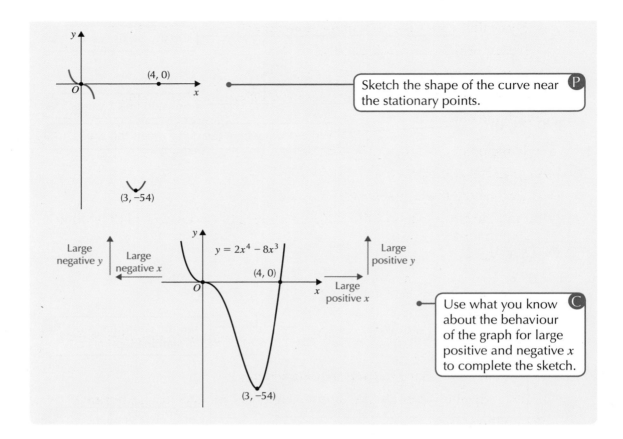

Exercise 10D

1 For each of these curves:

i find where the curve meets the x- and y-axes

ii find the stationary points and determine their nature

iii sketch the curve.

a $y = x^3 - 3x^2$ **b** $y = x^3 - 12x$ **c** $y = 3x - x^3$

d $y = (x + 2)(x - 1)^2$ **e** $y = (2x - 1)(x - 2)^2$ **f** $y = (x + 1)^2(x - 5)$

g $y = 3x^4 - 8x^3$ **h** $y = 6x^3 - x^4$ **i** $y = (9 - x^2)(x^2 + 4)$

Only attempt the remaining questions if you have already worked through Chapters 1 and 7.

2 A function is defined by $f(x) = x^3 + 3x^2 - 4$.

a **i** Show that $(x - 1)$ is a factor of $f(x)$.

    ii Hence fully factorise $f(x)$.

b Find the stationary points of the curve with equation $y = f(x)$ and determine their nature.

c Find where the curve meets the x- and y-axes.

d Sketch the curve.

3 a i Show that $x = -1$ is a root of the equation $x^3 - 7x^2 - 17x - 9 = 0$.

ii Hence find the remaining roots.

b A curve has equation $y = x^3 - 7x^2 - 17x - 9$.

Find the stationary points of the curve and determine their nature.

c Sketch the curve.

 4 a Fully factorise $x^3 - 10x^2 + 28x - 24$.

b A curve has equation $y = x^3 - 10x^2 + 28x - 24$.

i Find where the curve meets the x-axis.

ii Find the coordinates of the stationary points of the curve and determine their nature.

c Sketch the curve.

 5 a i Show that $(x + 2)$ is a factor of $x^3 - x^2 - x + 10$.

ii Fully factorise $x^3 - x^2 - x + 10$.

b i A curve has equation $y = x^3 - x^2 - x + 10$. Find the x-coordinates of the stationary points.

ii Determine the nature of the stationary points.

c Explain why the y-coordinates of both stationary points must be positive.

d Sketch the curve.

6 A function is given by $f(x) = x^4 + 2x^3 - 3x^2 - 4x + 4$.

a i Show that $(x + 2)$ is a factor of $f(x)$

ii Hence fully factorise $f(x)$

b Explain why the graph of $y = f(x)$ never lies below the x-axis.

c i Find the stationary points of the curve with equation $y = f(x)$

ii Determine their nature.

d Sketch the curve.

7 The function g is defined by $g(x) = x^3 + kx^2 + (k + 1)x + 18$.

a Given that $(x + 2)$ is a factor of $f(x)$ determine the value of k.

b i Fully factorise $g(x)$.

ii Explain why the x-axis is a tangent to the curve with equation $y = g(x)$.

c Find the stationary points of the curve with equation $y = g(x)$ and determine their nature.

d Sketch the curve.

- I can find the equation of a tangent to a curve given suitable information. ★ Exercise 10A Q3, Q4, Q6

- I can evaluate when a function is increasing or decreasing. ★ Exercise 10B Q1, Q2

- I can find the stationary points of a curve and determine their nature. ★ Exercise 10C Q1, Q2

- I can sketch a curve, showing all key points. ★ Exercise 10D Q1

For further assessment opportunities, see the Preparation for Assessment for Unit 2 on page 291–292.

11 Integrating functions

This chapter will show you how to:

- integrate an algebraic function which is, or can be, simplified to an expression in powers of x
- integrate functions of the form $f(x) = (x + q)^n$ and $f(x) = (px + q)^n$
- integrate functions of the form $f(x) = p\sin x$ and $f(x) = p\cos x$; $f(x) = p\sin(qx + r)$ and $f(x) = p\cos(qx + r)$
- solve differential equations of the form $\frac{dy}{dx} = f(x)$ to determine and use a function from a given rate of change and initial conditions.

You should already know:

- how to differentiate algebraic, trigonometric and simple composite functions.

Integrating a algebraic function which is already an expression in powers of x

If we differentiate $\frac{x^3}{3} + 5$ we get x^2. If we differentiate $\frac{x^3}{3} - 2$ we still get x^2. In general, if we differentiate $\frac{x^3}{3} + c$, where c is a constant, we will get x^2.

The inverse process – or **anti-differentiation** – which takes us from x^2 back to $\frac{x^3}{3} + c$ is called **integration**. $\frac{x^3}{3} + c$ is called an **anti-derivative** of x^2.

Similarly, differentiating $\frac{x^4}{4} + c$ gives x^3 and so $\frac{x^4}{4} + c$ is an anti–derivative of x^3.

More generally an anti-derivative of x^n is $\frac{x^{n+1}}{n+1} + c$.

We write the process of integration as follows:

$$\int x^n dx = \frac{x^{n+1}}{n+1} + c$$

The left–hand side is read "the integral of x^n with respect to x".

In the same way that when you differentiate x^n to obtain the derivative of x^n, you can also integrate x^n to obtain the integral of x^n.

x^{-1} cannot be integrated in this way. This would give $\frac{x^0}{0}$, which is undefined. x^{-1} can be integrated using techniques developed in Advanced Higher maths.

As with differentiation there are three main rules for integration:

1 $\int ax^n dx = a\int x^n dx = \frac{ax^{n+1}}{n+1} + c$ where $n \neq -1$

2 $\int (f(x) + g(x))\, dx = \int f(x)\, dx + \int g(x)\, dx$ \qquad This is the **sum rule**.

3 For any constant k, $\int k\, dx = kx + c$ \qquad If you differentiate $kx + c$ you get k.

Example 11.1

Integrate with respect to x.

a x^6 b x^{-2} c $x^{\frac{1}{4}}$ d $x^{-\frac{1}{2}}$

a $\displaystyle\int x^6\,dx = \frac{x^{6+1}}{6+1} + c$

$= \dfrac{x^7}{7} + c$ —— The answer could be written as $\frac{1}{7}x^7 + c$ Ⓟ

b $\displaystyle\int x^{-2}\,dx = \frac{x^{-1}}{-1} + c$

$= -\dfrac{1}{x} + c$ —— Write the final answer with a positive index. Ⓟ

c $\displaystyle\int x^{\frac{1}{4}}\,dx = \frac{x^{\frac{5}{4}}}{\frac{5}{4}} + c$

$= \dfrac{4}{5}x^{\frac{5}{4}} + c$ —— Dividing by $\frac{5}{4}$ is the same as multiplying by $\frac{4}{5}$. Ⓟ

$= \dfrac{4}{5}\left(\sqrt[4]{x^5}\right) + c$ —— Express $x^{\frac{5}{4}}$ in root form. Ⓒ

d $\displaystyle\int x^{-\frac{1}{2}}\,dx = \frac{x^{\frac{1}{2}}}{\frac{1}{2}} + c$

$= 2x^{\frac{1}{2}} + c$

$= 2\sqrt{x} + c$

Example 11.2

Integrate with respect to x: $\frac{1}{2}x^3 - 4x^{-2} + 3x^{-\frac{1}{2}} + 6$.

$\displaystyle\int \frac{1}{2}x^3 - 4x^{-2} + 3x^{-\frac{1}{2}} + 6\,dx = \frac{1}{2}\times\frac{x^4}{4} - 4\times\frac{x^{-1}}{-1} + 3\times\frac{x^{\frac{1}{2}}}{\frac{1}{2}} + 6x + c$ —— Integrate term by term. Ⓟ

$= \dfrac{1}{8}x^4 + 4x^{-1} + 6x^{\frac{1}{2}} + 6x + c$ —— Simplify the coefficient of each term in the answer. Ⓟ

$= \dfrac{1}{8}x^4 + \dfrac{4}{x} + 6\sqrt{x} + 6x + c$ —— Express all terms with positive indices and in root form where appropriate. Ⓒ

Example 11.3

Integrate with respect to the given variable.

a $4p^9$ b $u^3 - 2u^{-\frac{2}{3}}$

a $\displaystyle\int 4p^9\,dp = 4\times\frac{p^{10}}{10} + c$

$= \dfrac{2}{5}p^{10} + c$ —— We can integrate with respect to any variable we choose. Ⓟ

b $\displaystyle\int u^3 - 2u^{-\frac{2}{3}}\, du = \frac{u^4}{4} - 2 \times \frac{u^{\frac{1}{3}}}{\frac{1}{3}} + c$

$\displaystyle = \frac{u^4}{4} - 6\left(\sqrt[3]{u}\right) + c$

Exercise 11A

1 Integrate with respect to x.

a x^6 b x c x^{-3} d 5 e $4x^3$

f $-3x^{-7}$ g $\frac{3}{4}x^5$ h $8x^{-\frac{1}{4}}$ i $x^{\frac{3}{2}}$ j $x^{\frac{4}{3}}$

k $x^{-\frac{1}{2}}$ l $-x^{-\frac{2}{5}}$ m $6x^{\frac{1}{2}}$ n $-\frac{1}{6}x^{-5}$

2 Integrate with respect to x.

a $3x^2 + x - 1$ b $7 - 5x + x^4$ c $\frac{3}{2}x^5 - \frac{1}{4}x - 4$ d $\frac{4x^2}{3} - \frac{1}{5}x + 5x^{-6}$

3 Integrate with respect to x.

a $x^{\frac{1}{2}} - x^{-\frac{1}{2}}$ b $10x^{\frac{3}{2}} - 21x^{\frac{5}{2}}$ c $\frac{2}{3}x^{\frac{2}{3}} - 4x^{-\frac{1}{5}} - 8x$ d $\frac{15}{4}x^{\frac{1}{4}} - \frac{2}{25}x^{-\frac{1}{5}}$

4 Integrate with respect to the given variable.

a $k^4 + 3k - 5$ b $6p^8 - p^{-4}$ c $9t^{\frac{1}{2}} + 4t^{-\frac{1}{2}}$

★5 The table below contains eight functions and their integrals. Six of the eight integrals in the second column contain mistakes. Find all the mistakes and correct them.

	$f(x)$	$\int f(x)\, dx$
a	$2x^2 - 5$	$\frac{2}{3}x^3 - 5 + c$
b	$5x^{-4} + 2x^{\frac{1}{2}}$	$-x^{-3} - \frac{4}{\sqrt{x}} + c$
c	$\frac{1}{2}x^{-3} + 6x^{-\frac{3}{4}}$	$-\frac{1}{4x^2} + 24\left(\sqrt[4]{x}\right) + c$
d	$\frac{2}{5}x^{-\frac{4}{5}}$	$2\left(\sqrt{x^5}\right) + c$
e	$3x^{-5} + 4 - x^2$	$-\frac{3}{4x^3} + 4x - \frac{1}{3}x^3 + c$
f	$\frac{1}{2}x^{-\frac{3}{2}} - \frac{3}{2}x^{\frac{1}{2}}$	$-\frac{1}{\sqrt{x}} - \sqrt{x^3} + c$
g	$1 - 4x^{-\frac{1}{6}}$	$x + \frac{24}{5}\left(\sqrt[5]{x^6}\right) + c$
h	$\frac{1}{2}x^{-2} + \frac{1}{3}x^{-3} - 4x^{\frac{1}{4}}$	$-\frac{1}{2x} - \frac{2}{3x^{-2}} - \frac{16}{5}\left(\sqrt[4]{x^5}\right) + c$

6 Follow these instructions:

a Make up five functions to integrate. Include a mixture of positive, negative and fractional powers.

b Work out the correct integrals for your five functions (keep these secret).

c Create a 'wrong' answer for the integral of each function.

d Present your partner with the five original functions and the corresponding wrong answers. Have them correct your answers.

Integrating a function which is not already an expression in powers of x

When the function is not in integrable form, we proceed in the same as we did when differentiating.

Example 11.4

Find $\int \dfrac{1}{3x^4}\, dx$.

$$\frac{1}{3x^4} = \frac{1}{3}x^{-4}$$ — Write in integrable form. **P**

$$\int \frac{1}{3}x^{-4}\, dx = \frac{1}{3} \times \frac{x^{-3}}{-3} + c$$

$$= -\frac{1}{9x^3} + c$$ — Integrate and simplify. **P**

Example 11.5

Find $\int 9\sqrt{x} - 10\left(\sqrt{x^3}\right) dx$.

$$9\sqrt{x} - 10\left(\sqrt{x^3}\right) = 9x^{\frac{1}{2}} - 10x^{\frac{3}{2}}$$ — Write $9\sqrt{x} - 10\left(\sqrt{x^3}\right)$ in integrable form. **P**

$$\int 9x^{\frac{1}{2}} - 10x^{\frac{3}{2}}\, dx = 9 \times \frac{x^{\frac{3}{2}}}{\frac{3}{2}} - 10 \times \frac{x^{\frac{5}{2}}}{\frac{5}{2}} + c$$ — Integrate. **P**

$$= 6x^{\frac{3}{2}} - 4x^{\frac{5}{2}} + c$$ — Simplify coefficients. **P**

$$= 6\sqrt{x^3} - 4\sqrt{x^5} + c$$ — Express terms in root form. **C**

Example 11.6

$\int \dfrac{x^3 - 5}{x^2}\, dx$. Find $f'(x)$.

$$\frac{x^3 - 5}{x^2} = \frac{x^3}{x^2} - \frac{5}{x^2} = x - 5x^{-2}$$ — Write $\frac{x^3-5}{x^2}$ in integrable form. **P**

$$\int x - 5x^{-2}\, dx = \frac{1}{2}x^2 - 5 \times \frac{x^{-1}}{-1} + c$$

$$= \frac{1}{2}x^2 + \frac{5}{x} + c$$ — Integrate and simplify **P**

Exercise 11B

1 For each of these expressions

   i write the expression in integrable form

   ii integrate with respect to x.

     a $(2x - 1)(x - 3)$ **b** $x(x - 4)(x + 1)$ **c** $(x + 2)(x^2 + 3x - 4)$

     d $5x^2(x - 3)^2$ **e** $(x - 2)(x + 3)^2$

2 The first column of the table contains eight functions. In the second column, the functions have been written in integrable form. Some of them contain mistakes. Identify and correct all errors. Do not integrate the functions.

	$f(x)$	Integrable form
a	$\dfrac{1}{5x^3}$	$5x^{-3}$
b	$6\left(\sqrt[4]{x^3}\right)$	$6x^{\frac{3}{4}}$
c	$\sqrt{x}(2x - 5)$	$2x^{\frac{3}{2}} - 5x^{\frac{1}{2}}$
d	$\dfrac{2x^4 - 1}{3x^2}$	$\dfrac{2}{3}x^2 - 3x^{-2}$
e	$\dfrac{3}{5\left(\sqrt[3]{x^4}\right)}$	$\dfrac{3}{5}x^{-\frac{3}{4}}$
f	$(x - 2)\left(x^2 - 3x + 4\right)$	$x^3 - 5x^2 + 10x - 8$
g	$\dfrac{\sqrt{x} - 3x}{\sqrt{x}}$	$1 - 3x^{-\frac{1}{2}}$
h	$\dfrac{1}{6x^2} - \dfrac{5x^2}{3}$	$\dfrac{1}{6}x^{-2} - \dfrac{5}{2}x^2$

★3 For each of these expressions

   i write the expression in integrable form

   ii integrate with respect to x.

     a $\dfrac{6}{x^3}$ **b** $\dfrac{1}{5x^4}$ **c** $y = \dfrac{7}{3x^8}$ **d** $\dfrac{4}{x^2} - x^2 + 5$

★4 For each of these expressions

   i write the expression in integrable form

   ii integrate with respect to x.

     a $3\sqrt{x}$ **b** $\sqrt[3]{x^4}$ **c** $6\left(\sqrt[5]{x}\right)$ **d** $\dfrac{4}{\sqrt{x}}$

     e $\dfrac{1}{x\sqrt{x}}$ **f** $\dfrac{3}{\sqrt[4]{x}}$ **g** $\dfrac{10}{\sqrt{x^5}}$ **h** $\dfrac{1}{2\left(\sqrt[4]{x^3}\right)}$

★**5** For each of the expressions:

    i write the expression in integrable form

    ii integrate with respect to x.

    a $\dfrac{x^6 - 4}{x^2}$ **b** $\dfrac{9x - x^5}{x^4}$ **c** $\dfrac{x^4 - x - 3}{x^4}$

    d $y = \dfrac{5 - 2x^4}{3x^2}$ **e** $\dfrac{(x - 2)(x + 3)}{x^4}$ **f** $\dfrac{(x - 1)(3x + 2)^2}{3x^5}$

 6 Integrate with respect to x.

    a $x\left(\sqrt{x} - 4\right)$ **b** $\dfrac{2}{x}\left(x^2 - \dfrac{1}{x}\right)$ **c** $\left(1 - x^2\right)\left(2 + \dfrac{1}{\sqrt{x}}\right)$

    d $\left(\dfrac{5}{x} - \dfrac{x}{5}\right)^2$ **e** $\dfrac{1 - x^3}{\sqrt{x}}$ **f** $\dfrac{(x + 1)(2 - x)}{x\sqrt{x}}$

 7 Integrate with respect to the given variable.

    a $\dfrac{2}{x^8}$ **b** $\dfrac{1}{5t^2}$ **c** $(p - 1)(p + 2)(p - 4)$

    d $6\left(\sqrt[3]{x^2}\right)$ **e** $\dfrac{2}{3\sqrt{w}}$ **f** $\dfrac{3 - x^4}{x^3}$

    g $2t^3(4 - t)$ **h** $\dfrac{1}{u^2} - 3\sqrt{u} + 2$ **i** $\dfrac{3}{4\left(\sqrt[5]{x}\right)}$

 8 Follow these instructions:

    a Make up five functions to integrate. Choose from the types of functions covered in Questions 1 to 5.

    b Express your functions in integrable form but make sure that each contains a mistake (more than one if you want).

    c Present your partner with the five original functions and the corresponding wrong answers. Have them correct your answers.

 9 Follow these instructions:

    a Make up five functions to integrate. Choose from the types of functions covered in Questions 1 to 5.

    b Work out the correct integrals for your five functions (keep these secret).

    c Create a 'wrong' answer for the integral of each function.

    d Present your partner with the five original functions and the corresponding wrong answers. Have them correct your answers.

Integrate functions of the form $(x + q)^n$ and $(px + q)^n$

In Chapter 9 you learned how to differentiate functions of the form $(x + q)^n$ and $(px + q)^n$. This will help us to integrate such functions.

For example, when we differentiate $(x - 3)^5$ we get $5(x - 3)^4$ and so:

$$\int 5(x - 3)^4 \, dx = (x - 3)^5 + c.$$

It now follows that $\int (x-3)^4\, dx = \frac{(x-3)^5}{5} + c$.

More generally, the derivative of $(x+q)^{n+1}$ is $(n+1)(x+q)^n$ and so

$\int (n+1)(x+q)^n\, dx = (x+q)^{n+1} + c$

It follows that:

$$\int (x+q)^n\, dx = \frac{(x+q)^{n+1}}{n+1} + c$$

Differentiating $(3x+1)^7$ gives $21(3x+1)^6$ and so $\int 21(3x+1)^6\, dx = (3x+1)^7 + c$

It now follows that $\int (3x+1)^6\, dx = \frac{(3x+1)^7}{21} + c$

More generally, the derivative of $(px+q)^{n+1}$ is $p(n+1)(px+q)^n$ and so:

$\int p(n+1)(px+q)^n\, dx = (px+q)^n + c$

It follows that:

$$\int (px+q)^n\, dx = \frac{(px+q)^{n+1}}{p(n+1)} + c$$

Example 11.7

Find $\int (x+1)^5\, dx$.

$\int (x+1)^5\, dx = \frac{(x+1)^6}{6} + c$ ⚫────────────── Integrate. The answer can also be written as $\frac{1}{6}(x+1)^6 + c$. **P**

Example 11.8

Find $\int (3x+7)^4\, dx$

$p=3,\ q=7,\ n=4$ and so $(n+1) = 5$ ⚫──────── Write the values of p, q and n. **P**
Write the value of $(n+1)$.

$\int (px+q)^n\, dx = \frac{(px+q)^{n+1}}{p(n+1)} + c$ ⚫──────── Write the general integral. **C**

$\int (3x+7)^4\, dx = \frac{(3x+7)^5}{3 \times 5} + c$

$\qquad = \frac{1}{15}(3x+7)^5 + c$ ⚫──────── Substitute for p, q and $(n+1)$, **P**
integrate and simplify.

Example 11.9

Find $\int \frac{1}{(2x-1)^4}\, dx$

$\dfrac{1}{(2x-1)^4} = (2x-1)^{-4}$

> Write $\frac{1}{(2x-1)^4}$ in integrable form. **P**

$p = 2,\ q = -1,\ n = -4$ and so $(n+1) = -3$

$\int (px + q)^n\, dx = \dfrac{(px+q)^{n+1}}{p(n+1)} + c$

$\int \dfrac{1}{(2x-1)^4}\, dx = \dfrac{(2x-1)^{-3}}{2 \times (-3)} + c$

$\qquad\qquad = -\dfrac{1}{6(2x-1)^3} + c$

> Integrate and simplify. **P**

Example 11.10

Find $\int \sqrt[3]{2 - 5x}\, dx$

$\sqrt[3]{2 - 5x} = (2 - 5x)^{\frac{1}{3}}$

> Write $\sqrt[3]{2-5x}$ in integrable form. **P**

$p = -5,\ q = 2,\ n = \frac{1}{3}$ and so $(n+1) = \frac{4}{3}$

$\int (px + q)^n\, dx = \dfrac{(px+q)^{n+1}}{p(n+1)} + c$

$\int \sqrt[3]{2 - 5x}\, dx = \dfrac{(2 - 5x)^{\frac{4}{3}}}{-5 \times \frac{4}{3}} + c$

> Integrate. **P**

$\qquad\qquad = -\dfrac{3}{20}(2 - 5x)^{\frac{4}{3}} + c$

> Note that $\frac{1}{-5 \times \frac{4}{3}} = -\frac{1}{\frac{20}{3}} = -\frac{3}{20}$ **P**

$\qquad\qquad = -\dfrac{3}{20}\left(\sqrt[3]{(2 - 5x)^4}\right) + c$

> Express the final answer in root form. **C**

Example 11.11

Find $\int \dfrac{8}{\sqrt{1 - 4x}}\, dx$

$\dfrac{8}{\sqrt{1 - 4x}} = \dfrac{8}{(1 - 4x)^{\frac{1}{2}}} = 8(1 - 4x)^{-\frac{1}{2}}$

> Write $\frac{8}{\sqrt{1-4x}}$ in integrable form. **P**

$p = -4,\ q = 1,\ n = \frac{-1}{2}$ and so $(n+1) = \frac{1}{2}$

$$\int k(px + q)^n \, dx = k\int (px + q)^n = k \times \frac{(px + q)^{n+1}}{p(n + 1)} + c$$

$$\int \frac{8}{\sqrt{1 - 4x}} \, dx = 8 \times \frac{(1 - 4x)^{\frac{1}{2}}}{(-4) \times \frac{1}{2}} + c \quad \text{—————— Integrate. \textcircled{P}}$$

$$= -4\sqrt{1 - 4x} + c \quad \text{—————— Simplify your answer. \textcircled{C}}$$

Exercise 11C

 ★ **1** Integrate with respect to x.

a $(x + 1)^5$ b $(x - 3)^8$ c $10(x - 2)^4$ d $4(x + 6)$

e $(2x + 3)^4$ f $(5x - 2)^7$ g $3(4x + 1)^6$ h $6(3x - 4)^8$

 ★ **2** Integrate with respect to x.

a $(x + 2)^{-3}$ b $(x - 7)^{-2}$ c $8(x - 5)^{-5}$ d $3(x + 8)^{-6}$

e $(3x - 1)^{-4}$ f $(2x + 1)^{-2}$ g $2(4x - 3)^{-5}$ h $7(6x - 5)^{-8}$

 ★ **3** Integrate with respect to x.

a $(x + 1)^{\frac{1}{2}}$ b $(x - 4)^{\frac{1}{3}}$ c $12(x + 6)^{\frac{1}{5}}$ d $10(x + 4)^{\frac{3}{2}}$

e $(x - 2)^{-\frac{1}{2}}$ f $(x + 1)^{-\frac{1}{3}}$ g $4(x - 6)^{-\frac{3}{2}}$ h $\frac{3}{4}(x + 4)^{-\frac{3}{4}}$

i $(3x + 2)^{\frac{1}{2}}$ j $(5x - 3)^{-\frac{1}{2}}$ k $\frac{5}{8}(6x + 1)^{\frac{1}{4}}$ l $\frac{3}{4}(7x - 4)^{-\frac{5}{2}}$

m $(2x - 5)^{-\frac{2}{3}}$ n $\frac{1}{2}(4x - 1)^{-\frac{3}{4}}$ o $\frac{2(8x + 3)^{-\frac{5}{4}}}{3}$ p $(7x - 1)^{\frac{5}{3}}$

4 Integrate with respect to x.

a $\dfrac{1}{(x - 3)^2}$ b $\dfrac{3}{(x + 1)^3}$ c $\dfrac{1}{5(x - 2)^4}$ d $\dfrac{4}{3(x + 8)^5}$

e $\dfrac{1}{(3x - 1)^4}$ f $\dfrac{9}{(5x - 3)^2}$ g $\dfrac{1}{3(2x - 7)^5}$ h $\dfrac{3}{2(x + 4)^6}$

5 Integrate with respect to x.

a $\sqrt{x + 4}$ b $8(\sqrt{x - 1})$ c $\dfrac{1}{\sqrt{x - 1}}$ d $\dfrac{6}{\sqrt{x + 3}}$

e $\sqrt[3]{x + 4}$ f $\sqrt[4]{(x - 1)^3}$ g $\sqrt{(x + 5)^3}$ h $\dfrac{1}{\sqrt{(x - 2)^3}}$

i $8(\sqrt[4]{x - 6})$ j $\frac{2}{3}(\sqrt[5]{x + 1})$ k $\dfrac{5}{(\sqrt[3]{x - 6})}$ l $\dfrac{4}{5(\sqrt[6]{(x - 1)^5})}$

 6 Integrate with respect to x.

a $\sqrt{3x + 2}$ b $\sqrt{(4x - 1)^3}$ c $3(\sqrt[5]{(2x - 3)^2})$ d $\dfrac{4}{3(\sqrt[6]{(7x - 1)^5})}$

7 Integrate with respect to x.

a $(3x - 1)^5$

b $\dfrac{1}{(x - 2)^3}$

c $(3 - x)^4$

d $\sqrt{2 - x}$

e $(1 - 5x)^8$

f $4(3x - 2)^5$

g $\dfrac{1}{(6x - 5)^4}$

h $\dfrac{1}{\left(\sqrt{x} - 2\right)^3}$

i $\dfrac{3}{2}(4 - x)^3$

j $\sqrt[4]{(5 - x)^3}$

k $\dfrac{3}{2(1 - 3x)^5}$

l $\dfrac{3}{7}(4x - 5)^4$

Integrate functions of the form $f(x) = p\sin x$, $f(x) = p\cos x$; $f(x) = p\sin(qx + r)$, $f(x) = p\cos(qx + r)$

If we differentiate $5\sin x$ we get $5\cos x$. Remember this is only true when x is measured in radians.

It follows that $\displaystyle\int 5\cos x\, dx = 5\sin x + c$

In general, $\displaystyle\int p\cos x\, dx = p\sin x + c$. In a similar way it is straightforward to show that $\displaystyle\int p\sin x\, dx = -p\cos x + c$

Exercise 11D

 1 Complete the table below, which asks you to find the derivatives of some trigonometric functions.

$f(x)$	$f'(x)$
$\sin 3x$	
$\sin 2x$	
$\sin\left(x - \dfrac{\pi}{6}\right)$	
$\sin(4x - \pi)$	
$\sin(6x + 5)$	
$\sin(qx + r)$	

 2 Using your answers to Question 1 as a guide, complete the table below.

$f(x)$	$\int f(x)\, dx$	$f(x)$	$\int f(x)\, dx$
$3\cos 3x$	$\sin 3x$	$\cos 3x$	
$2\cos 2x$		$\cos 2x$	
$\cos\left(x - \dfrac{\pi}{6}\right)$		$3\cos\left(x - \dfrac{\pi}{6}\right)$	
$4\cos(4x - \pi)$		$\cos(4x - \pi)$	

$f(x)$	$\int f(x)\,dx$	$f(x)$	$\int f(x)\,dx$
$6\cos(6x+5)$		$\cos(6x+5)$	
$q\cos(qx+r)$		$\cos(qx+r)$	

3 Using your answers from Questions 1 and 2 as a guide, complete the table below.

$f(x)$	$\int f(x)\,dx$
$6\sin 3x$	
$5\sin 2x$	
$3\sin\left(x-\dfrac{\pi}{6}\right)$	
$\dfrac{1}{2}\sin(4x-\pi)$	
$-3\sin(6x+5)$	
$p\sin(qx+r)$	

4 Try to complete the table below

$f(x)$	$\int f(x)\,dx$
$\cos 2x$	
$5\cos 2x$	
$2\cos\left(x+\dfrac{\pi}{2}\right)$	
$\dfrac{3}{4}\cos\left(\dfrac{1}{2}x-2\pi\right)$	
$-2\sin(4x-1)$	
$p\cos(qx+r)$	

Exercise 11D directed you towards two key results in integration.

$$\int \cos(qx+r) = \frac{1}{q}\sin(qx+r)$$

$$\int \sin(qx+r) = -\frac{1}{q}\cos(qx+r)$$

More generally,

$$\int p\cos(qx+r) = p\times\frac{1}{q}\sin(qx+r)$$

$$\int p\sin(qx+r) = p\times-\frac{1}{q}\cos(qx+r)$$

Example 11.12
Find $\int 2\cos x\,dx$

$$\int 2\cos x\,dx = 2\sin x + c$$

Example 11.13

Find $\int -\frac{2}{7}\sin x \, dx$

$\int -\frac{2}{7}\sin x \, dx = \frac{2}{7}\cos x + c$ •———————— The answer contains $\frac{2}{7}$ and not $-\frac{2}{7}$.

Example 11.14

Find $\int 3\cos 2x \, dx$

$\int 3\cos 2x \, dx = 3 \times \frac{1}{2}\sin 2x + c = \frac{3}{2}\sin 2x + c$

Example 11.15

Find $\int \frac{3}{5}\sin(4x - 1) \, dx$

$\int \frac{3}{5}\sin(4x - 1) \, dx = \frac{3}{5} \times -\frac{1}{4}\cos(4x - 1) + c$

$\qquad\qquad = -\frac{3}{20}\cos(4x - 1) + c$

Exercise 11E

 ★1 Integrate with respect to x.

   a $8\cos x$ **b** $3\sin x$ **c** $-4\sin x$

   d $2\cos x$ **e** $\frac{3}{2}\cos x$ **f** $-\frac{5\sin x}{4}$

   g $4\cos\left(x - \frac{\pi}{3}\right)$ **h** $5\sin(x - 2)$ **i** $\cos 5x$

   j $\sin 4x$ **k** $8\cos 2x$ **l** $\frac{1}{2}\cos(3x - \pi)$

   m $\sin\frac{1}{2}x$ **n** $6\cos\frac{3}{2}x$ **o** $9\sin 5x$

   p $-2\sin\left(4x + \frac{\pi}{2}\right)$

 2 Integrate with respect to x.

   a $3\sin x + 5x^2$ **b** $2\cos x + \frac{3}{x^2}$ **c** $4\sqrt{x} - \sin 2x$

   d $\sin\left(x - \frac{\pi}{6}\right) + (x - 3)^5$ **e** $(4x + 1)^5 - \cos 3x$ **f** $\frac{x - 1}{x^3} + 4\sin(x - 1)$

   g $\frac{2}{\sqrt{x}} - 5\cos 3x$ **h** $\frac{5}{4x^3} - 4\sin\frac{1}{2}x$ **i** $\frac{1}{(x - 5)^4} + \cos(8x - \pi)$

   j $\sqrt{1 - 4x} + 2\sin\left(3x + \frac{\pi}{4}\right)$

3 **a** By using the fact that $\cos 2x = 2\cos^2 x - 1$, show that $\cos^2 x = \frac{1}{2} + \frac{1}{2}\cos 2x$.

  b Using your answer from part **a** show that $\int \cos^2 x\, dx = \frac{1}{2}x + \frac{1}{4}\sin 2x + c$.

 4 **a** By using the fact that $\cos 2x = 1 - 2\sin^2 x$, show that $\sin^2 x = \frac{1}{2} - \frac{1}{2}\cos 2x$.

  b Hence show that $\int \sin^2 x\, dx = \frac{1}{2}x - \frac{1}{4}\sin 2x + c$.

 5 Integrate with respect to x.

  a $4\sin^2 x$ **b** $\dfrac{1}{2}\cos^2 x$ **c** $\sin 2x + \sin^2 x$

  d $\cos^2 x - \dfrac{2}{3}x^2$ **e** $\sin^2 x - \cos^2 x$ **f** $\dfrac{2}{5}\cos^2 x$

 6 Find two different ways to find
$\int \sin^2 x + \cos^2 x\, dx$.

> One method should use the
> answers to Questions 3 and 4.

⛰ Challenge

a Show that $(x + y)$ is a factor of $x^3 + y^3$ and that $(x - y)$ is a factor of $x^3 - y^3$

b Hence, or otherwise, show that:

 i $\sin^3 x + \cos^3 x = (\sin x + \cos x)\left(1 - \dfrac{1}{2}\sin 2x\right).$

 ii $\sin^3 x - \cos^3 x = (\sin x - \cos x)\left(1 + \dfrac{1}{2}\sin 2x\right).$

c Show that $\sin^6 x - \cos^6 x = -\cos 2x\left(1 - \dfrac{1}{4}\sin^2 2x\right)$ and hence deduce that

  $\sin^6 x - \cos^6 x = -\dfrac{7}{8}\cos 2x - \dfrac{1}{8}\cos 2x \cos 4x.$

d **i** Show that $\cos(A + B) + \cos(A - B) = 2\cos A \cos B.$

 ii Deduce that $\sin^6 x - \cos^6 x = -\dfrac{1}{16}\cos 6x - \dfrac{15}{16}\cos 2x.$

e Find $\int \sin^6 x - \cos^6 x\, dx.$

Solve differential equations of the form $\frac{dy}{dx} = f(x)$ to determine and use a function from a given rate of change and initial conditions

Differential equations are used to model and describe many real-life processes, and their solutions allow scientists and others to make predictions and test theories.

For example, the differential equation $v\frac{dv}{ds} = -\frac{gr^2}{s^2}$ describes the equation of motion of a rocket at a distance s metres from the centre of the Earth.

A differential equation is any equation which contains at least one derivative. The equation of motion of the rocket contained the derivative of v with respect to s.

The solution to a differential equation will be a function, not a number.

Exercise 11F

The use of a graphics calculator or computer software is strongly encouraged throughout this exercise.

1 Consider the differential equation $\frac{dy}{dx} = 2x$. By inspection, a solution to this differential equation is $y = x^2$. Explain why the phrase 'a solution' is used rather than 'the solution'.

The diagram might help.

2 The general solution to the differential equation $\frac{dy}{dx} = 2x$ is $y = x^2 + c$.

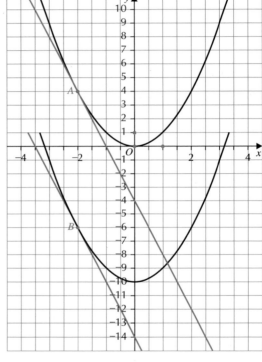

   a Starting with the graph of $y = x^2$, describe how to obtain the graph of $y = x^2 + c$.

   Suppose that P is on the graph of $y = x^2$ and Q is on the graph of $y = x^2 + c$, where $x_P = x_q = a$.

   b Explain why the tangents at P and Q have equal gradients.

3 Choosing $c = 5$ we have the solution $y = x^2 + 5$.

   a Why do you think $y = x^2 + 5$ is called a particular solution of the differential equation $\frac{dy}{dx} = 2x$?

   b Explain why $y = x^2 + x + 5$ is **not** a particular solution of the equation $\frac{dy}{dx} = 2x$.

4 Craig says: "I'm thinking of a function, $f(x)$, and I know that $f'(x) = 2x$."

   Andy says: "That's easy. $f(x) = x^2 + c$, where c is a constant".

   Craig says: "Not so fast Andy boy. I'm thinking of a particular solution."

   Andy says: "Well how I am supposed to know which one?"

   Craig says: "It's really quite easy. For my particular $f(x)$, the point $(3, 4)$ is on the graph with equation $y = f(x)$."

   Andy says: "How is that supposed to help me?"

   Help Andy out and explain to him why Craig has given enough information to be sure of a particular solution.

5 In general, what information will you need to be able to find a particular solution to a differential equation?

Example 11.16

For a given function it is known that $\frac{dy}{dx} = x^2 - 6$ and when $x = -3$, $y = 7$. Express y in terms of x.

$y = \int x^2 - 6 \, dx$ — Given $\frac{dy}{dx}$, integrate to find y **S**

$y = \frac{1}{3}x^3 - 6x + c$ — Integrate. Remember the constant. **P**

$7 = \frac{1}{3}(-3)^3 - 6(-3) + c$ — Substitute for x and y. **S**

$7 = -9 + 18 + c$

$c = -2$ — Solve to find c. The value of c is needed to obtain a particular solution. **P**

$y = \frac{1}{3}x^3 - 6x - 2$ — Write the particular solution by expressing y in terms of x. **C**

Example 11.17

A curve has equation $y = f(x)$. It is known that $f'(x) = \frac{2}{\sqrt{x}}$ and $f\left(\frac{1}{4}\right) = 3$. Find the equation of the curve.

$y = \int \frac{2}{\sqrt{x}} \, dx$

$\frac{2}{\sqrt{x}} = 2x^{-\frac{1}{2}}$ — Write $\frac{2}{\sqrt{x}}$ in integrable form. **P**

$y = 2 \times \frac{x^{\frac{1}{2}}}{\frac{1}{2}} + c = 4x^{\frac{1}{2}} + c = 4\sqrt{x} + c$ — Integrate, expressing the answer in root form. **P**

$3 = 4\sqrt{\frac{1}{4}} + c$ — Substitute for x and y. When $x = \frac{1}{4}$, $y = 3$. **S**

$3 = 4 \times \frac{1}{2} + c$

$c = 1$ — Solve to find c. **P**

$y = 4\sqrt{x} + 1$

Example 11.18

A curve has equation $y = g(x)$. For this curve it is known that $\frac{dy}{dx} = 6\cos 3x$. The point $\left(\frac{\pi}{2}, 3\right)$ lies on the curve. Express $g(x)$ in terms of x.

$y = \int 6\cos 3x \, dx$

$y = 6 \times \frac{1}{3}\sin 3x + c = 2\sin 3x + c$

$3 = 2\sin\frac{3\pi}{2} + c$

$3 = 2(-1) + c$

$c = 5$

$g(x) = 2\sin 3x + 5$ — Complete the answer by writing an expression for $g(x)$. **C**

Exercise 11G

★1 For a given function it is known that $\frac{dy}{dx} = 3x^2 - 2x + 4$ and when $x = -2$, $y = 5$. Express y in terms of x.

★2 A curve has equation $y = f(x)$. It is known that $f'(x) = 6\sqrt{x}$ and $f(4) = 8$. Find the equation of the curve.

★3 For the function $g(x)$ it is known that $g'(x) = 4\sin 2x$. The point $\left(\frac{\pi}{3}, 4\right)$ lies on the curve with equation $y = g(x)$. Express $g(x)$ in terms of x.

4 A curve has equation $y = f(x)$.

$(1, -20)$ lies on the curve and it is known that $\frac{dy}{dx} = 3x^2 + 10x - 2$.

   a   Find the equation of the curve.

   b   i   Show that $(x + 3)$ is a factor of $f(x)$

       ii   Fully factorise $f(x)$.

   c   Find the coordinates of the points where the curve intersects the x-axis.

5 The function $f(x)$ is such that $f(0) = 7$ and $f'(x) = 4(x - 1)$.

   a   Find an expression for $f(x)$.

   b   Show that the graph of $y = f(x)$ does not intersect the x-axis.

6 For the function $h(x)$ it is known that $h'(x) = 3x^2 + 2x - 5$ and $h(2) = 5$.

   a   Express $h(x)$ in terms of x.

   b   i   Show that -3 is a root of the equation $h(x) = 0$.

       ii   Hence find the remaining roots of $h(x) = 0$.

   c   i   Find the coordinates of the stationary points of $h(x)$.

       ii   Determine the nature of the stationary points.

   d   Sketch the graph of $y = h(x)$.

7 The function $f(x)$ is such that $f'(x) = 3x^2 + 4x - 4$.
$(-1, -3)$ lies on the graph with equation $y = f(x)$.

   a   Find an expression for $f(x)$.

   b   i   Show that $(x - 2)$ is a factor of $f(x)$.

       ii   Fully factorise $f(x)$.

   c   Show that the x-axis is a tangent to the graph of $y = f(x)$ and state the coordinates of the point of contact.

8 A straight line has equation $y = f'(x)$. It passes through A $(0, -4)$ and makes an angle of $a°$ with the positive direction of the x-axis.

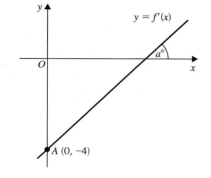

  a Given that $\sin a° = \frac{2}{\sqrt{13}}$ find the exact value of $\tan a°$.

  b Hence express $f'(x)$ in terms of x.

    $(6, 3)$ lies on the graph of $y = f(x)$.

  c Obtain an expression for $f(x)$.

9 A straight line has equation $2y - x + 8 = 0$.

It intersects the x-axis at A and the y-axis at B.

The perpendicular bisector of AB has equation $y = f(x)$.

  a Express $f(x)$ in terms of x.

    The function $h(x)$ is defined by $h'(x) = f(x)$.

  b Given that the x-axis is a tangent to the graph of $y = h(x)$, express $h(x)$ in terms of x.

10 The diagram shows part of the graph of a quadratic function $f(x)$.

The graph passes through $(0, 0)$ and $(4, 0)$. The line $y = -12$ is a tangent to the curve at A.

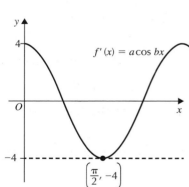

  a Given that $f(x) = ax(x - b)$, determine the values of a and b.

    The function $h(x)$ is defined by $h'(x) = f(x)$.

  b Given that $h(1) = 11$, express $h(x)$ in terms of x.

  c Fully factorise $h(x)$.

  d Sketch the graph of $y = h(x)$, showing clearly the coordinates of any turning points and the x- and y-axis intercepts.

11 The diagram shows the graph of $y = f'(x)$, where $f'(x) = a\cos bx$.

$\left(\frac{\pi}{2}, -4\right)$ is a minimum turning point.

  a Determine the values of a and b.

    $\left(\frac{\pi}{4}, 3\right)$ lies on the graph with equation $y = f(x)$

  b Express $f(x)$ in terms of x.

  c Find the coordinates of the points where the graph of $y = f(x)$ intersects the x-axis in the range $0 \le x \le \pi$.

12 Functions f and g are defined on suitable domains by $f(x) = 2x - 1$ and $g(x) = 3x^5$.

    a   Write down an expression for $g(f(x))$.

         The function h is such that $h'(x) = g(f(x))$ and $h\left(\frac{3}{2}\right) = -20$.

    b   Express $h(x)$ in terms of x.

         The function k is defined by $k(x) = \frac{1}{3}f(x - 2) + \frac{20}{3}$.

    c **i**   Write down an expression for $k(x)$.

        ii   Determine the x-coordinates of the points where the graph of $y = k(x)$ meets the x-axis.

13 The function f is such that $f'(x) = 2px + 12$. The graph of $y = f(x)$ passes through $(0, p - 5)$.

    a   Determine algebraically the least positive integer value of p for which the graph of $y = f(x)$ does not touch or intersect the x-axis.

    b   Show that when $p \neq 0$, the line $y = \frac{p^2 - 5p - 36}{p}$ is a tangent to the graph of $y = f(x)$.

14 The following information is known about the function $f(x)$:

    • $f'(x) = ax + b$, where a and b are constants

    • $f'(4) = 2$

    •   When $x = 2$, the tangent to the graph of $y = f(x)$ makes an angle of $135°$ with the positive direction of the x-axis.

    a   Express $f'(x)$ in terms of x.

         $f(x)$ is of the form $f(x) = px^2 + qx + r$.

    b   Determine the range of values of r for which the graph of $y = f(x)$ lies entirely above the x-axis.

•   I can integrate an expression reducible to powers of x. ★ Exercise 11A Q5, Exercise 11B Q3–5

•   I can integrate functions of the form $(x + q)^n$ and $(px + q)^n$. ★ Exercise 11C Q1–3

•   I can integrate trigonometric functions. ★ Exercise 11E Q1

•   I can solve differential equations of the form $\frac{dy}{dx} = f(x)$. ★ Exercise 11G Q1–3

For further assessment opportunities, see the Preparation for Assessment for Unit 2 on pages 291–292.

12 Using integration to calculate definite integrals

This chapter will show you how to:

- evaluate definite integrals for functions which can be expressed in powers of x
- evaluate definite integrals for functions of the form $f(x) = (x + q)^n$ and $f(x) = (px + q)^n$
- evaluate definite integrals for trigonometric functions of the form $p\sin(qx + r)$ and $f(x) = p\cos(qx + r)$.

You should already know:

- how to integrate an algebraic function which is, or can be, simplified to an expression in powers of x
- how to integrate functions of the form $f(x) = (x + q)^n$ and $f(x) = (px + q)^n$
- how to integrate functions of the form $f(x) = p\sin x$ and $f(x) = p\cos x$; $f(x) = p\sin(qx + r)$ and $f(x) = p\cos(qx + r)$
- how to solve differential equations of the form $\frac{dy}{dx} = f(x)$.

Evaluating definite integrals

Chapter 11 introduced **indefinite integration**. Indefinite integration produces a function called the **anti-derivative**.

A second type of integration is **definite integration**. Definite integration is defined by:

$$\int_{a}^{b} f(x)\,dx = F(b) - F(a)$$

- In the expression above, a and b are the **limits** of the integral. They are the numbers representing the endpoints of a closed interval $a \leq x \leq b$
- a is called the **lower limit** and b the **upper limit** (since $a < b$)
- $F(x)$ is an anti-derivative of $f(x)$, so $F(x) = \int f(x)\,dx$
- the calculation $F(b) - F(a)$ produces a **number**.

Evaluating definite integrals for non-trigonometric functions

Example 12.1

Evaluate $\int_1^3 3x^2 - 2x \, dx$

$a = 1$ (lower limit), $b = 3$ (upper limit) and $f(x) = 3x^2 - 2x$

$$\int_1^3 3x^2 - 2x \, dx = \left[3 \times \frac{x^3}{3} - 2 \times \frac{x^2}{2} \right]_1^3 = \left[x^3 - x^2 \right]_1^3$$

> Integrate to get $F(x) = x^3 - x^2$. Ⓟ The limits appear in the 'lower' and 'upper' positions at the end of the square brackets.

$$= \left(3^3 - 3^2 \right) - \left(1^3 - 1^2 \right)$$

> Substitute the upper and lower limits to begin the process of evaluating $F(b) - F(a)$ Ⓟ

$$= (27 - 9) - (1 - 1) = 18 - 0 = 18$$

> Evaluate $F(b) - F(a)$. The answer is a number, not a function. Ⓟ

Example 12.2

Evaluate $\int_{-3}^0 (2x - 1)(3x - 1) \, dx$

$a = -3$ (lower limit), $b = 0$ (upper limit) and $f(x) = (2x - 1)(3x - 1)$

$f(x) = 6x^2 - 5x + 1$

> Express $f(x)$ in integrable form. Ⓟ

$$\int_{-3}^0 6x^2 - 5x + 1 \, dx = \left[6 \times \frac{x^3}{3} - 5 \times \frac{x^2}{2} + x \right]_{-3}^0 = \left[2x^3 - \frac{5}{2}x^2 + x \right]_{-3}^0$$

> Integrate to obtain $F(x)$. Ⓟ

$$= \left(2(0)^3 - \frac{5}{2}(0)^2 + (0) \right) - \left(2(-3)^3 - \frac{5}{2}(-3)^2 + (-3) \right)$$

$$= (0) - \left(2(-27) - \frac{5}{2}(9) - 3 \right)$$

> Substitute the upper and lower limits to evaluate $F(b) - F(a)$. Ⓟ

$$= 0 - \left(-54 - \frac{45}{2} - 3 \right)$$

$$= 0 - \left(-\frac{108}{2} - \frac{45}{2} - \frac{6}{2} \right)$$

$$= \frac{159}{2} = 79\frac{1}{2}$$

> Evaluate $F(b) - F(a)$. Don't try to evaluate everything at once. Follow conventions for order of operations. Ⓟ

Example 12.3

Evaluate $\displaystyle\int_{-1}^{1}(3x-1)^4\,dx$

$$\int_{-1}^{1}(3x-1)^4\,dx = \left[\frac{(3x-1)^5}{15}\right]_{-1}^{1} = \frac{1}{15}\left[(3x-1)^5\right]_{-1}^{1}$$

> Integrate and take the $\frac{1}{15}$ outside the square brackets. **P**

$$= \frac{1}{15}\left[\left(3(1)-1\right)^5 - \left(3(-1)-1\right)^5\right]$$

$$= \frac{1}{15}\left[(2)^5 - (-4)^5\right]$$

$$= \frac{1}{15}\left[32 - (-1024)\right]$$

$$= \frac{1056}{15} = 70\frac{2}{5}$$

Example 12.4

Evaluate $\displaystyle\int_{-2}^{-\frac{3}{4}}\sqrt{1-4x}\,dx$

$$\sqrt{1-4x} = (1-4x)^{\frac{1}{2}}$$

$$\int_{-2}^{-\frac{3}{4}}(1-4x)^{\frac{1}{2}}\,dx = \left[\frac{(1-4x)^{\frac{3}{2}}}{-4\times\frac{3}{2}}\right]_{-2}^{-\frac{3}{4}} = -\frac{1}{6}\left[(1-4x)^{\frac{3}{2}}\right]_{-2}^{-\frac{3}{4}} = -\frac{1}{6}\left[\left(\sqrt{1-4x}\right)^3\right]_{-2}^{-\frac{3}{4}}$$

$$= -\frac{1}{6}\left[\left(\sqrt{1-4\left(-\frac{3}{4}\right)}\right)^3 - \left(\sqrt{1-4(-2)}\right)^3\right]$$

> Integrate, expressing $F(x)$ in root form. It is easier to evaluate $\left(\sqrt{1-4x}\right)^3$ than the equivalent $\sqrt{(1-4x)^3}$. **P**

$$= -\frac{1}{6}\left[\left(\sqrt{1-(-3)}\right)^3 - \left(\sqrt{1-(-8)}\right)^3\right]$$

$$= -\frac{1}{6}\left[\left(\sqrt{4}\right)^3 - \left(\sqrt{9}\right)^3\right]$$

$$= -\frac{1}{6}\left[(2)^3 - (3)^3\right]$$

$$= -\frac{1}{6}[8-27] = -\frac{1}{6}[-19] = \frac{19}{6} = 3\frac{1}{6}$$

Exercise 12A

⊞★ **1** Evaluate these integrals.

a $\displaystyle\int_1^3 2x + 5 \ dx$

b $\displaystyle\int_0^4 6 - 2x \ dx$

c $\displaystyle\int_{-2}^1 3x^2 + 4x - 1 \ dx$

d $\displaystyle\int_{-1}^2 2x^4 \ dx$

e $\displaystyle\int_{-2}^2 4x^3 + x^2 - 3x + 2 \ dx$

f $\displaystyle\int_0^6 5 - 2x^2 \ dx$

⊞★ **2** Evaluate these integrals.

a $\displaystyle\int_1^4 (x+2)(3x+2) \ dx$

b $\displaystyle\int_{-2}^1 x(2x-1)^2 \ dx$

c $\displaystyle\int_{-1}^2 (x+1)(x^2-4) \ dx$

d $\displaystyle\int_2^4 (x+1)(x^2+2x-3) \ dx$

e $\displaystyle\int_{-3}^3 2x^2(1-x) - 3(x^2-2) \ dx$

⊞★ **3** Evaluate these integrals.

a $\displaystyle\int_1^3 \frac{4}{x^2} \ dx$

b $\displaystyle\int_3^4 \frac{1}{2x^3} \ dx$

c $\displaystyle\int_{-2}^{-1} \frac{4}{3x^5} - 1 \ dx$

d $\displaystyle\int_4^9 3\sqrt{x} \ dx$

e $\displaystyle\int_1^4 \frac{2}{\sqrt{x}} \ dx$

f $\displaystyle\int_1^4 6(\sqrt[3]{x}) \ dx$

g $\displaystyle\int_1^{27} \frac{2}{3(\sqrt[3]{x^2})} \ dx$

h $\displaystyle\int_1^9 10(\sqrt{x^3}) \ dx$

i $\displaystyle\int_4^9 \frac{3}{(\sqrt{x^5})} \ dx$

⊞★ **4** Evaluate these integrals.

a $\displaystyle\int_1^3 (x-2)^4 \ dx$

b $\displaystyle\int_{-2}^0 (x+3)^5 \ dx$

c $\displaystyle\int_7^8 8(x-5)^3 \ dx$

d $\displaystyle\int_{-3}^{-1} (2x+3)^5 \ dx$

e $\displaystyle\int_0^1 (5x-4)^3 \ dx$

f $\displaystyle\int_1^2 \frac{1}{8}(3x-5)^8 \ dx$

⊞ **5** Evaluate these integrals.

a $\displaystyle\int_1^3 \frac{x^3-2}{x^2} \ dx$

b $\displaystyle\int_4^9 \sqrt{x}\left(2 - \frac{1}{x}\right) \ dx$

c $\displaystyle\int_{-2}^{-1} \frac{1}{(1-x)^3} \ dx$

d $\displaystyle\int_1^6 \sqrt{x+3} \ dx$

e $\displaystyle\int_4^{16} \frac{2}{x^3} - \frac{3}{4}\sqrt{x} \ dx$

f $\displaystyle\int_6^8 \left(5 - \frac{1}{2}x\right)^3 \ dx$

g $\displaystyle\int_{-3}^{-1} 6 - \frac{1}{2}x - \frac{1}{3x^2} \ dx$

h $\displaystyle\int_{\frac{1}{2}}^{\frac{5}{3}} 8(x-1)^3 \ dx$

i $\displaystyle\int_{\sqrt{2}}^{\sqrt{3}} x(x^2-1) \ dx$

⊞ **6** Show that $\displaystyle\int_8^{18} \frac{1}{4}\sqrt{x} \ dx = \frac{19\sqrt{2}}{3}$

7 Determine algebraically the values of k such that $\int\limits_{-2}^{k} 4x - 3\ dx = -15$.

 8 a Given that $\int\limits_{-2}^{p} 3x^2 - 4x\ dx = 48$, show that $p^3 - 2p^2 - 32 = 0$.

    b **i** Show that $(p - 4)$ is a factor of $p^3 - 2p^2 - 32$.

        ii Hence show that 4 is the only real value of p for which $\int\limits_{-2}^{p} 3x^2 - 4x\ dx = 48$.

 9 Determine algebraically the negative value of p for which $\int\limits_{0}^{3} \dfrac{7}{3(2x + p)^2}\ dx = 1$.

 10 Determine algebraically the values of t, where $t \neq 0$, for which

$$\int\limits_{\frac{1}{4}}^{1} \frac{1}{(tx + 1)^2} - \frac{1}{\sqrt{x}}\ dx = -\frac{5}{6}.$$

Evaluating definite integrals for trigonometric functions

Recall from Chapter 11 these two important integration facts:

$$\int \cos(qx + r) = \frac{1}{q}\sin(qx + r)$$

$$\int \sin(qx + r) = -\frac{1}{q}\cos(qx + r)$$

These results are only true when x is measured in radians. (See Chapters 8 and 9.)

In this section, where a calculator is permitted, you must ensure that it is in radian mode. Most basic scientific calculators now let you enter expressions exactly as you've written them down. Be careful not to miss out any brackets, numbers or other symbols if you evaluate in this way. Always check that what you have entered into the calculator matches what you have written down.

Example 12.5

Evaluate $\int\limits_{0}^{\frac{\pi}{2}} 4\cos x\ dx$

$$\int\limits_{0}^{\frac{\pi}{2}} 4\cos x\ dx = \left[4\sin x\right]_{0}^{\frac{\pi}{2}}$$

$$= \left(4\sin\frac{\pi}{2}\right) - (4\sin 0)$$

$$= 4 - 0 = 4$$

Example 12.6

Evaluate $\int_{\frac{\pi}{4}}^{\frac{\pi}{2}} 2\sin 2x \; dx$

$$\int_{\frac{\pi}{4}}^{\frac{\pi}{2}} 2\sin 2x \; dx = \left[2 \times -\frac{1}{2}\cos 2x \right]_{\frac{\pi}{4}}^{\frac{\pi}{2}} = \left[-\cos 2x \right]_{\frac{\pi}{4}}^{\frac{\pi}{2}}$$

$$= \left(-\cos\left(2 \times \frac{\pi}{2} \right) \right) - \left(-\cos\left(2 \times \frac{\pi}{4} \right) \right)$$

$$= \left(-\cos(\pi) \right) - \left(-\cos\left(\frac{\pi}{2} \right) \right)$$

$$= (-(-1)) - (0) = 1$$

Example 12.7

Evaluate $\int_{2}^{4} \frac{1}{2}\cos(3x - 1) \; dx$, giving your answer correct to 2 significant figures.

$$\int_{2}^{4} \frac{1}{2}\cos(3x - 1) \; dx = \left[\frac{1}{2} \times \frac{1}{3}\sin(3x - 1) \right]_{2}^{4} = \left[\frac{1}{6}\sin(3x - 1) \right]_{2}^{4} = \frac{1}{6}\left[\sin(3x - 1) \right]_{2}^{4}$$

$$= \frac{1}{6}\left[\sin(3(4) - 1) - \sin(3(2) - 1) \right]$$

Integrate in the usual way, extracting the $\frac{1}{6}$ to make working easier. **P**

$$= \frac{1}{6}\left[\sin(11) - \sin(5) \right]$$

Substitute the limits as normal. **C**

$$= -0.0068$$

Set your calculator to radian mode and evaluate, rounding to 2 s.f. **P**

Example 12.8

Evaluate $\int_{-2}^{\frac{\pi}{12}} 5 + \sin x \; dx$, giving your answer correct to 2 decimal places.

$$\int_{-2}^{\frac{\pi}{12}} 5 + \sin x \; dx = \left[5x - \cos x \right]_{-2}^{\frac{\pi}{12}}$$

$$= \left(5 \times \frac{\pi}{12} - \cos\frac{\pi}{12} \right) - \left(5 \times (-2) - \cos(-2) \right)$$

Substitute the limits as normal. **P**

$$= 9.93$$

Set your calculator to radian mode and evaluate, rounding to 2 decimal places. **P**

Exercise 12B

Do not use a calculator in this Exercise except where instructed.

★ **1** Evaluate these integrals.

a $\displaystyle\int_{0}^{\frac{\pi}{6}} 2\cos x \, dx$

b $\displaystyle\int_{0}^{\frac{\pi}{3}} 5\sin x \, dx$

c $\displaystyle\int_{\frac{\pi}{6}}^{\frac{\pi}{2}} 8\cos x \, dx$

d $\displaystyle\int_{0}^{\frac{\pi}{6}} 3\cos 2x \, dx$

e $\displaystyle\int_{\frac{\pi}{6}}^{\frac{\pi}{3}} \sin 3x \, dx$

f $\displaystyle\int_{\frac{\pi}{3}}^{\frac{2\pi}{3}} \cos \frac{1}{2}x \, dx$

★ **2** Evaluate these integrals.

a $\displaystyle\int_{\frac{\pi}{6}}^{\frac{\pi}{3}} 3\cos\left(x - \frac{\pi}{6}\right) dx$

b $\displaystyle\int_{0}^{\frac{\pi}{4}} 6\sin\left(x + \frac{\pi}{4}\right) dx$

c $\displaystyle\int_{\frac{\pi}{6}}^{\frac{\pi}{2}} 8\cos x \, dx$

d $\displaystyle\int_{0}^{\frac{\pi}{6}} 3\cos 2x \, dx$

e $\displaystyle\int_{\frac{\pi}{6}}^{\frac{\pi}{3}} \sin 3x \, dx$

f $\displaystyle\int_{\frac{\pi}{3}}^{\frac{2\pi}{3}} \cos \frac{1}{2}x \, dx$

★ **3** Evaluate these integrals.

a $\displaystyle\int_{\frac{\pi}{4}}^{\pi} 6\cos\left(x + \frac{\pi}{2}\right) dx$

b $\displaystyle\int_{3\pi}^{5\pi} \sin 4x \, dx$

c $\displaystyle\int_{0}^{\frac{5\pi}{4}} 2\cos\left(2x - \frac{\pi}{4}\right) dx$

d $\displaystyle\int_{\frac{\pi}{12}}^{\pi} 4\sin(3x - \pi) \, dx$

e $\displaystyle\int_{-\frac{5\pi}{3}}^{-\frac{\pi}{3}} \cos\frac{1}{2}x \, dx$

f $\displaystyle\int_{-\pi}^{\pi} \sin\frac{2}{3}x \, dx$

★ **4** Evaluate these integrals, giving your answers to 2 decimal places.

a $\displaystyle\int_{1}^{5} 2\sin x \, dx$

b $\displaystyle\int_{2.3}^{3.6} -4\cos x \, dx$

c $\displaystyle\int_{2}^{4} \sin(4x - 1) \, dx$

d $\displaystyle\int_{-1.2}^{3.4} \frac{2}{3}\cos(2 - x) \, dx$

e $\displaystyle\int_{3}^{5} \cos\left(2x - \frac{\pi}{2}\right) + 3 \, dx$

f $\displaystyle\int_{-1.3}^{0.5} 4 - \sin(5 - 2x) \, dx$

5 Find the values of t, $0 \le t < 2\pi$, for which $\displaystyle\int_{\frac{\pi}{6}}^{t} 3\cos x \, dx = \frac{3\sqrt{2}}{2} - \frac{3}{2}$.

6 Find the values of p, $0 \le p < \pi$, for which $\displaystyle\int_{0}^{p} \sin 2x \, dx = \frac{1}{4}$.

7 Find the value of k, where $\pi < k < \frac{3\pi}{2}$, such that $\displaystyle\int_{\pi}^{k} 2\cos 2x - \sin x \, dx = 1$.

8 a Show that $\sin^2 x = \frac{1}{2}(1 - \cos 2x)$.

b Hence show that $\int_{\frac{\pi}{4}}^{\frac{\pi}{3}} \sin^2 x \, dx = \frac{1}{24}\left(\pi + 6 - 3\sqrt{3}\right)$.

9 a By expressing $\frac{\pi}{12}$ as $\left(\frac{\pi}{4} - \frac{\pi}{6}\right)$ show that $\cos\frac{\pi}{12} = \frac{\sqrt{3}+1}{2\sqrt{2}}$.

b Similarly, show that $\sin\frac{\pi}{12} = \frac{\sqrt{3}-1}{2\sqrt{2}}$.

c Deduce that $\int_{-\frac{\pi}{12}}^{\frac{\pi}{12}} \cos x - \sin x \, dx = \frac{\sqrt{6}-\sqrt{2}}{2}$.

- I can evaluate definite integrals involving non-trigonometric functions.
  ★ Exercise 12A Q1-4
- I can evaluate definite integrals involving trigonometric functions.
  ★ Exercise 12B Q1-4

For further assessment opportunities, see the Preparation for Assessment for Unit 2 on pages 291–292.

Unit 2 Relationships and Calculus

The questions in this section cover the minimum competence for the content of the course in Unit 2. They are a good preparation for your unit assessment. In an assessment you will get full credit only if your solution includes the appropriate steps of process and accuracy, so make sure you show your thinking when writing your answers.

Remember that reasoning questions marked with the symbol ⚙ expect you to interpret the situation, select an appropriate strategy to solve the problem and then clearly explain your solution. If a context is given you must relate your answer to this context.

The use of a calculator is allowed.

Applying algebraic skills to solve equations (Chapter 7)

1 Factorise these cubic expressions fully.

   a $x^3 + x^2 - 22x - 40$

   b $x^3 - 6x^2 + 11x - 6$

2 Solve the equation $x^3 + 5x^2 - 3x - 18 = 0$

⚙ 3 Find the value of the constant k for which the equation $x^2 - kx + (2k - 3)$ has two equal roots.

Applying trigonometric skills to solve equations (Chapter 8)

4 Solve $\sqrt{3} \cos x = 2,\ 0 \le x \le 180$

5 Solve $\sqrt{2 \sin 2x - \cos x} = 1,\ 0 \le x \le 180$

6 Solve $3\sin(x + 30) = 2,\ 0 \le x \le 90$

7 Solve $\cos(2x) + 4\cos x + 5 = 2,\ 0 \le x \le 180$

Applying calculus skills of differentiation (Chapters 9 and 10)

8 A curve has equation $y = 3\sqrt{x}$. Find the equation of the tangent to the curve at the point where $x = 4$.

9 Find the gradient of the tangent to the curve $y = 3\sin\Phi$ at the point where $\Phi = \pi$.

10 Differentiate the expression $y = \dfrac{1 - x^3}{x}$

11 A function f is defined at $f(x) = 2x^3 - 3x^2 + x + 4$. Find $f'(x)$.

12 Differentiate

    a $-\frac{1}{2}\sin x$ **b** $8\cos x$

13 Find $\frac{dy}{dx}$ for the function $y = \sqrt{x(x^3 - x)}$.

14 The altitude of a two-stage rocket gets to after take-off is given by the formula $s = 10t^2 - t^3$. The velocity, in m/s of the rocket at time, t, is given by $\frac{ds}{dt}$.

    a   Find the velocity of the rocket after 5 seconds.

    b   At what time should the second-stage rockets fire in order to avoid the rocket coming to a standstill?

Applying calculus skills of integration (Chapters 11 and 12)

15 Integrate with respect to u: $u^3 + 2u^2 - 6u + 4$

16 If $\frac{dy}{dx} = 6x^{\frac{1}{2}}$, find y.

17 Find $\int \dfrac{2x^3 + 5}{\sqrt{x}}\,dx$

18 Evaluate $\displaystyle\int_{0}^{3}\left(t^4 + 2t^2 - 3t + 2\right)dt$

19 Find $\int (2x + 4)^3\,dx$

20 Find $\int (6\cos\alpha)\,d\alpha$

21 Evaluate $\displaystyle\int_{0}^{\frac{\Pi}{2}}(3\sin x)\,dx$

13 Applying algebraic skills to rectilinear shapes

This chapter will show you how to:

- find the equation of a line parallel to a given line
- find the equation of a line perpendicular to a given line
- use $m = \tan\theta$ to calculate a gradient or angle
- use gradients to show that points are collinear
- find equations of medians, altitudes and perpendicular bisectors
- solve problems involving medians, altitudes and perpendicular bisectors.

You should already know:

- how to use the gradient formula $m = \dfrac{y_2 - y_1}{x_2 - x_1}$
- how to use the formula $y - b = m(x - a)$ or equivalent to find the equation of a straight line
- how to identify the gradient and y-intercept from various forms of the equation of a straight line
- how to find the midpoint of a line segment
- how to calculate the distance between two points
- how to solve two linear equations simultaneously and that the solutions represent the point of intersection of the two lines.

Parallel lines

Parallel lines never meet, no matter how far they are extended. Parallel lines have **equal** gradients. The mathematical way to indicate that two lines are parallel is to write $PQ \parallel RS$, where the symbol $\parallel$ means 'is parallel to'. This means the line joining points P and Q is parallel to the line joining points R and S. In this case the gradients will be equal, this is written $m_{PQ} = m_{RS}$. To show two lines are not parallel we write $PQ \nparallel RS$.

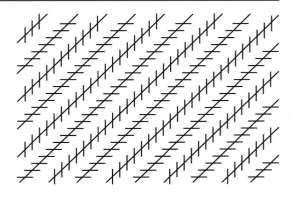

Are the diagonal lines in this image parallel? Use your ruler to check.

Example 13.1

Consider the points $S(2, 5)$, $T(5, 9)$, $U(-3, -4)$ and $V(3, 4)$. Show that ST and UV are parallel.

• $T(5, 9)$

$S(2, 5)$ •

• $V(3, 4)$

O •

• $U(-3, -4)$

⚠ You may find it helpful to make a quick sketch for questions like this. There is no need to draw axes and accurately plot the points; just chose a sensible point to be your origin (shown here in red), then estimate where the points will be.

$$m_{ST} = \frac{y_2 - y_1}{x_2 - x_1}$$

Comparing the gradients of ST and UV will allow us to determine if the lines are parallel Ⓢ

$$m_{ST} = \frac{9 - 5}{5 - 2}$$

$$m_{ST} = \frac{4}{3}$$

$$m_{UV} = \frac{y_2 - y_1}{x_2 - x_1}$$

$$m_{UV} = \frac{4 - (-4)}{3 - (-3)}$$

$$m_{UV} = \frac{8}{6} = \frac{4}{3}$$

$$m_{ST} = m_{UV} = \frac{4}{3} \therefore ST \parallel UV$$

This could also be written as "$m_{ST} = m_{UV} = \frac{4}{3}$ therefore ST is parallel to UV". Ⓒ

Example 13.2

Find the equation of the line GH which passes through the point $(4, -2)$ and is parallel to the line RS with equation $3x + 2y = 7$.

$$3x + 2y = 7$$

Rearrange the equation of RS into the form $y = mx + c$ to find the gradient of RS. Ⓢ

$$2y = -3x + 7$$

$$y = -\frac{3}{2}x + \frac{7}{2}$$

$$m_{GH} = m_{RS} = -\frac{3}{2}$$

Since GH is parallel to RS, the gradients are the same. Ⓢ

$$y - b = m(x - a)$$

$$y - (-2) = -\frac{3}{2}(x - 4)$$

$$y + 2 = -\frac{3}{2}x + 6$$

$$y = -\frac{3}{2}x + 4 \text{ or } 2y + 3x = 8$$

It is good practice to simplify linear equations to either of these forms. The equation is often required in a later part of the question. Ⓢ

Exercise 13A

Do not use a calculator for this Exercise.

1 Consider the points $K(-2, 5), L(4, 7), M(-7, -2), N(-1, 0)$.

   a Find m_{KL}

   b Find m_{MN}

   c Explain why lines KL and MN are parallel.

★ ⚙ 2 The line AB has equation $2x + 3y = 7$.

   a Find the gradient of line AB.

   b Write down the coordinates of the point at which AB crosses the y-axis.

   c Determine the equation of the line CD which is parallel to AB, given that CD passes through the point $(9, -3)$.

★ ⚙ 3 The line EF has equation $3x + 2y = 5$. Determine the equation of the line GH which is parallel to EF, given that GH passes through the point $(-2, 2)$.

4 Consider the points $T\left(-\frac{3}{2}, -5\right), U\left(\frac{5}{2}, 3\right), V\left(\frac{7}{2}, \frac{2}{5}\right), W\left(\frac{15}{2}, \frac{42}{5}\right)$.

   Show that TU and VW are parallel.

⚙ 5 The points $P(-5, -5)$, $Q(-2, -1)$, $R(2, 2)$, $S(-1, -2)$ are vertices of a quadrilateral. Show that $PQRS$ is a rhombus.

> You should first make a sketch here. Then you will need to consider the properties of a rhombus: opposite sides are parallel *and* all sides are equal in length.

⚙ 6 The line $ax - 2y + 7 = 0$ is parallel to the line $3x + 4y = 9$. Find the value of a.

⚙ 7 The points $A(-1, -3)$, $B(2, 5)$, $C(7, 6)$, $D(a, -2)$ are vertices of a quadrilateral. Given that AD is parallel to BC, find the value of a.

Collinearity

A set of points which lie on the same straight line are said to be **collinear**. We have already seen that if the gradients of two lines are equal, then the lines are parallel. If these lines also share a common point, then the set of points will be collinear.

In the first diagram AB and CD are parallel, so $m_{AB} = m_{CD}$. However, points A, B, C and D are not collinear since AB and CD don't share a common point.

In the second diagram AB and BC are parallel (their gradients are equal) and, since B is a common point, we can also say that the points A, B and C are collinear.

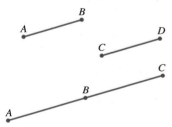

Example 13.3

Decide whether P lies on the line passing through AB where $A(-5, -4)$, $B(1, 4)$ and $P(4, 8)$.

(continued)

$$m_{AB} = \frac{y_2 - y_1}{x_2 - x_1} = \frac{4 - (-4)}{1 - (-5)} = \frac{8}{6} = \frac{4}{3}$$

$$m_{BP} = \frac{y_2 - y_1}{x_2 - x_1} = \frac{8 - 4}{4 - 1} = \frac{4}{3}$$

$m_{AB} = m_{BP}$ ∴ $AB \parallel BP$ and since B is a common point, A, B and P are collinear.

> If the gradients are not the same you should write "$m_{AB} \neq m_{BP}$ ∴ A, B and P are not collinear" Ⓒ

Exercise 13B

Do not use a calculator for this Exercise.

★ **1** Use gradients to decide whether each set of points below is collinear.

  a $A(-3, 0), B(0, 3), C(3, 7)$

  b $D(-3, 3), E(-2, 1), F(0, -3)$

  c $G(4, 0), H(-2, -3), J(-4, -4)$

  d $K(1, -5), L(-1, 1), M(-3, 5)$

2 Points $P(-5, -3), Q(1, 1)$ and $R(10, k)$ are collinear. Find the value of k.

3 A fly lands at the point (5, −5) on coordinate diagram and walks in a straight line to point (−1, 10). On its journey, did it walk over point (3, −1) or point (3, 0)?

4 Two teams of explorers are independently crossing the Arctic. Relative to a suitable set of coordinate axes, Team 1 passes through the points A (6, 13) and B (−3, −2). Team 2 passes through the points C (−9, −17) and D (0, −11). Both teams are heading for a research station at point S (−6, −7). Assuming both teams continue travelling in a straight line, which team(s), if any, will make it to the station?

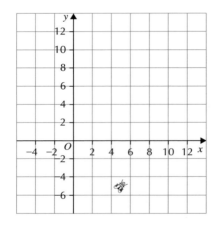

Perpendicular lines

Any two straight lines which meet at right angles are **perpendicular** to one another.

$m_\perp$ is the notation for a perpendicular gradient.

> **If two lines are perpendicular then their gradients will multiply to give −1, so $m_1 \times m_2 = -1$.**

> **The converse is also true, so if two gradients multiply to give −1 then the lines are perpendicular.**

If we divide −1 by the gradient of a line we can find the gradient of any line perpendicular to it.

For example:

a If $m = \frac{2}{5}$ then $m_\perp = -1 \div \frac{2}{5} = -\frac{5}{2}$ **b** If $m = \frac{4}{3}$ then $m_\perp = -1 \div \frac{4}{3} = -\frac{3}{4}$

c If $m = -3$ then $m_\perp = -1 \div -3 = \frac{1}{3}$ **d** If $m = \frac{1}{2}$ then $m_\perp = -1 \div \frac{1}{2} = -2$

Dividing -1 like this can be awkward. An easier way to determine a perpendicular gradient is to take the **negative reciprocal** of the first gradient. Swapping numerator and denominator creates a reciprocal. When finding reciprocals it can help to think of a whole number, n as $\frac{n}{1}$ giving a reciprocal of $\frac{1}{n}$.

Example 13.4

Consider the points $A(-1, 2)$, $B(-5, 7)$ and $C(4, 6)$. Show that AB is perpendicular to AC.

$$m_{AB} = \frac{y_2 - y_1}{x_2 - x_1} = \frac{7 - 2}{-5 - (-1)} = -\frac{5}{4}$$

$$m_{AC} = \frac{y_2 - y_1}{x_2 - x_1} = \frac{6 - 2}{4 - (-1)} = \frac{4}{5}$$

Since $m_{AB} \times m_{AC} = -1$ then AB is perpendicular to AC.

Using suitable notation, this can be written as $m_{AB} \times m_{AC} = -1 \therefore AB \perp AC$

A short way of writing AB is perpendicular to AC is $AB \perp AC$.

Example 13.5

Find the gradient of a line perpendicular to the line with gradient $\frac{3}{5}$.

$m \times m_\perp = -1$

$\frac{3}{5} \times m_\perp = -1$

$\therefore m_\perp = -\frac{5}{3}$

Since $\frac{3}{5} \times -\frac{5}{3} = -1$

Example 13.6

Find the gradient of a line perpendicular to the line with equation $2x + 3y = 12$.

$2x + 3y = 12$

$3y = -2x + 12$

$y = -\frac{2}{3}x + 4$

Rearranging the equation into the form $y = mx + c$ gives the gradient.

$m \times m_\perp = -1$

$-\frac{2}{3} \times m_\perp = -1$

$\therefore m_\perp = \frac{3}{2}$

Example 13.7

Find the equation of the line passing through the point $(3, 5)$ which is perpendicular to the line with equation $y = -4$

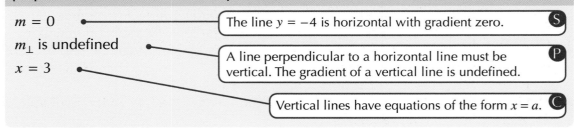

$m = 0$ — The line $y = -4$ is horizontal with gradient zero. Ⓢ

$m_\perp$ is undefined —

$x = 3$ — A line perpendicular to a horizontal line must be vertical. The gradient of a vertical line is undefined. Ⓟ

Vertical lines have equations of the form $x = a$. Ⓒ

Example 13.8

Triangle ABC has vertices $A(-3, -1)$, $B(-1, 2)$ and $C(5, -2)$. Use gradients to show that it is right-angled.

⚠ To show that a triangle is right-angled using gradients, we must prove that two sides are perpendicular (meet at 90°).

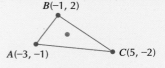

$B(-1, 2)$

$A(-3, -1)$ $C(5, -2)$

Make a quick sketch. The right angle appears to be at vertex B. Ⓢ

$m_{AB} = \dfrac{y_2 - y_1}{x_2 - x_1} = \dfrac{2 - (-1)}{-1 - (-3)} = \dfrac{3}{2}$

$m_{BC} = \dfrac{y_2 - y_1}{x_2 - x_1} = \dfrac{-2 - 2}{5 - (-1)} = -\dfrac{4}{6} = -\dfrac{2}{3}$

$m_{AB} \times m_{BC} = \dfrac{3}{2} \times -\dfrac{2}{3} = -1$

$\therefore AB \perp BC$ and triangle ABC is right angled at B

Exercise 13C

Do not use a calculator for this Exercise.

1   Find the gradient of a line that is perpendicular to lines with these gradients.

   a $m = \dfrac{2}{3}$ **b** $m = -\dfrac{4}{3}$ **c** $m = \dfrac{1}{2}$ **d** $m = 7$

   e $m = 1$ **f** $m = -3$ **g** $m = -\dfrac{1}{5}$ **h** $m = 0$

2   Find the gradient of a line perpendicular to the line joining the points $F(-4, 3)$ and $G(-2, 12)$.

★ 3   Find the equation of the line which is perpendicular to the line $3x - 2y = -4$ and which passes through the point $(12, -6)$.

⚙ 4   Consider the points $S(2, 1)$ and $T(-6, -5)$. Show that the equation of the line which is perpendicular to ST and which passes through the midpoint of ST is $3y + 4x = -14$.

5 Make a sketch showing the points $C(-2, 4)$, $D(5, 5)$ and $E(2, 1)$. Show that CDE is a right-angled triangle.

6 The points $P(-5, 3)$, $Q(0, 15)$, $R(12, 10)$ and $S(7, -2)$ are the vertices of a quadrilateral. Show that $PQRS$ is a square.

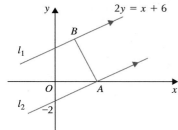

Making a sketch will help here. Consider the properties of a square: opposite sides are parallel, all sides are equal in length and all sides meet at 90°.

7 Show that the lines with equations $5y - 2x = 4$ and $2y + 5x = 4$ are perpendicular.

8 The line with equation $7y - ax = 13$ is perpendicular to the line with equation $5y - 7x = 24$. Find the value of a.

9 A kite has vertices $A(-7, 11)$, $B(2, 13)$, $C(8, 8)$ and $D(-2, -7)$. Show that the diagonals are perpendicular.

10 Part of the line l_1, with equation $2y = x + 6$ is shown in the diagram. The line l_2 is parallel to l_1 and passes through the point $(0, -2)$. Point A lies on the x-axis.

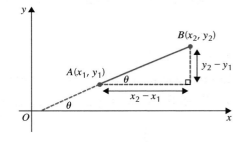

a Show that the equation of l_2 is $2y = x - 4$ and hence write down the coordinates of A.

b Given that the line AB is perpendicular to both lines, find, algebraically, the coordinates of B.

c Calculate the exact shortest distance between the lines l_1 and l_2.

11 Line l_1 passes through the points $A(2, 1)$ and $B(4, 7)$. Line l_2 passes through the points $C(3, 1)$ and $D(-6, y)$. Given that the lines are perpendicular, find the value of y.

Gradients and angles

The diagram shows points A and B. If the line AB is extended to the left, it meets the positive direction of the x-axis at an angle θ. This corresponds to an angle of θ inside the triangle (since corresponding angles are equal).

From the definition of the tangent ratio we can see that $\tan\theta = \dfrac{\text{opposite}}{\text{adjacent}} = \dfrac{y_2 - y_1}{x_2 - x_1}$

Since $m = \dfrac{y_2 - y_1}{x_2 - x_1}$ so the gradient of the line is equal to the tangent of the angle it makes with the positive direction of the x-axis. This is represented by:

$$m = \tan\theta$$

The converse is also true. The tangent of the angle the line makes with the positive direction of the x-axis equals the gradient. Given a gradient m, we can use $\tan\theta = m$ to find an angle.

Example 13.9

Calculate the gradient of the straight line *ST* which makes an angle of 60° with the positive direction of the *x*-axis.

$$m_{ST} = \tan\theta$$
$$m_{ST} = \tan 60°$$
$$m_{ST} = \sqrt{3}$$ •────────(Leave answers as exact values where possible. Ⓒ)

Example 13.10

Calculate the angle θ that the line with gradient $-\frac{5}{2}$ makes with the positive direction of the *x*-axis.

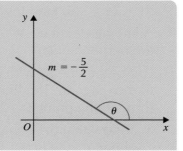

$$\tan\theta = m$$
$$\tan\theta = -\frac{5}{2}$$ •────(The gradient is negative, so we should expect an obtuse angle. Ⓢ)
$$\theta = 180° - 68.2° = 111.8°$$

Exercise 13D

★ **1** Find the gradient of the line which meets the positive direction of the *x*-axis at these angles.

    a 45°   **b** 30°   **c** 135°   **d** 120°

    e 90°   **f** 150°   **g** 180°   **h** 60°

★ **2** Calculate the angle θ which is made with the positive direction of the *x*-axis by the lines with these gradients.

    a $m = 5$ **b** $m = \dfrac{1}{3}$ **c** $m = -2$

    d $m = -\dfrac{2}{5}$ **e** $m = \dfrac{7}{4}$ **f** $m = -\dfrac{11}{5}$

3 Find the size of the angle that the line joining the points $A(-5, 1)$ and $B(4, 4)$ makes with the positive direction of the *x*-axis.

4 Find the size of the angle that the line joining the points $C(-4, 2)$ and $D(4, -2)$ makes with the positive direction of the *x*-axis.

5 The line *OA* connects the origin to point $A(3, 7)$.

Line *OB* connects the origin to point $B(5, 2)$

Calculate the size of angle $\angle AOB$.

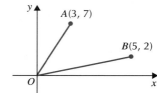

6 Calculate the size of the acute angle between the lines *JK* and *PQ*.

7 Line *AB* has equation $y - 7x + 11 = 0$. Line *AC* has equation $7y + x = 23$. Calculate the size of angle *BAC*.

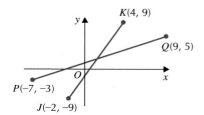

8 Lines l_1, l_2 and the *x*-axis form a triangle. Given that l_1 has equation $2y - 3x + 11 = 0$ and l_2 has equation $3y + 5x = 31$, calculate the sizes of all the angles in the triangle.

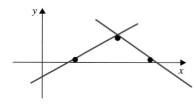

Special lines in a triangle

Thousands of special lines and points associated with triangles have been discovered, each satisfying some unique property. Often they are found by finding the equations of lines associated with each vertex and proving the lines meet at a single point. Believe it or not, there is an encyclopaedia of triangle centres, listing over 5000 points or 'centres' associated with the geometry of a triangle. Some of these points are more interesting than others!

> Do an internet search for 'encyclopaedia of triangle centres'.

Medians

A **median** joins a vertex to the **midpoint** of the opposite side. Every triangle has three medians. Each median divides the triangle into two smaller triangles of equal area. The three medians pass through the same point and are said to be **concurrent**. This point is called the **centroid** of the triangle. The centroid is the centre of gravity of the triangle and is always inside the triangle. The centroid is exactly two-thirds of the way along each median.

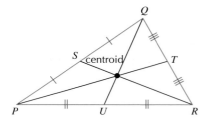

Example 13.11

Triangle *ABC* has coordinates $A(3, 2)$, $B(5, 8)$ and $C(11, 2)$.

Find the equation of the median *BN*.

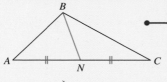

> Drawing a quick sketch will ensure you calculate the correct median.

Coordinates of *N* are $\left(\dfrac{x_1 + x_2}{2}, \dfrac{y_1 + y_2}{2}\right)$

$N = \left(\dfrac{3 + 11}{2}, \dfrac{2 + 2}{2}\right)$

> Find *N*, the midpoint of *AC*.

(continued)

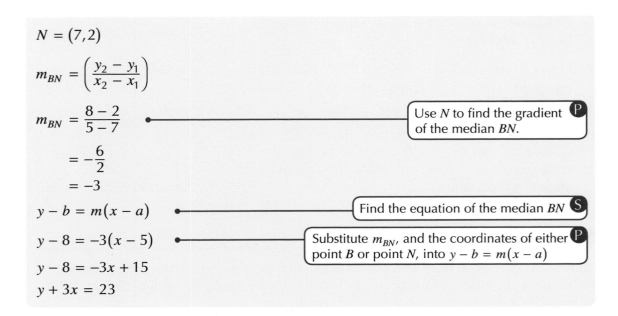

$N = (7, 2)$

$m_{BN} = \left(\dfrac{y_2 - y_1}{x_2 - x_1} \right)$

$m_{BN} = \dfrac{8 - 2}{5 - 7}$ ●————————————— Use N to find the gradient of the median BN. Ⓟ

$\quad = -\dfrac{6}{2}$

$\quad = -3$

$y - b = m(x - a)$ ●————————————— Find the equation of the median BN Ⓢ

$y - 8 = -3(x - 5)$ ●————————————— Substitute m_{BN}, and the coordinates of either point B or point N, into $y - b = m(x - a)$ Ⓟ

$y - 8 = -3x + 15$

$y + 3x = 23$

Exercise 13E

Cut any size of triangle from heavy cardboard. Mark a point half-way along each side. Draw a line from each midpoint to the opposite corner. These are the medians of the triangle. They should meet at the centroid.

Make a small hole at the centroid and thread a knotted string through it.

When held up and suspended by the string it should hang level (it might be tricky to get it exactly balanced, but you should get close). Explain why.

Altitudes

An **altitude** is a line through a vertex perpendicular to the opposite side. Every triangle has three altitudes. The three altitudes are concurrent and meet at the **orthocentre** of the triangle. The orthocentre is not always inside the triangle. If the triangle is obtuse, it will be outside.

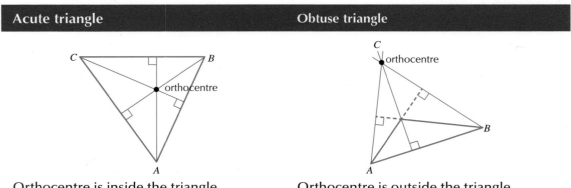

Acute triangle	Obtuse triangle
Orthocentre is inside the triangle.	Orthocentre is outside the triangle.

Example 13.12

Triangle ABC has coordinates $A(2,7)$, $B(4,-1)$ and $C(-4,3)$.

Find the equation of the altitude AD.

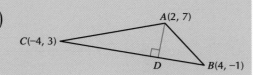

$m_{BC} = \left(\dfrac{y_2 - y_1}{x_2 - x_1}\right)$

$m_{BC} = \dfrac{-1-3}{4--4}$ — Find the gradient of BC. **P**

$= -\dfrac{4}{8}$

$m_{BC} = -\dfrac{1}{2}$

$m_{BC} \times m_{AD} = -1$ — Altitude AD is perpendicular to side BC, so $m_{BC} \times m_{AD} = -1$. **S**

$\therefore m_{AD} = 2$

$y - b = m(x - a)$ — Find the equation of the altitude. **P**

$y - 7 = 2(x - 2)$

$y - 7 = 2x - 4$

$y - 2x = 3$

Perpendicular bisectors

A line which cuts another line in half at right angles is a **perpendicular bisector**. In a triangle, the three perpendicular bisectors are concurrent. They meet at a point called the **circumcentre** of the triangle. This point is the centre of a circle which passes through the three vertices of the triangle. The circumcentre is not always inside the triangle. If the triangle is obtuse, it will be outside.

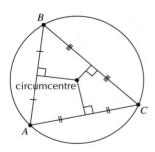

Example 13.13

Find the equation of the perpendicular bisector of the line joining $A(-2,0)$ and $B(2,6)$.

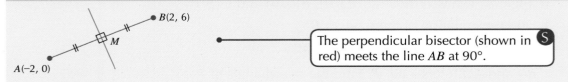

The perpendicular bisector (shown in red) meets the line AB at 90°. **S**

(continued)

$$M\left(\frac{x_1 + x_2}{2}, \frac{y_1 + y_2}{2}\right)$$

$$M\left(\frac{-2 + 2}{2}, \frac{0 + 6}{2}\right)$$ ●————————— Find the midpoint of AB. Ⓟ

$$M(0, 3)$$ ●————————— The perpendicular bisector passes Ⓢ through the midpoint of AB.

$$m_{AB} = \frac{y_2 - y_1}{x_2 - x_1}$$

$$m_{AB} = \frac{6 - 0}{2 - (-2)} = \frac{3}{2}$$ ●————————— Find the gradient of AB. Ⓟ

$$m_{AB} \times m_\perp = -1$$

$$\frac{3}{2} \times \frac{-2}{3} = -1$$ ●————————— Find the perpendicular gradient. Ⓟ

$$m_\perp = -\frac{2}{3}$$

$$y - b = m(x - a)$$ ●————————— Find the equation of the Ⓟ perpendicular bisector.

$$y - 3 = -\frac{2}{3}(x - 0)$$

$$y - 3 = -\frac{2}{3}x$$

$$3y + 2x = 9$$

Angle bisectors

An **angle bisector** is a line that cuts an angle in half. Every triangle has three angle bisectors. They are concurrent and meet at the **incentre** of the triangle. The incentre is the centre of the largest circle that will fit inside the triangle and touch all three sides. The incentre is always inside the triangle. It turns out that the orthocentre, centroid and circumcentre of any triangle are collinear, lying on the same straight line called the Euler line, named after Swiss mathematician Leonhard Euler who discovered it.

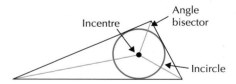

Exercise 13F

Do not use a calculator for this Exercise

★ **1** Triangle ABC has vertices at $A(1, 0)$, $B(3, 6)$ and $C(9, 0)$.

   a Find the equation of the median BN.

   b Find the equation of the altitude AM.

   c BN and AM intersect at P. Find the coordiantes of P.

   d The point Q lies on AC and has coordinates $(7, 0)$. Show that PQ is parallel to BC.

★ **2** Find the equation of the perpendicular bisector of the line joining $A(-1, 1)$ and $B(3, 7)$.

3 Triangle *JKL* has vertices at $J(-7, -1)$, $K(3, 9)$ and $L(9, -9)$.

   a Find the equation of side *JL*.

   b Find the equation of altitude *KP*.

   c Hence find the coordinates of *P*, the point where the altitude *KP* meets side *JL*.

4 Triangle *ABC* has vertices at $A(-4, 2)$, $B(0, 6)$ and $C(6, -4)$.

   a Find the equations of the medians *AP* and *BQ*.

   b Find the point of intersection of *AP* and *BQ*.

   c Find the equation of the third median, *CR* and show it passes through the point you found in part **b**.

5 A triangle has vertices at *A*, *B* and *C*. The altitude *BN* has equation $y = 3x - 2$. Given that *N* is the point $(6, 16)$ and *BC* has equation $8y = 14x - 6$, find the coordinates of *C*.

6 A triangle has vertices $A(-3, -4)$, $B(-1, 2)$ and $C(7, -4)$.

   Find the equations of all three medians, show that the medians are concurrent and state the coordinates of the centroid.

7 There is a quick way of finding the centroid of a triangle. Examine the vertices given in Question 6, and your answer for the centroid. What is the quick method?

★ **8** Find the orthocentre of a triangle with vertices at $A(-2, -1)$, $B(-1, -4)$ and $C(0, -5)$

9 A triangle has vertices $A(-1, -3)$, $B(-4, 0)$ and $C(-7, -3)$.

   Show that triangle *ABC* is right-angled at *B*.

   Find the orthocentre of triangle *ABC*. What do you notice?

- I can find the equation of a line parallel to a given line. ★ Exercise 13A Q2, 3
- I can find the equation of a line perpendicular to a given line. ★ Exercise 13C Q3
- I can use $m = \tan \theta$ to calculate a gradient or angle. ★ Exercise 13D Q1, 2
- I can use gradients to show points are collinear. ★ Exercise 13B Q1
- I can find equations of medians, altitudes and perpendicular bisectors. ★ Exercise 13F Q1, 2
- I can solve problems involving medians, altitudes and perpendicular bisectors. ★ Exercise 13F Q8

For further assessment opportunities, see the Preparation for Assessment for Unit 3 on page 370–372.

14 Applying algebraic skills to circles

This chapter will show you how to:

- determine and use the equation of a circle $(x - a)^2 + (y - b)^2 = r^2$
- determine and use the general equation of a circle $x^2 + y^2 + 2gx + 2fy + c = 0$
- use properties of tangency when solving problems
- determine the intersection of circles or a line and a circle.

You should already know:

- the terminology used to describe features of a circle
- how to complete the square in a quadratic expression
- how to solve inequalities
- how to factorise and solve quadratic equations
- how to determine the equation of a straight line
- how to use simultaneous equations to find where two lines intersect
- how to find the mid-point of a line segment
- how to use the discriminant to examine the roots of a quadratic equation
- how to use Pythagoras' theorem (distance formula) to calculate the distance between two points.

The equation of a circle $(x - a)^2 + (y - b)^2 = r^2$

A circle is a simple and elegant shape which has been known to humankind since before the beginning of recorded history and has been the subject of much scholarly debate. One definition of a circle treats a circle as a type of line. If a straight line segment is bent around until its ends meet and positioned so that all the points that make up the line are the same distance from a central point, the resulting shape is defined as a circle.

An alternative definition describes a circle as the **locus** of points at a fixed distance from a given centre point. A locus is a shape created by a set of points whose position satisfies a given rule.

The term circle is commonly used to mean either the boundary of the shape, or the whole shape including its interior. Strictly speaking a circle is a line and, as such, has no area. The area enclosed by a circle is called a disk.

A useful property of the circle is that for any closed shape with a fixed perimeter, a circle will enclose the largest area.

A circle is defined by the position of its centre, and by the length of its radius. This is enough information from which to create a circle. The diagram shows a circle with its centre fixed at point (a, b). Any point on the circumference of the circle can be represented by the general point (x, y). Pythagoras' theorem can be used to establish a connection between the centre, the radius and any point on the circle.

The equation of a circle with centre (a, b) and radius r is :

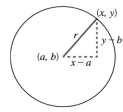

$$r^2 = (x - a)^2 + (y - b)^2$$

Notice that the circle equation is essentially a version of the distance formula which states: $d^2 = (x_2 - x_1)^2 + (y_2 - y_1)^2$

If the centre is at the origin, so (a, b) is $(0, 0)$, then the equation is $x^2 + y^2 = r^2$.

The circle which is centred at the origin with radius 1 is called the **unit circle**.

The three circles in the diagram are congruent (identical).

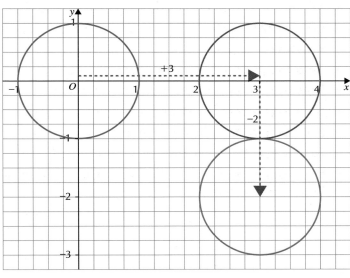

- The green circle is the unit circle, with centre $(0, 0)$ and radius 1.

- The blue circle is the result of a translation of 3 units to the right, with equation $(x - 3)^2 + y^2 = 1$. The translation to the right corresponds to the $(x - 3)$ component in the equation.

- The red circle is the result of a second translation of 2 units down, with equation $(x - 3)^2 + (y + 2)^2 = 1$. The translation down corresponds to the $(y + 2)$ component in the equation.

It follows that a circle with equation $(x - a)^2 + (y - b)^2 = 1$ is the unit circle following a translation of a units right and b units up. It will have its centre at the point (a, b).

Example 14.1

Find the equation of the circle with centre $(3, -4)$ and radius 6.

$$(x - a)^2 + (y - b)^2 = r^2$$
$$(x - 3)^2 + (y - (-4))^2 = 6^2$$
$$(x - 3)^2 + (y + 4)^2 = 36$$

> **S** Substitute $a = 3$, $b = -4$ and $r = 6$ into the circle equation.

> **P** You don't have to square the brackets but you must square the radius.

307

Example 14.2

State the centre and radius of the circle with equation $(x + 2)^2 + (y - 3)^2 = 25$.

Centre is $(-2, 3)$ and radius is $\sqrt{25} = 5$

Example 14.3

Find the equation of the circle which passes through the point $(6, 3)$ and has centre $(3, 1)$.

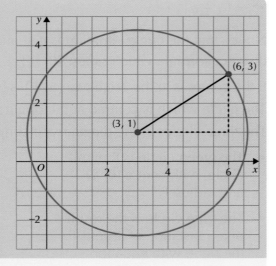

$$\overset{x, y}{\text{Point } (6, 3)} \quad \overset{a, b}{\text{Centre } (3, 1)}$$

$r^2 = (x - a)^2 + (y - b)^2$ — Substitute the point (x, y) and centre (a, b) into the circle equation. Labelling the coordinates can be helpful. **P**

$r^2 = (6 - 3)^2 + (3 - 1)^2 = 13$ — Use Pythagoras' theorem to find r^2. **S**

$(x - 3)^2 + (y - 1)^2 = 13$

Exercise 14A

Do not use a calculator for this Exercise.

1   Find the equation of the circle with the given centre and radius.

   a $(4, 6)$; 5   **b** $(7, -2)$; 4   **c** $(-3, 6)$; 2   **d** $(-3, -5)$; 6

   e $(0, -3)$; 7   **f** $(-4, 0)$; 3   **g** $(0, 0)$; $\sqrt{5}$ **h** $(-7, -1)$; $\sqrt{13}$

2   Write down the centre and radius of each of these circles.

   a $(x - 2)^2 + (y - 3)^2 = 36$ **b** $(x + 4)^2 + (y - 1)^2 = 9$

   c $(x - 7)^2 + (y + 5)^2 = 100$ **d** $x^2 + (y - 9)^2 = 16$

   e $(x + 3)^2 + y^2 = 84$ **f** $(x - 1)^2 + (y + 1)^2 = 1$

   g $x^2 + y^2 = 17$ **h** $(x + 4)^2 + (y - 5)^2 = 7$

3 Find the equation of this circle which passes through the points (0, 5) and (5, 0).

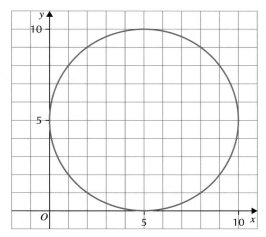

4 Find the equation of the circle that has a diameter with end-points at (0, 2) and (0, 12).

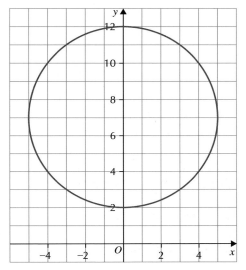

5 Find the equation of the circle that passes through the point (2, 4) and has centre (−2, 1).

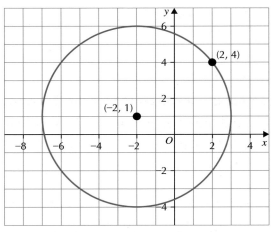

★ **6** Find the equation of the circle that passes through the point (−3, 3) and has centre (−5, −1).

7 Find the equation of the circle that has a diameter with end-points at (−6, −9) and (2, −3).

8 The circle with centre $(4, a)$ passes through the points $(-1, 3)$ and $(1, 1)$. Find the value of a and hence find the equation of the circle.

9 Find the shortest distance from the point $(10, 0)$ to the circle with equation $(x + 2)^2 + (y + 5)^2 = 25$.

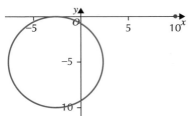

The general equation of a circle $x^2 + y^2 + 2gx + 2fy + c = 0$

The equation $(x - 2)^2 + (y - 1)^2 = 3^2$ represents a circle with centre $(2, 1)$ and radius 3.

Expanding gives: $x^2 - 4x + 4 + y^2 - 2y + 1 - 9 = 0$

Rewriting gives: $x^2 + y^2 - 4x - 2y - 4 = 0$

The equation is now in general form.

Conversely, if an equation is expressed in general form, it is possible to work back to standard form.

Start with: $x^2 + y^2 - 4x - 2y - 4 = 0$

Rewrite as: $x^2 - 4x + y^2 - 2y = 4$

Complete the square for x and y: $(x - 2)^2 - 4 + (y - 1)^2 - 1 = 4$

Rewrite as: $(x - 2)^2 + (y - 1)^2 = 9$

This represents a circle with centre $(2, 1)$ and radius 3.

The general equation of a circle is $x^2 + y^2 + 2gx + 2fy + c = 0$.

If we follow the above process using this version of the general equation, we can establish a quick way to obtain the centre and radius from an equation in general form.

Start with: $x^2 + y^2 + 2gx + 2fy + c = 0$

Rewrite as: $x^2 + 2gx + y^2 + 2fy = -c$

Complete the square for x and y: $(x + g)^2 - g^2 + (y + f)^2 - f^2 = -c$

Rewrite as: $(x - (-g))^2 + (y - (-f))^2 = g^2 + f^2 - c$

We can see that $x^2 + y^2 + 2gx + 2fy + c = 0$ represents a circle with centre $(-g, -f)$ and radius $\sqrt{g^2 + f^2 - c}$

For a circle to exist, $g^2 + f^2 - c > 0$

As you will see in Example 14.6, some of the questions which follow can be solved using quite different approaches. The stepping-out method is visual and appeals to some people more than others. If you have experience of working with vectors then you might prefer the vector approach. You should choose the method you are most comfortable with.

Example 14.4

Does the equation $x^2 + y^2 - 4x + 2y + 8 = 0$ represent a circle?

For a circle to exist, $g^2 + f^2 - c > 0$ State the condition required. **S**

$2g = -4 \Rightarrow g = -2$

$2f = 2 \Rightarrow f = 1$

$c = 8$

$g^2 + f^2 - c = (-2)^2 + 1^2 - 8 = -3$

The equation $x^2 + y^2 - 4x + 2y + 8 = 0$
does not represent a circle since $g^2 + f^2 - c < 0$

> Write a statement to explain your solution. Make sure you include a reason for your answer. **C**

Example 14.5

Find the centre and radius of the circle with equation $x^2 + y^2 - 4x + 2y + 1 = 0$

$2g = -4 \Rightarrow g = -2$

$2f = 2 \Rightarrow f = 1$

Centre is $(2, -1)$

> Remember the centre is $(-g, -f)$. **S**

$c = 1$

$r = \sqrt{g^2 + f^2 - c} = \sqrt{(-2)^2 + (1)^2 - 1} = 2$

> Substitute g and f (not $-g$ and $-f$) and c into the formula to find the radius. **P**

Example 14.6

A circle $x^2 + y^2 + 4x - 2y - 84 = 0$ has diameter AB. If A is the point $(-10, -4)$, find the coordinates of B.

Coordinates of centre: $C(-2, 1)$

> Identify the coordinates of the centre from the circle equation. **P**

Coordinates of point B: $B(a, b)$

> Introduce unknowns to represent the coordinates of B. **S**

$\dfrac{a + (-10)}{2} = -2$

$a - 10 = -4$

$a = 6$

> Use the mid-point formula with the known point and the centre to set up an equation. **S**

$\dfrac{b + (-4)}{2} = 1$

$b - 4 = 2$

$b = 6$

> Repeat the process for the y-coordinate. **P**

$\therefore B(6, 6)$

Stepping-out method

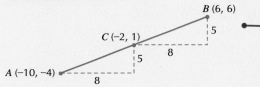

> A neat and well-annotated sketch is essential if using this method. This method uses what you are told about the point A and what you calculate about the centre. **S**

(continued)

Vectors method

$$\overrightarrow{AC} = \begin{pmatrix} -2 \\ 1 \end{pmatrix} - \begin{pmatrix} -10 \\ -4 \end{pmatrix} = \begin{pmatrix} 8 \\ 5 \end{pmatrix}$$

$$\overrightarrow{CB} = \begin{pmatrix} a \\ b \end{pmatrix} - \begin{pmatrix} -2 \\ 1 \end{pmatrix} = \begin{pmatrix} a + 2 \\ b - 1 \end{pmatrix}$$

The journey from A to C is the same as that from C to B. Use this fact, along with the point (a, b) to create two equations.

$$\begin{pmatrix} a + 2 \\ b - 1 \end{pmatrix} = \begin{pmatrix} 8 \\ 5 \end{pmatrix} \Rightarrow a = 6, b = 6$$

Solving gives the coordinates of the point B. (P)

Exercise 14B

Do not use a calculator for this Exercise.

1 Which of these equations represent circles? Explain your answers.

    a $x^2 + y^2 + 4x + 4y + 3 = 0$ **b** $x^2 + y^2 - 6x + 2y + 6 = 0$

    c $x^2 + y^2 - 2x + 2y + 6 = 0$ **d** $x^2 + y^2 + 4x + 6y + 13 = 0$

    e $x^2 + y^2 + 4x - 5 = 0$ **f** $2x^2 + 2y^2 + 20x - 16y = 16$

2 Find the centre and radius of each circle.

    a $x^2 + y^2 + 4x - 2y + 1 = 0$ **b** $x^2 + y^2 - 6x + 4y + 4 = 0$

    c $x^2 + y^2 + 6x - 4y - 3 = 0$ **d** $x^2 + y^2 - 6x - 10y + 17 = 0$

    e $x^2 + y^2 - 6y - 2 = 0$ **f** $\dfrac{x^2}{2} + \dfrac{y^2}{2} + x - \dfrac{y}{2} - \dfrac{1}{2} = 0$

3 Find the centre and radius of the circle $x^2 + y^2 + 4x - 8y - 5 = 0$. Use your answers to help decide if each of the points S (1, 5), T (1, 0), U (–6, –3) and V (2, –2) lies inside, outside or on the circle.

★ **4** The point (2, k) lies on the circle with equation $x^2 + y^2 + 4x - 8y - 5 = 0$. Find the possible value(s) of k.

5 The point (m, 4) lies on the circle with equation $x^2 + y^2 + 4x - 8y - 5 = 0$. Find the possible value(s) of m.

6 The point (9, 10) lies on the circle with equation $x^2 + y^2 + 6x - 10y + c = 0$. Find the value of c.

7 Two concentric circles are shown. The smaller circle has equation $x^2 + y^2 + 14x - 4y + 17 = 0$. Given that the radius of the larger circle is three times that of the smaller circle, find the equation of the larger circle.

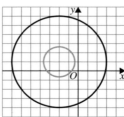

8 The circle with equation $x^2 + y^2 - 12x - 10y + 24 = 0$ cuts the y-axis at points S and T. Find the length of the line ST.

9 The circle with equation $x^2 + y^2 - 8x + 4y + 15 = 0$ cuts the x-axis at points U and V. Find the length of the line UV.

10 These equations represent circles. Find the possible range of values of k.

   a $x^2 + y^2 + 4x - 2y - k = 0$ **b** $x^2 + y^2 + 4x + k = 0$

   c $x^2 + y^2 + 6x - 2ky + 13 = 0$ **d** $x^2 + y^2 + kx + ky + 8 = 0$

11 The circles with equations $x^2 + y^2 + 20x - 16y - 5 = 0$ and
$x^2 + y^2 - 52x + 14y + 49 = 0$ touch externally as shown.
The two centres and the point of contact are collinear. Find the
coordinates of the point of contact.

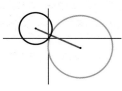

12 A computer animator is designing a snowman for a greetings card. The snowman
is to be 20 cm high with a circular body and head.

   The equation of the body is $x^2 + y^2 - 8x - 6y - 11 = 0$ and the line between both
centres is vertical. Find the equation of the head.

13 These three circles have collinear centres. The equations of
the outer circles are $x^2 + y^2 + 10x - 55 = 0$ and
$x^2 + y^2 - 26x - 18y + 205 = 0$. Find the equation of the
middle circle.

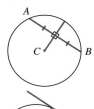

14 The line $3y + 4x + 15 = 0$ is a common tangent to two circles.
The larger circle has equation $x^2 + y^2 + 22x + 14y + 70 = 0$.
The radius of the smaller circle is half that of the larger circle.
Find the equation of the smaller circle.

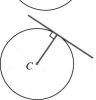

Use properties of tangency when solving problems

A chord is a line (AB) which joins any two points on the circumference
of circle. A chord divides a circle into two segments. The perpendicular
bisector of a chord passes through the centre, C, of the circle.

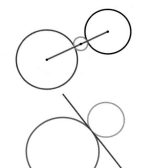

A tangent to a circle is a straight line which touches the circle at only
one point. The angle between a tangent and a radius is 90°. The radius
and tangent are perpendicular.

To determine whether or not a line is a tangent to a circle we need to
examine how (if at all) the circle and straight line meet. There are
three possibilities:

 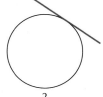 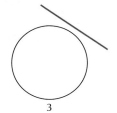

1   the circle and the line intersect at two points

2   the line touches the circle at one point

3   the line does not meet the circle.

The three possibilities here are similar to the three possibilities for quadratic roots: two distinct real roots; one repeated real root or no real roots. The discriminant can be used here to investigate the intersection of a circle and a straight line.

Example 14.7

Find the equation of the tangent to the circle $x^2 + y^2 - 4x - 2y - 3 = 0$ at the point $(4, 3)$.

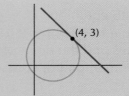
(4, 3)

Centre $(2, 1)$ — Use the circle equation to establish the centre of the circle. **P**

$m_{radius} = \dfrac{y_2 - y_1}{x_2 - x_1}$

$m_{radius} = \dfrac{3 - 1}{4 - 2} = 1$ — Use the centre and the given point to determine the gradient of the radius. **P**

$m_{rad} \times m_{tgt} = -1$

$m_\perp = -1$ — The radius and tangent are perpendicular. Use this fact to determine the gradient of the tangent. **P**

$y - b = m(x - a)$

$y - 3 = -(x - 4)$ — Substitute into $y - b = m(x - a)$ **P**

$y + x = 7$

Example 14.8

Show that the line $x + 3y - 11 = 0$ is a tangent to the circle $x^2 + y^2 + 2x + 12y - 53 = 0$ and find the point of contact.

$x = 11 - 3y$ — Rearrange the linear equation to make either x or y the subject. In this case it is simpler to rearrange for x. **P**

$(11 - 3y)^2 + y^2 + 2(11 - 3y) + 12y - 53 = 0$ — Substitute into the circle equation to create a new equation in only one variable. **S**

$121 - 66y + 9y^2 + y^2 + 22 - 6y + 12y - 53 = 0$ — Expand both brackets. **P**

$10y^2 - 60y + 90 = 0$
$y^2 - 6y + 9 = 0$ — Rearrange into standard quadratic form. Simplifying the equation by dividing both sides by a number does not affect the solutions. **P**

$a = 1, b = -6, c = 9$ — Use the discriminant to determine the nature of the roots. **S**

$b^2 - 4ac = (-6)^2 - 4(1)(9) = 0$ — Evaluate the discriminant. **P**

$b^2 - 4ac = 0$ hence there is one point of contact and the line is a tangent to the circle. — Write a statement to explain your solution. **C**

$(y - 3)(y - 3) = 0$ — Factorise the quadratic. **P**

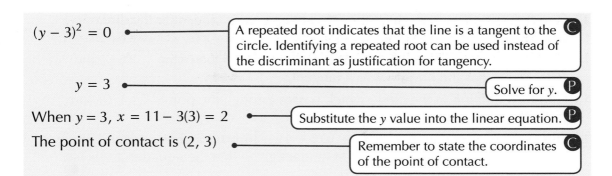

$(y - 3)^2 = 0$ — A repeated root indicates that the line is a tangent to the circle. Identifying a repeated root can be used instead of the discriminant as justification for tangency. **C**

$y = 3$ — Solve for y. **P**

When $y = 3$, $x = 11 - 3(3) = 2$ — Substitute the y value into the linear equation. **P**

The point of contact is $(2, 3)$ — Remember to state the coordinates of the point of contact. **C**

Exercise 14C

1 Find the equation of the tangent to the circle $x^2 + y^2 - 4x - 10y - 24 = 0$ at the point $(9, 3)$.

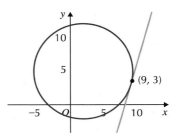

2 Find the equation of the tangent to the circle $x^2 + y^2 + 6x - 14y - 27 = 0$ at the point $(6, 9)$.

3 Find the equation of the tangent to the circle $x^2 + y^2 = 25$ at the point $(-4, -3)$.

4 The line $x = 4$ meets the circle $x^2 + y^2 - 2x - 4y - 20 = 0$ at P and Q. Find the coordinates of P and Q and hence show the tangent to the circle at P has equation $4y + 3x = 36$.

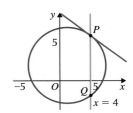

5 Show that the line $2y + x = 31$ is a tangent to the circle $x^2 + y^2 - 6x - 8y - 55 = 0$ and determine the point of contact.

6 Determine whether the line with equation $3y - 5x = 11$ is a chord, a tangent, or does not touch the circle with equation $x^2 + y^2 - 14x - 8y + 31 = 0$.

7 Determine whether the line with equation $5y - 8x = 22$ is a chord, a tangent, or does not touch the circle with equation $x^2 + y^2 - 10x + 14y - 15 = 0$.

★ **8** Determine whether the line with equation $5y - 7x + 6 = 0$ is a chord, a tangent, or does not touch the circle with equation $x^2 + y^2 + 4x - 18y + 11 = 0$.

9 The circle with equation $x^2 + y^2 - 10x - 14y + 29 = 0$ has two tangents with gradient 2. Find the points of contact of the circle and these tangents and hence find the equations of both tangents.

10 Find the values of k for which the line $y = x + k$ is a tangent to the circle $x^2 + y^2 = 5$.

$\textcircled{\tiny{\$}}$ **11** Find the values of k for which the line $y = x - k$ is a tangent to the circle $(x - 1)^2 + y^2 = 2$.

★ $\textcircled{\tiny{\$}}$ **12** Find the equation of the circle which passes through the points A (0, 1) and B (0, 7) and has the line $x = 1$ as a tangent.

$\textcircled{\tiny{\$}}$ **13** Find the equation of the circle which passes through the points $C(-1, 0)$ and $D(5, 0)$ and has the line $y = 1$ as a tangent.

$\textcircled{\tiny{\$}}$ **14** The circle with equation $x^2 + y^2 + 10x - 4y - 11 = 0$ touches a second circle at exactly one point. Find this point of contact given that the equation of the second circle is $x^2 + y^2 - 8x + 2y + 7 = 0$.

Intersecting circles

To determine whether or not two circles intersect we compare the distance between the two centres and the sum of the radii. There are five possibilities:

1 The circles meet externally at one point. The distance, d, between the centres equals the sum of the radii: $d = r_1 + r_2$

2 The circles meet at only one point and that one circle is positioned inside the other. The distance, d, between the centres will equal the **difference** between the radii: $d = r_1 - r_2$

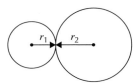

3 The circles intersect at two points. The distance, d, between the centres is less than the sum of the radii: $d < r_1 + r_2$

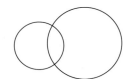

4 The circles do not touch. The distance, d, between the centres is greater than the sum of the radii: $d > r_1 + r_2$

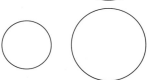

5 The circles do not touch and one circle is contained in the other. The distance, d, between the centres is less than the difference between the radii: $d < r_1 - r_2$

> ### Example 14.9
> Determine how, if at all, the circles with equations $x^2 + y^2 - 8x + 6y - 11 = 0$ and $x^2 + y^2 + 4x - 10y + 4 = 0$ intersect.

Centres are at $c_1(4, -3)$ and $c_2(-2, 5)$ —•———— Identify the coordinates of the centre from the circle equations. **P**

$d = \sqrt{(4 - (-2))^2 + (-3 - 5)^2} = 10$ —•———— Use the distance formula to calculate the distance between the centres. **P**

$r_1 = \sqrt{(-4)^2 + 3^2 + 11} = 6$

$r_2 = \sqrt{2^2 + (-5)^2 - 4} = 5$

$r_1 + r_2 = 11, d = 10$

The circles intersect at two points as $d < r_1 + r_2$

> Calculate both radii. **P**

> Compare the sum of the radii with the distance between the centres and write a statement to explain your answer. **C**

Exercise 14D

Do not use a calculator for this Exercise.

★ ⚙ **1** Determine how, if at all, the two circles intersect.

a $x^2 + y^2 + 4y - 32 = 0$ and $x^2 + y^2 - 16x - 12y + 75 = 0$

b $x^2 + y^2 + 8x + 4y - 61 = 0$ and $x^2 + y^2 - 16x - 6y - 57 = 0$

c $x^2 + y^2 + 8x - 4y + 11 = 0$ and $x^2 + y^2 - 10y + 16 = 0$

d $x^2 + y^2 + 10x + 4y - 115 = 0$ and $x^2 + y^2 + 6x - 8y + 21 = 0$

e $x^2 + y^2 + 16x + 8y - 320 = 0$ and $x^2 + y^2 + 8x + 2y - 47 = 0$

⚙ **2** Circle C_1 has equation $x^2 + y^2 - 18x - 22y - 23 = 0$. Circle C_2 has equation $x^2 + y^2 + 14x + 2y + p = 0$.

a Find the value of p if the circles touch at exactly one point.

b Find the point of contact.

- I can determine and use the equation $(x - a)^2 + (y - b)^2 = r^2$.
  ★ Exercise 14A Q6

- I can determine and use the general equation of a circle $x^2 + y^2 + 2gx + 2fy + c = 0$. ★ Exercise 14B Q4

- I can use properties of tangency when solving problems.
  ★ Exercise 14C Q12

- I can determine the intersection of a line and a circle and of two circles.
  ★ Exercise 14C Q8, Exercise 14D Q1

For further assessment opportunities, see the Preparation for Assessment for Unit 3 on page 370–372.

15 Modelling situations using sequences

This chapter will show you how to:

- use the terminology and notation associated with sequences
- use and determine nth term formulae
- determine a recurrence relation from given information
- use a recurrence relation to calculate a required term
- find and interpret a limit of a sequence, where it exists.

You should already know:

- commonly used sequences such as the square and cube numbers.

Sequences

A sequence is an ordered list of numbers generated by a set rule. Each number in the sequence is called a **term**. The set rule must show a relationship between any given term and the term following it.

Sequences such as (2, 4, 6, ..., 70) are called **finite** sequences. The ellipsis (three dots) in the sequence followed by the final number indicates that we have omitted some of the terms.

Some sequences are **infinite** where the terms go on for ever. (3, 6, 9, 12, ...) is an example of an infinite sequence, the three dots with nothing after them indicate the sequence never ends.

Terms are often denoted by $u_1, u_2, u_3, ..., u_n, u_{n+1}$ where u_1 is the first term, u_2 is the second term and u_n is the nth term. Where appropriate, the final term is denoted by l.

Nth term formulae

One way of defining a sequence is to give a formula for the nth term of the sequence, for example $u_n = 2n - 1$. By substituting values for n ($n = 1, 2, 3, ...$) we can generate the sequence, as shown in the table.

n	1	2	3	4	...
$u_n = 2n - 1$	1	3	5	7	...

The value of any particular term in the sequence can be found by substituting its position number (n) in to the formula. In the example given, the 50th term is $u_{50} = 2(50) - 1 = 99$.

The nth term formulae are useful because they let us calculate the value of the term at any position in the sequence, without knowing any other terms in the sequence. However, some nth term formulae can be difficult to determine.

Example 15.1

Given the nth term formula $u_n = 3n + 2$, calculate the first five terms of the sequence. Calculate the value of the 20th term in the sequence.

$u_1 = 3(1) + 2 = 5$

$u_2 = 3(2) + 2 = 8$

$u_3 = 3(3) + 2 = 11$

$u_4 = 3(4) + 2 = 14$

$u_5 = 3(5) + 2 = 17$

$u_{20} = 3(20) + 2 = 62$

> Substituting any value (n) into an nth term formula will give the value of the term at position (n) in the sequence. **S**

Example 15.2

Find a formula for the nth term of the sequence 3, 8, 13, 18, 23, ...

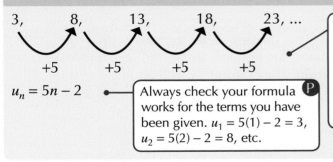

$u_n = 5n - 2$

> Always check your formula works for the terms you have been given. $u_1 = 5(1) - 2 = 3$, $u_2 = 5(2) - 2 = 8$, etc. **P**

> Try to find a relationship between any two consecutive terms which is true throughout the sequence. In this case, the difference between consecutive terms is equal to 5. This sequence contains the multiples of 5, minus 2. **S**

Example 15.3

Find a formula for the nth term of the sequence 3, 5, 9, 17, 33, ...

3, 5, 9, 17, 33, ...

$u_n = 2^n + 1$

> The terms in this sequence are one greater than the powers of 2 ($2^2 = 4$, $2^3 = 8$, $2^4 = 16$, and so on.) **S**

Exercise 15A

★ 1 Given these nth term formulae, calculate the first three terms and the value of the 30th term.

  a $u_n = 5n$ b $u_n = 2n + 6$ c $u_n = -3n + 2$ d $u_n = n^2 - 4$

  e $u_n = 3^n - 1$ f $u_n = 3^{n-1}$ g $u_n = \dfrac{1}{2^n}$

★ **2** Find a formula for the nth term of the following sequences and use it to find u_{25}

    a 4, 8, 12, 16, 20, ...   **b** 6, 11, 16, 21, 26, ...   **c** 7, 10, 13, 16, 19, ...

    d −7, −4, −1, 2, 5, ...   **e** 17, 13, 9, 5, 1, ...   **f** 2, 4, 8, 16, 32, ...

    g 2, 8, 26, 80, 242, ...   **h** 66, 34, 18, 10, 6, ...

3 Examine this pattern of blocks and find a formula to calculate the number of blocks required to build the nth structure in the sequence.

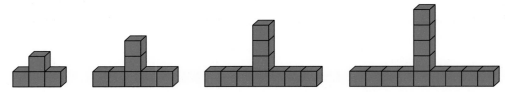

4 A pentagon has 5 sides and 5 diagonals. A hexagon has 6 sides and 9 diagonals. By considering the number of sides and diagonals in other polygons, find a formula that calculates the number of diagonals in a polygon with n sides.

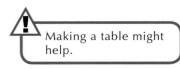

Making a table might help.

Determining a recurrence relation from given information

A sequence can also be defined by giving a **starting value** together with a rule, that shows the connection between a term and the one following it in the sequence. This is sometimes called a **recursive definition**. For example, $u_1 = 5$ and $u_{n+1} = 2u_n - 1$ defines the sequence 5, 9, 17, 33, ... A recursive definition must always include a starting value.

The rule $u_{n+1} = 2u_n - 1$ is an example of a **recurrence relation**.

A recurrence relation will generate different sequences for different starting values.

Applications of recurrence relations include: economics; biology (in particular population changes); digital signal processing (DSP), where recurrence relations are used to model feedback in a system where present time outputs become future inputs; and computer science.

For a given sequence it is possible to find an nth term formulae. However, finding nth term formulae can be quite tricky. Instead of finding an nth term formula, it is often simpler to find a recurrence relation that connects consecutive terms.

Linear recurrence relations

A recurrence relation of the form $u_{n+1} = mu_n + c$, where m and c are constant, is a **linear** recurrence relation. Special sequences can be formed if $m = 1$ or $c = 0$.

If $m = 1$ the recurrence relation becomes $u_{n+1} = u_n + c$. This gives **arithmetic** sequences like 3, 6, 9, 12, … where the difference between successive terms is constant.

If $c = 0$ we get $u_{n+1} = mu_n$. This gives **geometric** sequences like 2, 4, 8, 16, 32, … where the **ratio** of successive terms is constant.

If you go on study Mathematics at Advanced Higher level you will learn how to find the sum (total) of n terms of arithmetic and geometric sequences. This will allow you to quickly answer questions like "Find the sum of the first 50 natural numbers" or "Find the sum of the first 20 powers of 2", without having to add them all together.

Example 15.4

Find a recurrence relation, including the first term u_1 for the sequence 7, 11, 15, 19, …

$u_{n+1} = u_n + 4$ with $u_1 = 7$

The recurrence relation generates the next term by adding 4 to the previous term. The nth term formula is straightforward: $u_n = 4n + 3$. The 4 which was **added** in the recurrence relation to generate the next term becomes the **multiplier** in the nth term formula. Ⓢ

Example 15.5

In an isolated village 30 people are struck with a mysterious illness from which no-one recovers. Each day the number of people affected increases by 15% from the previous day's total. Find a recurrence relation to model this situation.

$u_{n+1} = 1.15u_n$

$u_0 = 30$

Every day the number infected is 115% of the previous day's total. Ⓢ

If a recurrence is used to model a real-life situation it is often sensible to call the starting value u_0. Doing this means u_1 will represent the first term actually generated by the recurrence relation. Ⓟ

Exercise 15B

1 Find a recurrence relation (including the first term u_1) for each of the following sequences.

  a 4, 8, 12, 16, 20, …   **b** 6, 11, 16, 21, 26, …   **c** 7, 10, 13, 16, 19, …
  d −7, −4, −1, 2, 5, …   **e** 17, 13, 9, 5, 1, …   **f** 2, 4, 8, 16, 32, …
  g 80, −40, 20, −10, 5, …   **h** 1, 2, 5, 14, 41, …

2 A biologist is studying a colony of termites. The number of termites doubles during the day. Initially there are 150 termites in the colony. Form a recurrence relation (including the first term u_0) to model this situation.

3 A patient is given a single 40 ml dose of antibiotics. Every hour the amount of antibiotic in the patient's bloodstream has dropped by 18%. Form a recurrence relation (including the first term u_0) to model this situation.

4 A generous relative puts £100 into a bank account when her granddaughter is born on 2nd April. On every subsequent birthday, the relative puts another £50 into the account. Interest at a rate of 0.5% per annum is credited to the account on 1st April. Form a recurrence relation (including the first term u_0) to model this situation.

★ 5 At a big game conservation park in Africa an elephant needs treatment and is injected with 500 ml of anaesthetic. It is known that 15% of the anaesthetic wears off every hour, so the elephant is given a booster of 40 ml every hour. Form a recurrence relation (including the first term u_0) to model this situation.

Using a recurrence relation to calculate a required term

Starting with the first term, we can use the recurrence relation to calculate successive terms in the sequence until we reach any specified term. The next example shows how each term is fed back into the recurrence relation to generate the next term.

Example 15.6
Given the recurrence relation $u_{n+1} = u_n + 3$ and the starting value $u_1 = 7$ calculate u_4 the fourth term of the sequence.

$u_1 = 7$
$u_2 = u_1 + 3 = 7 + 3 = 10$
$u_3 = u_2 + 3 = 10 + 3 = 13$
$u_4 = u_3 + 3 = 13 + 3 = 16$

Example 15.7
240 ml of a drug is administered to a patient. Every 3 hours 30% of the drug passes out of her bloodstream. To compensate, a further 10 ml dose is given every 3 hours.

a Form a recurrence relation to model this situation.

b Use your answer to calculate the amount of drug remaining after 12 hours

a $u_{n+1} = 0.7u_n + 10$

We are concerned with how much of the drug remains in the patient's system. If 30% or 0.3 leaves the system, that means that 70% or 0.7 remains.

S

b $u_0 = 240$

$u_1 = 0.7(240) + 10 = 178\,\text{ml}$

$u_2 = 0.7(178) + 10 = 134.6\,\text{ml}$

$u_3 = 0.7(134.6) + 10 = 104.22\,\text{ml}$

$u_4 = 0.7(104.22) + 10 = 82.954\,\text{ml}$

> **S** Since this recurrence relation takes place over a 3 hour period, u_4 will be the term that represents the amount of drug remaining after 12 hours

Using a calculator to generate terms

Some calculators can be used to generate successive terms in a recurrence relation. To find the terms generated by $u_{n+1} = 0.7u_n + 10$ with $u_0 = 240$, simply key in the starting value **240** and press = then type **0.7Ans + 10**. Repeatedly pressing = will generate successive terms in the sequence. Try this yourself and check that the values match the answers in Example 15.7. Continue pressing = and what do you notice about the sequence?

Example 15.8

A sequence is defined by the recurrence relation $u_{n+1} = mu_n + c$.

Find the values of m and c if $u_1 = 32$, $u_2 = 20$, $u_3 = 14$

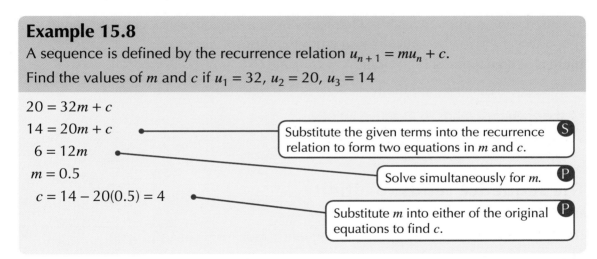

$20 = 32m + c$

$14 = 20m + c$

$6 = 12m$

$m = 0.5$

$c = 14 - 20(0.5) = 4$

> **S** Substitute the given terms into the recurrence relation to form two equations in m and c.

> **P** Solve simultaneously for m.

> **P** Substitute m into either of the original equations to find c.

Exercise 15C

1 Given the recurrence relation $u_{n+1} = 2u_n - 7$ and the starting value $u_1 = 9$ calculate u_4 the fourth term of the sequence.

2 Given the recurrence relation $u_{n+1} = 1.2u_n + 2$ and the starting value $u_1 = 10$ calculate u_4 the fourth term of the sequence.

★ 3 During heavy rainfall the depth of a river increases by 40 units per hour. Anti-flood drains reduce the level by 20% each hour. The initial depth was 80 units. Find a recurrence relation to model this situation and use it to calculate the depth of the river after 3 hours.

4 A sequence is defined by the recurrence relation $u_{n+1} = 1.5u_n + 2k$ and term $u_1 = 4k$. Find an expression for u_3 in terms of k.

⚙ **5** A sequence is defined by the recurrence relation $u_{n+1} = mu_n + c$ where m and c are constants. If $u_0 = 5$ and $u_1 = 10$ find an expression for m in terms of c.

⚙ **6** A sequence is defined by the recurrence relation $u_{n+1} = mu_n + c$ where m and c are constants. Given $u_0 = 2$, $u_1 = 5$ and $u_2 = 23$, find the values of m and c.

⚙ **7** For a sequence of the form $u_{n+1} = au_n + b$ the terms $u_0 = 10$, $u_1 = 30$ and $u_2 = 46$. Use this information to determine the values of a and b and hence calculate u_5.

⚙ **8** On February 1, a bank lends a customer £4000 at a fixed interest rate of 3.5% per month. The interest is added on the last day of every month and is calculated on the amount due on the first day of that month. The customer agrees to make repayments on the first day of each subsequent month. Each repayment is £400 except for the smaller final amount which will pay off the loan. The customer has calculated that he will pay off the loan on March 1 the following year, with a final payment of £203.49. Is the customer correct? Explain your answer.

The limit of a sequence

It is often very useful to examine what happens to a sequence as n gets very large. This is described mathematically as n tends to infinity $n \to \infty$

There are three types of behaviour:

- the sequence tends to infinity

- the sequence tends to negative infinity

- the sequence **converges** to a real limit.

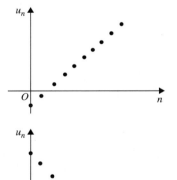

The sequence tends to infinity

$u_{n+1} = 4u_n - 10$ tends to infinity as $n \to \infty$
The graph shows the behaviour of the sequence as $n \to \infty$

The sequence tends to negative infinity

$u_{n+1} = -2u_n + 6$ tends to negative infinity as $n \to \infty$
The graph shows the behaviour of the sequence as $n \to \infty$

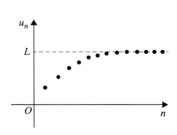

The sequence converges to a real limit

The recurrence relation in Example 15.7 ($u_{n+1} = 0.7u_n + 10$) is an example of a sequence that converges to a real limit.

The graph below shows the behaviour of the sequence as $n \to \infty$

For a linear recurrence relation of the form $u_{n+1} = mu_n + c$, if $-1 < m < 1$ the sequence will eventually settle around some final limiting value, L. At this point, the term generated by the recurrence relation is fed back in to generate the same term. In other words $u_{n+1} = u_n = L$ so:

$$L = mL + c$$

Changing the subject of this relationship to L gives:

$$L = mL + c$$
$$L - mL = c$$
$$L(1 - m) = c$$
$$L = \frac{c}{1 - m}$$

This formula can be used to calculate the limit of a sequence (if one exists) using the values of m and c in the recurrence relation. Note that the limit of a sequence depends on the values of m and c, but not on the starting value.

Example 15.9

Explain why the sequence defined by the recurrence relation $u_{n+1} = 0.7u_n + 15$ converges to a limit as $n \to \infty$ and find the value of this limit.

A limit exists as $-1 < 0.7 < 1$

$$L = \frac{c}{1 - m}$$

Substitute m and c from the recurrence relation into the limit formula. **S**

$$L = \frac{15}{1 - 0.7}$$

$$L = \frac{15}{0.3}$$

Be prepared to evaluate limits without the use of a calculator. It can be easier to think of the denominator as a fraction, in this case $L = 15 \div \frac{3}{10} = 15 \times \frac{10}{3} = 50$ **S**

$$L = 50$$

Example 15.10

A chemical factory wants to dump 200 kg of waste into a sea loch every week. It is estimated that the tide will remove 65% of this waste each week. Environmental groups claim that more than 350 kg of waste in the loch will be harmful to the local seal colony. Is it safe for the factory to dump the waste?

$u_{n+1} = 0.35u_n + 200$, $u_0 = 0$

Define a recurrence relation for this situation. **S**

A limit exists since $-1 < 0.35 < 1$

State why this recurrence relation tends to a limit. **C**

$$L = \frac{200}{1 - 0.35}$$

$L = 307.69$ kg

307.69 kg < 350 kg so it is safe for the factory to dump the waste.

Always include a numerical comparison and relate your explanation to the context of the question. Consider what this limit means in this question. At any particular time, the level of waste in the loch will not exceed 307.69 kg. However, the level of waste will not become fixed at 307.69 kg. As the week progresses the waste will reduce by 65%, dropping to a low of 107.69 kg. The weekly waste dump of 200 kg will take the level back to its maximum before gradually falling again. This cycle will continue to repeat. **C**

Exercise 15D

 1 For each recurrence relation below, explain whether the sequence formed has a limit as $n \to \infty$. If a limit exists, find its value.

    a $u_{n+1} = 0.3u_n + 6$ **b** $u_{n+1} = 0.9u_n - 4$ **c** $u_{n+1} = -0.4u_n + 1$

    d $u_{n+1} = -4u_n + 0.5$ **e** $2u_{n+1} = 1.4u_n - 12$ **f** $u_{n+1} = 8 - \frac{3}{2}u_n$

    g $3u_{n+1} = -1u_n + 0.6$

2 A sequence is defined by $u_{n+1} = 0.8u_n + 4$ and term $u_1 = 5$

    a What is the smallest value of n for which $u_n > 15$?

    b Find the limit of this sequence.

3 A sequence is defined by $u_{n+1} = mu_n + 6$ and term $u_0 = 2$, with $-1 < m < 1$. Find the value of m for which the sequence has a limit of 20.

4 Two sequences defined by $u_{n+1} = ku_n + 2$ and $v_{n+1} = 0.5v_n + 3$ have the same limit. Find the value of k.

5 A farmer sprays his field weekly with a pesticide called X-Pest, whose manufacturers claim will remove 60% of all pests. Between treatments 450 new pests appear in the field. A new product called Pest-Away is produced. The manufacturers claim it will remove 75% of all pests, although 550 new pests will appear in the fields.

Which pesticide will be more effective in the long term?

★ **6** A patient is given a 50 ml dose of medication. It is known that 16% of the medication leaves the bloodstream every day. To compensate, the patient is given daily booster injections of 25 ml. It could be harmful if the level of medication in the bloodstream exceeds 150 ml. Is it safe to continue this course of treatment over a long period of time?

7 A gardener plants a number of fast-growing trees between his and his neighbour's garden. The trees are expected to grow 0.7 metres taller each year. The gardener decides to trim the top 20% off the height of the trees at the start of every year.

    a To what height will the trees grow to in the long run?

    b His neighbour is worried that the trees are growing at an alarming rate and wants the height of the trees restricted to 2.5 metres. What is the minimum percentage that will need to be trimmed from the trees each year to meet this condition?

8 A woman opens a new savings account by depositing £100. The account receives 2.4% interest per annum. Each year, after the interest is paid in, the woman plans to add £k to her savings, where k is a constant.

    a Find a recurrence relation for u_n for the amount in pounds in the account after n years.

    b The woman would like to have exactly £500 in her account after three years. Find the required value of k, correct to two decimal places.

- I can use the terminology and notation associated with sequences. ★ Exercise 15A Q1

- I can use and determine nth term formulae. ★ Exercise 15A Q2

- I can determine a recurrence relation from given information. ★ Exercise 15B Q5

- I can use a recurrence relation to calculate a required term. ★ Exercise 15C Q3

- I can find and interpret a limit of a sequence, where it exists. ★ Exercise 15D Q6

For further assessment opportunities, see the Preparation for Assessment for Unit 3 on page 370–372.

16 Applying differential calculus

This chapter will show you how to:

- determine the greatest and least values of an algebraic function on a closed interval
- construct a model to represent a situation and use differentiation to obtain an optimal solution
- solve problems using rates of change.

You should already know:

- how to differentiate algebraic, trigonometric and simple composite functions
- how to determine the stationary points of a curve.

Determine the greatest and least values of an algebraic function on a given interval

Knowing how to find the stationary points of a curve and determine their nature (Chapter 10) can be used to analyse the graphs of algebraic functions in greater detail.

Intervals are used to specify the part of the graph under consideration and to identify solutions. For example, the quadratic inequality $x^2 - 3x - 18 < 0$ has solution $-3 < x < 6$. To satisfy the inequality, x must take values between -3 and 6.

$-3 < x < 6$ is an example of a **open interval**. The numbers -3 and 6 are the **endpoints** of the interval. The interval is described as open because the endpoints are being excluded, so we are not allowed to choose x to be either -3 or 6.

If, however, the inequality was $x^2 - 3x - 18 \leq 0$ our solution would be $-3 \leq x \leq 6$. $-3 \leq x \leq 6$ is a ***closed interval*** because the endpoints are now included, so we are allowed to choose x to be either -3 or 6.

Real-life problems invariably involve intervals. Consider the following:

A firm manufactures water tanks from rectangular sheets of metal 3 m long by 2 m wide.

Squares of side x m are cut out of each corner as shown in the diagram.

$PQRS$ forms the base of the tank and the remaining parts are folded upwards and sealed to form the open tank. Given suitable information, it is possible to derive a

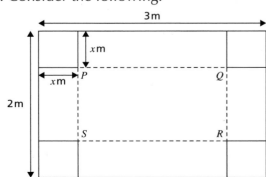

function which represents the profit obtained on an individual tank. This will depend on the value of x (as well as a number of other factors).

Note, however, that $0 < x < 1$ (why?) and our analysis will need to take account of this. This question is explored further in Exercise 16B Question 12.

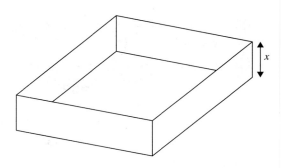

Maximum and minimum values and stationary points

Consider the graph of the function $f(x) = (x - 3)^2 + 2$. The green line shows the interval $0 \leq x \leq 4$ and the portion of the graph corresponding to this interval is shown in red.

On the interval $0 \leq x \leq 4$ the minimum value of the function is 2. This occurs when $x = 3$.

Evaluating the function at the endpoints of the interval gives:

$f(0) = (0 - 3)^2 + 2 = 11$ and

$f(4) = (4 - 3)^2 + 2 = 3$

On the interval $0 \leq x \leq 4$ we can therefore say that the maximum value of $f(x)$ is 11 and the minimum value is 2.

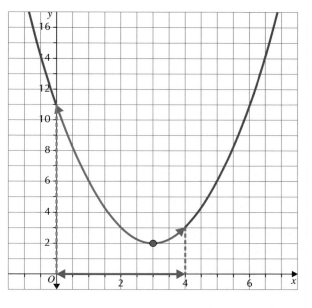

Suppose now we look at the interval $4 \leq x \leq 6$.

For $4 \leq x \leq 6$ the minimum turning point of the graph is not included in the red part of the graph. However, it is easily seen that on the interval $4 \leq x \leq 6$ the maximum value of the function occurs when $x = 6$ and the minimum value when $x = 4$.

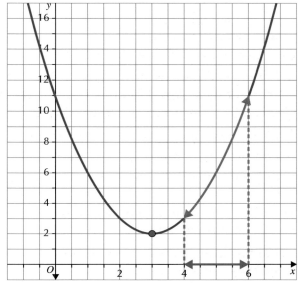

Consider now the graph of the cubic given by $g(x) = x^3 - 4x^2 - 3x + 1$ on the interval $-1 \leq x \leq 5$.

It can be shown that the graph has stationary points when $x = -\frac{1}{3}$ and $x = 3$.

From the graph it is clear that the stationary points occur within the given interval. On the interval $-1 \leq x \leq 5$ the minimum value of the function does occur at the minimum stationary point but the maximum value occurs at the endpoint where $x = 5$.

It can be seen that for a given closed interval, the greatest and least values of a function will occur either at stationary points or at the end points of the interval.

To determine the greatest and least values of a function $f(x)$ on a closed interval $a \leq x \leq b$:

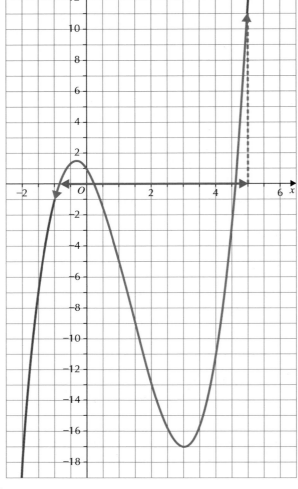

- find the stationary points and determine their nature
- find the value of the function at the endpoints of the interval (determine $f(a)$ and $f(b)$).
- make a sketch of the graph of $y = f(x)$, showing the stationary points and the endpoints (do not worry about finding intersections with the axes).
- clearly communicate the greatest and least values.

Example 16.1

The function f is defined by $f(x) = 2x^3 - 5x^2 - 4x + 3$. Find the greatest and least values of $f(x)$ on the interval $-1 \leq x \leq 4$.

$$f(x) = 2x^3 - 5x^2 - 4x + 3$$
$$f'(x) = 6x^2 - 10x - 4$$
$$6x^2 - 10x - 4 = 0$$
$$2(3x + 1)(x - 2) = 0$$

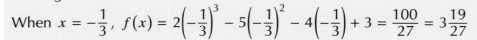

Differentiate and factorise to find the values of x at the stationary points. **S**

$$x = -\frac{1}{3} \text{ or } x = 2$$

When $x = -\frac{1}{3}$, $f(x) = 2\left(-\frac{1}{3}\right)^3 - 5\left(-\frac{1}{3}\right)^2 - 4\left(-\frac{1}{3}\right) + 3 = \frac{100}{27} = 3\frac{19}{27}$

When $x = 2$, $f(x) = 2(2)^3 - 5(2)^2 - 4(2) + 3 = -9$

Stationary points at $\left(-\frac{1}{3},\ 3\frac{19}{27}\right)$ and $(2, -9)$.

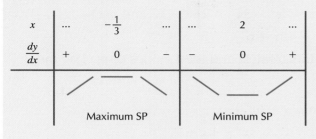

x	...	$-\frac{1}{3}$	...	...	2	...
$\frac{dy}{dx}$	+	0	–	–	0	+

Maximum SP   Minimum SP

> Use a nature table to determine the nature of the stationary points. **S**

Maximum SP at $\left(-\frac{1}{3},\ 3\frac{19}{27}\right)$, minimum SP at $(2, -9)$.

$f(-1) = 2(-1)^3 - 5(-1)^2 - 4(-1) + 3 = 0$

$f(4) = 2(4)^3 - 5(4)^2 - 4(4) + 3 = 35$

> Find the value of the function at the endpoints. **S**

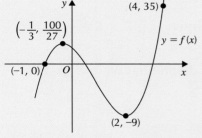

> Make a sketch of the graph of $y = f(x)$, showing the stationary points and the endpoints. **S**

On the interval $-1 \le x \le 4$, the greatest value of $f(x)$ is 35 and the least value of $f(x)$ is -9.

> Clearly communicate the greatest and least values. **C**

Exercise 16A

1 For each of the following functions, determine the greatest and least values of the function on the given closed interval.

a $f(x) = x^3 - 2x^2 + x - 5$, $-2 \le x \le 1$

b $g(x) = x^3 - 3x^2 - 24x + 10$, $-1 \le x \le 6$

c $h(x) = x^3 - 6x^2 + 3x + 2$, $2 \le x \le 5$

d $y = 4 + 9x - 3x^2 - x^3$, $-1 \le x \le 6$

e $y = 3x^4 + 8x^3 + 6x^2 - 6$, $-3 \le x \le -1$

f $y = x^3(x - 2)$, $1 \le x \le 4$

2 A curve has equation $y = f(x)$, where $f(x) = (x + 3)(x - 1)^2$

a Find where the curve crosses the x- and y-axes.

b Find the stationary points of the curve and determine their nature.

c i Sketch the curve, showing the stationary points and intersections with the axes.

ii Determine the greatest and least values of $f(x)$ on the interval $0 \leq x \leq 2$.

3 Functions f and g are defined by $f(x) = x^3 - 3x^2 - 24x - 12$ and $g(x) = 2x - 3$.

a Show that $f(g(x)) = 8x^3 - 48x^2 + 42x + 6$.

b Given $h(x) = f(g(x))$ determine the greatest and least values of $h(x)$ on the interval $0 \leq x \leq 3$.

4 a Express $3x^2 - 6x + 8$ in the form $a(x + b)^2 + c$.

b The function f is defined by $f(x) = x^3 - 3x^2 + 8x + 2$.

Andy claims that for any interval, $p \leq x \leq q$, the greatest and least values of $f(x)$ will always occur when $x = q$ or $x = p$. Is Andy right? Justify your answer.

5 A curve has equation $f(x) = 3\sin x + \sqrt{3}\cos x$, $0 \leq x \leq \frac{3\pi}{2}$.

a i Find the stationary points of the curve.

ii Determine their nature.

b Find the greatest and least values of $f(x)$ on the interval $\frac{\pi}{2} \leq x \leq \frac{3\pi}{2}$.

c Show that $2\sqrt{3}\cos\left(x - \frac{\pi}{3}\right) = 3\sin x + \sqrt{3}\cos x$.

d Using your answer to part **c**, find, without using differentiation, the greatest and least values of $f(x)$ on the interval $\frac{\pi}{2} \leq x \leq \frac{3\pi}{2}$.

▲ Challenge

A function is given by $f(x) = (x - a)^2(x - b)^2$, where $a < b$.

a Explain why the graph of $y = f(x)$ never lies below the x-axis.

b Show that the line $x = k$, where $k = \frac{a+b}{2}$, is an axis of symmetry of the graph.

c Without differentiating, show that the maximum value of $f(x)$ on the interval $a \leq x \leq b$ is $\frac{(b-a)^4}{4}$.

Determining the optimal solution to a given problem

Differentiation gives us a very powerful tool for solving real-life problems which involve maximising or minimising a given quantity. For the water tank on page xx, we can derive an expression for the profit, P, obtained in terms of x. In order to find the maximum profit we need to solve the equation $\frac{dP}{dx} = 0$. A nature table is then used to justify this.

As a further example, suppose you've come up with an idea for a Higher maths app. Your friends suggests that you should get someone to develop it for you, so you can sell it and make lots of money. But, you realise that you don't have the money to pay a developer, or know how much you would charge for the app. However, being a good problem-solver you are on the case immediately and start thinking about how to maximise your profit. Question 5 in Exercise 16B explores this further.

Example 16.2

The fuel consumption, in gallons per hour, of a bulldozer is given by the formula $y = 30 - 6x + 0.42x^2 - 0.008x^3$, where x is the steady speed of the bulldozer in miles per hour and $5 \le x \le 25$.

Determine the maximum and minimum fuel consumption.

$$y = 30 - 6x + 0.42x^2 - 0.008x^3$$

> The question asks about minimum and maximum values so think differentiation. **S**

$$\frac{dy}{dx} = -6 + 0.84x - 0.024x^2$$

$$-6 + 0.84x - 0.024x^2 = 0$$

> Maximum and minimum values **might** occur at stationary points so we solve $\frac{dy}{dx} = 0$. **P**

$$-6000 + 840x - 24x^2 = 0$$

$$x^2 - 35x + 250 = 0$$

$$(x - 10)(x - 25) = 0$$

> Rearrange the equation into standard form and divide by 24. Note that for this equation $b^2 - 4ac = 225$, a square number. This tells us that we should be able to solve by factorising. Alternatively, solve the equation $-6 + 0.84x - 0.024x^2 = 0$ using the quadratic formula. **P**

$$x - 10 = 0 \text{ or } x - 25 = 0$$

$$x = 10 \text{ or } x = 25$$

> Solve to find the x-coordinates of the stationary points, to find the speeds where a maximum or minimum fuel consumption may occur. **P**

When $x = 10$, $y = 30 - 6(10) + 0.42(10)^2 - 0.008(10)^3 = 4$

When $x = 25$, $y = 30 - 5(25) + 0.4(25)^2 - 0.008(5)^3 = 17.5$

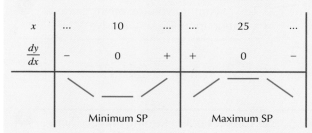

x	...	10	...	...	25	...
$\frac{dy}{dx}$	–	0	+	+	0	–

Minimum SP   Maximum SP

> Find the y-coordinates of the stationary points, to find the fuel consumption at speeds of 10 mph and 25 mph. **P**

> Check the endpoints of the interval, noting that we have already found the fuel consumption when $x = 25$. **S**

When $x = 5$, $y = 30 - 6(5) + 0.42(5)^2 - 0.008(5)^3 = 9.5$

Minimum fuel consumption is 4 gallons per hour at a steady speed of 10 mph.

Maximum fuel consumption is 17.5 gallons per hour at a steady speed of 25 mph.

> Explain your answer fully and communicate clearly. **C**

Example 16.3

Plastic storage containers are made from rectangular sheets measuring 2.5 m by 4 m. Two congruent rectangles of width x cm are removed from a sheet.

The remaining plastic is the net of a cuboid and this net is used to create a storage container of height x cm.

a Show that the volume, V cm^3, of a container is given by $V(x) = 2x^3 - 650x^2 + 50\,000x$.

b Find the value of x which maximises the volume of the tank.

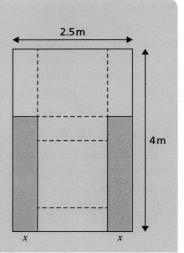

a Let l and b be the length and breadth of the cuboid.

> $V = lbh$. The height is x. Find expressions for the length and breadth in terms of x.

It follows that:
$$2x + l = 250$$

and so:
$$l = 250 - 2x$$

We also have:
$$2x + 2b = 400$$
$$2b = 400 - 2x$$
$$b = 200 - x$$

> Write down an expression **P** for the volume.

$$V = x(200 - x)(250 - 2x)$$
$$V = x(50\,000 - 650x + 2x^2)$$
$$V = 2x^3 - 650x^2 + 50\,000x$$

> Multiply out the brackets and **P** complete the proof.

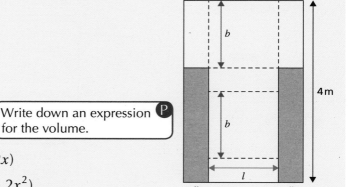

b $V'(x) = 6x^2 - 1300x + 50\,000$

> The maximum volume **might** occur at **P** a stationary point so we solve $\frac{dy}{dx} = 0$.

$$6x^2 - 1300x + 50\,000 = 0$$
$$(3x - 500)(2x - 100) = 0$$
$$3x - 500 = 0 \text{ or } 2x - 100 = 0$$
$$x = \frac{500}{3} \text{ or } x = 50$$

> Find the x-coordinates of the turning points. These **S** are the values of x which might give the maximum volume.

x	...	50	...	...	$\frac{500}{3}$	...
$V'(x)$	+	0	−	−	0	+

| | Maximum SP | | Minimum SP | |

In this problem, $x > 0$ and also since $l > 0$ we have $250 - 2x > 0$, so $x < 125$

When $x = 0$ or 125, $V(x) = 0$ and so the maximum volume will occur when $x = 50$.

Example 16.4

A small baking tin is in the shape of an open-topped, square-based cuboid with volume 500 cm^3.

The length of the base is x cm.

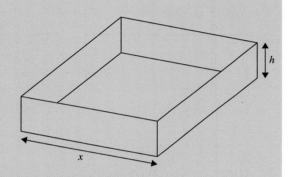

a Show that the surface area, S, of metal needed to make the tin is given by

$$S = x^2 + \frac{2000}{x}.$$

b Find the dimensions of the tin which minimise the surface area.

a Let h be the height of the tin.

$$S = 4xh + x^2$$

> The surface area consists of the square base and the 4 rectangular faces. **S**

$$V = l \times b \times h$$

$$500 = x \times x \times h = x^2 h$$

$$h = \frac{500}{x^2}$$

> The volume is 500 cm^3 so use this to express h in terms of x. **P**

$$S = 4x\left(\frac{500}{x^2}\right) + x^2$$

$$S = \frac{2000x}{x^2} + x^2$$

$$S = \frac{2000}{x} + x^2$$

> Substitute for h and complete the proof. Note that the middle line is required: it clearly shows that you know how to get to the final answer.

(continued)

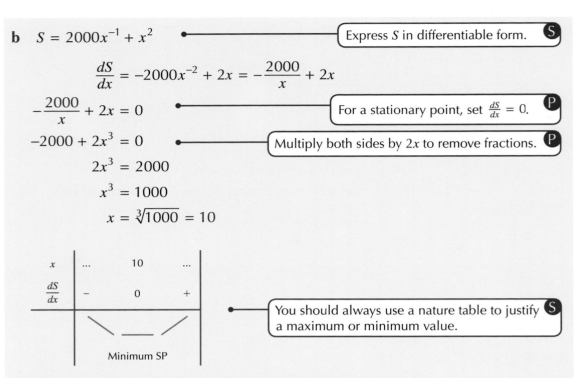

b $S = 2000x^{-1} + x^2$ — Express S in differentiable form. **S**

$$\frac{dS}{dx} = -2000x^{-2} + 2x = -\frac{2000}{x} + 2x$$

$-\frac{2000}{x} + 2x = 0$ — For a stationary point, set $\frac{dS}{dx} = 0$. **P**

$-2000 + 2x^3 = 0$ — Multiply both sides by $2x$ to remove fractions. **P**

$$2x^3 = 2000$$

$$x^3 = 1000$$

$$x = \sqrt[3]{1000} = 10$$

x	...	10	...
$\frac{dS}{dx}$	$-$	0	$+$

Minimum SP

You should always use a nature table to justify a maximum or minimum value. **S**

The minimum surface area will occur when $x = 10$.

When $x = 10$, $h = \frac{500}{10^2} = 5$ and so the dimensions of the tin are 10 cm by 10 cm by 5 cm.

The question asks for the dimensions of the tin. You need to find h and then communicate your final answer. **C**

Example 16.5

An oil drum is in the shape of a closed cylinder. The surface area of the oil drum is $9\,\text{m}^2$.

The oil drum has height $h\,$m and radius $r\,$m.

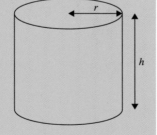

a Show that the volume, V, of the drum (in cubic metres) is given by:

$$V = \frac{9}{2}r - \pi r^3$$

b Show that the exact value of r for which the volume of the oil drum is a maximum is $\sqrt{\frac{3}{2\pi}}$.

c Show that the maximum volume of the oil drum is equal to three times its radius.

a $2\pi rh + 2\pi r^2$ ●————— Write an expression for the surface area. **S**

$2\pi rh + 2\pi r^2 = 9$ ●————— Use the fact that you know the surface area is 9. **S**

$2\pi rh = 9 - 2\pi r^2$

$h = \dfrac{9 - 2\pi r^2}{2\pi r}$ ●————— Solve the equation for h. **P**

$V = \pi r^2 h$ ●————— Write the formula for a cylinder. **S**

$V = \pi r^2 \left(\dfrac{9 - 2\pi r^2}{2\pi r} \right)$ ●————— Substitute for h. **P**

$V = \dfrac{9\pi r^2 - 2\pi^2 r^4}{2\pi r}$

$V = \dfrac{9\pi r^2}{2\pi r} - \dfrac{2\pi^2 r^4}{2\pi r}$

$V = \dfrac{9}{2} r - \pi r^3$ ●————— Complete the proof. **C**

b $\dfrac{dV}{dr} = \dfrac{9}{2} - 3\pi r^2$

$\dfrac{dV}{dr} = 0$

$\dfrac{9}{2} - 3\pi r^2 = 0$

$3\pi r^2 = \dfrac{9}{2}$

$r^2 = \dfrac{9}{6\pi} = \dfrac{3}{2\pi}$

$r = \pm\sqrt{\dfrac{3}{2\pi}}$

Since $r > 0$ we can disregard the negative solution. Hence $r = \sqrt{\dfrac{3}{2\pi}}$

Now test for a maximum in the usual way.

r	...	$\sqrt{\dfrac{3}{2\pi}}$	...
$\dfrac{dV}{dr}$	$+$	0	$-$
	╱	▔	╲
		Maximum SP	

Maximum turning point is at $r = \sqrt{\dfrac{3}{2\pi}}$.

(continued)

c $V = \frac{9}{2}r - \pi r^3 = r\left(\frac{9}{2} - \pi r^2\right)$

When $r = \sqrt{\frac{3}{2\pi}}$, $V = \sqrt{\frac{3}{2\pi}}\left(\frac{9}{2} - \pi\left(\sqrt{\frac{3}{2\pi}}\right)^2\right)$

$V = \sqrt{\frac{3}{2\pi}}\left(\frac{9}{2} - \pi\left(\frac{3}{2\pi}\right)\right)$

$V = \sqrt{\frac{3}{2\pi}}\left(\frac{9}{2} - \frac{3}{2}\right) = \sqrt{\frac{3}{2\pi}} \times 3 = 3r$

This now shows that the maximum volume is three times the radius, so $V = 3r$.

Exercise 16B

You may use a calculator throughout this exercise unless otherwise stated.

★ 1 Claire owns some shares and decides to keep track of their value.

The value, £V, of Claire's shares x days since the start of January is given by:

$V(x) = x^3 - 105x^2 + 3000x + 2000$, where $0 \le x \le 60$.

Determine the maximum and minimum value of Claire's shares during this period.

★ 2 A piece of wire 60 cm long is bent into the shape of a rectangle.

Let x cm be the length of the rectangle.

a Write down an expression for the breadth of the rectangle.

b Show that the area, A cm², of the rectangle is given by $A(x) = 30x - x^2$.

c Find the maximum area the rectangle could enclose.

★ 3 An open-topped box is formed by cutting squares of length x from the corners of a rectangular piece of card and folding up the remaining area.

The dimensions of the rectangle are 15 inches by 8 inches.

a Show that the volume, V, of the box formed is given by:

$V(x) = 4x^3 - 46x^2 + 120x$

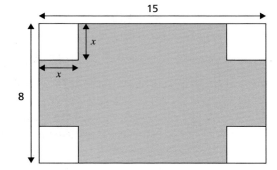

b Determine the volume of the box which has maximum volume.

★ 4 A large coffee tin is in the shape of a closed cuboid. It has a volume of 3375 ml.

The square base has length x cm and height h cm.

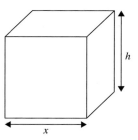

a Show that the surface area of metal needed to make the tin is given by $S = 2x^2 + \frac{13500}{x}$

b Find the value of x which minimises the surface area.

c Determine the minimum surface area.

5 After doing some research you come up with the following information about the Higher maths app you want to develop:

- The person who will design and develop your app for the lowest price estimates the cost of development at £30 000
- If the app is priced at £1, you can expect to sell 80 000
- If the app is priced at £6 you can expect no sales.

a Assume that demand for your app is linear, so an increase in the price of your app leads to a proportional decrease in demand.

  i If T represents the total sales of your app, explain why $0 \le T \le 80\,000$.

  ii Show that if your app costs £x, where $1 \le x \le 6$, then $T = -16\,000x + 96\,000$.

b Show that when the selling price of the app is £x the profit, £P, is given by $P(x) = -16\,000x^2 + 96\,000x - 30\,000$.

c Find the value of x for which the profit is maximised and hence find this maximum profit.

d Suppose now that you want to borrow the £30 000 necessary for product design and development. Write a short report which will convince a bank to lend you the money.

6 A large industrial water tank is made of galvanised steel. It is in the shape of a closed cylinder and has a volume of 216 000 litres.

The tank has radius r metres and height h metres.

a Show that the surface area, $S\,m^2$, of the tank is given by:

$$S = 2\pi r^2 + \frac{432}{r}$$

b i Find the value of r which minimises the surface area of the cylinder.

  ii Determine the minimum surface area.

7 Many medicines are prescribed in capsule form. Capsules are usually coated to help ensure they break up in the intestines and not the stomach.

A capsule of a certain medicine consists of a cylinder and two hemispheres.

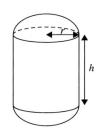

The volume of the cylinder is $0.6\,\text{cm}^3$.

a Show that the surface area, $S\,\text{cm}^2$, of a capsule is given by:

$$S = \frac{6}{5r} + \frac{4}{3}\pi r^2$$

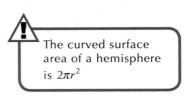

The curved surface area of a hemisphere is $2\pi r^2$

b **i** Find the value of r which minimises the surface area.

ii Determine the minimum surface area of a capsule.

8 A closed prism has a square base of length $x\,\text{cm}$ and height $h\,\text{cm}$. It has a surface area of $A\,\text{cm}^2$.

a Show that $h = \dfrac{A - 2x^2}{4x}$

b Show that for a fixed surface area, A, the maximum volume occurs when the prism is a cube.

9 The diagram shows part of the curve with equation $y = 8 - x^3$.

A rectangle is formed according to these conditions:

- one of the vertices of the rectangle is P where P is a point on the curve

- the rectangle is bounded by the x- and y-axes

- P has coordinates (x, y) where $x > 0$ and $y > 0$.

Determine the maximum area of the rectangle.

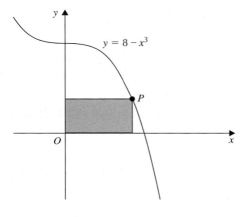

10 A company manufactures transparent plastic 3D objects and sells them to schools. One of the shapes is a triangular prism with volume 0.25 litres.

The height of the prism is $h\,\text{cm}$ and the cross-section of the prism is an equilateral triangle with sides of length x.

a Express h in terms of x.

b Show that the surface area, $S\,\text{cm}^2$, of the prism is given by:

$$S = \frac{\sqrt{3}}{2}\left(x^2 + \frac{2000}{x}\right)$$

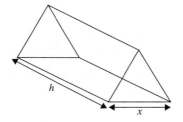

c Determine the minimum amount of plastic needed to make the prism.

11 A firm manufactures water tanks from rectangular sheets of metal 3 m long by 2 m wide.

Squares of side x m are cut out of each corner as shown in the diagram.

$PQRS$ forms the base of the tank and the remaining parts are folded upwards and sealed to form the open tank.

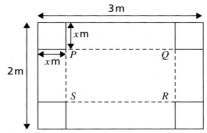

- Each sheet of metal purchased by the manufacturer costs £30.
- The internal surface area is to be coated with an anti-corrosive agent which costs £22.50 per square metre.
- When selling a tank, the manufacturer charges £350 for every cubic metre of capacity as well as a deposit of £165.
   a Neglecting all other costs, find and simplify an expression for the **profit** made on each tank.

   b Determine the maximum possible profit per tank.

12 At midday, car Q is 40 miles due east of car P.

Car Q is travelling due west at a steady speed of 80 mph. Car P is travelling due north at 60 mph.

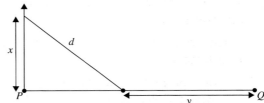

t hours after leaving, car P and car Q have travelled distances of x miles and y miles respectively. The distance between them is d miles.

a Write down expressions for x and y in terms of t.

b Show that $d(t) = 20\sqrt{25t^2 - 16t + 4}$.

c At what time (to the nearest minute) will the distance between the cars be at a minimum? Justify your answer.

13 In a race, Katie wants to get from X, on one side of a river, to a checkpoint Z in the shortest possible time.

The river is 300 m wide. XY is perpendicular to the riverbank. The distance from Y to Z is 1 km.

Katie can row at a speed of 2 ms^{-1} and run at 5 ms^{-1}.

Determine, to the nearest minute, the minimum time Katie could take to get from X to Z.

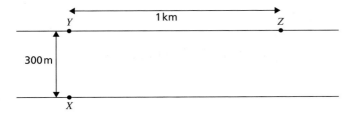

14 A right-angled triangle has a hypotenuse of length 30 cm.

    a Show that the perimeter, P cm, is given by $P(x) = 30 + x + \sqrt{900 - x^2}$, where x cm represents the length of one of the shorter sides.

    b Hence determine the maximum possible perimeter of the right-angled triangle.

    c Suppose that one of the angles in the triangle is a radians. Show now that:

        $P = 30(1 + \sin a + \cos a)$.

    d Determine the value of a for which P is a maximum and confirm that the maximum possible perimeter is equivalent to your answer in part **b**.

 15 A parabola has equation $f(x) = x^2 + 1$.

    From $R(0, 2)$ a straight line is drawn to the parabola. This line intersects the parabola at P.

    Given that PR is the shortest possible distance from R to the parabola, determine possible coordinates for P.

16 A rectangle is enclosed by a circle of radius r. Determine the area of the largest such rectangle.

> ⚠ Let the rectangle have area A. Maximising A^2 will also maximise A.

Challenge 1

A cylinder sits inside a cone of volume V. Given that this cylinder is to have maximum volume, determine in its simplest form the ratio $\frac{C}{V}$, where C is the volume of the cylinder.

Challenge 2

You have a piece of wire k cm long. You cut it into two pieces, one of which is bent into an equilateral triangle with the other forming a circle.

Determine the areas of the two shapes, given that the total area is to be a minimum.

Solving problems involving rates of change

Chapter 10 introduced the concept of **rate of change** and showed how this linked together distance (displacement) and speed. Displacement, velocity and acceleration are all inter-related.

Given a function which describes **displacement** with respect to **time**, we can obtain a function for **velocity** by differentiating.

Similarly, given a function for **velocity** with respect to **time**, we can obtain a function for **acceleration** by differentiating. Remember that acceleration is the rate of change of velocity with respect to time.

Rates of change underpin any real-life problem where quantities are changing, usually with respect to time.

Example 16.6

An aid package is dropped from a helicopter. The distance the package has dropped, after ts, is given by $h(t) = 5t^2$

Find the velocity of the package 2 s after it has been dropped.

$h(t) = 5t^2$

$h'(t) = 10t$ — Differentiate to obtain a function for velocity. P

$h'(2) = 10 \times 2 = 20$ — Substitute for t to obtain the velocity, remembering to use appropriate units. P

After 2 seconds the speed is $20\,\text{ms}^{-1}$

Example 16.7

The velocity of a particle in metres per second, at time t seconds, moving in a straight line along the x-axis is given by $v(t) = t - \sqrt{t}$, $t > 0$

Find the acceleration of the particle after 4 s.

$x(t) = t - \sqrt{t}$

$x(t) = t - t^{\frac{1}{2}}$

$x'(t) = 1 - \dfrac{1}{2}t^{-\frac{1}{2}} = 1 - \dfrac{1}{2\sqrt{t}}$ — Differentiate to obtain a function for acceleration. P

$x'(4) = 1 - \dfrac{1}{2\sqrt{4}} = 1 - \dfrac{1}{4} = \dfrac{3}{4}$

After 4 seconds the acceleration is $0.75\,\text{ms}^{-2}$.

Exercise 16C

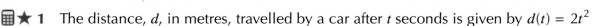 **1** The distance, d, in metres, travelled by a car after t seconds is given by $d(t) = 2t^2$

    a Write down an expression for the velocity, v, of the car after t seconds.

    b Find the velocity of the car after 2 seconds.

    c The acceleration of the car is given by $\frac{dv}{dt}$. Find the acceleration of the car.

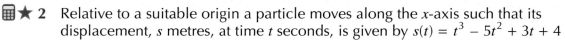 **2** Relative to a suitable origin a particle moves along the x-axis such that its displacement, s metres, at time t seconds, is given by $s(t) = t^3 - 5t^2 + 3t + 4$

    a How far from the origin is the particle when $t = 0$?

    b Find the velocity of the particle when $t = 4$.

    c At what times will the velocity of the particle be zero?

    d Evaluate $s(3)$ and explain your answer.

    e When $\frac{1}{3} < t < 3$ what can you say about the direction of motion of the particle?

3 The velocity, $v\,\text{ms}^{-1}$, of a particle after t seconds is given by $v(t) = 8\sqrt{t}$

    a Find the acceleration of the particle after 9 seconds.

    b Find the time for which the acceleration of the particle is $\frac{9}{16}\,\text{ms}^{-2}$

4 A stone is dropped into a pond. The circular ripple formed moves outwards at a constant rate of 2 metres per second.

    Find the rate at which the area enclosed by the ripple is changing (with respect to time) at the instant when the radius is 6 m.

5 The volume, V, of a sphere is increasing at a constant rate of $150\,\text{cm}^3$ per second. Find $\frac{dr}{dt}$ when the radius is 8 cm.

6 A paper cup is in the shape of a cone having radius 3 cm and height 10 cm.

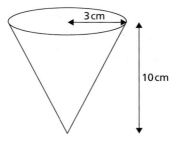

    Water is leaking from the bottom at a rate of $2\,\text{cm}^3$ per second.

    a Show that when the water inside the cup is at a height h cm from the bottom of the cone, the volume of water, V, is given by:

$$V = \frac{3}{100}\pi h^3$$

    b Find $\frac{dh}{dt}$ when $h = 2$.

7 A supertanker holds 840 million litres of oil. The tanker develops a leak causing it to spill oil into the sea. The spillage is modelled as follows:

- the oil 'slick' is a cylinder of constant height 15 centimetres
- the oil is leaking at a rate of 12 000 litres per second
- the oil slick is stationary (in practice, winds and tides would impact significantly).

    a Show that the volume, $V\,\text{m}^3$, of the oil slick is given by $V = 0.15\pi r^2$

    As the tanker releases more oil, the radius of the oil slick will increase as will the volume. It should be obvious that V is a function of r, so the volume of the oil slick depends on the radius. However, the radius is a function of time and therefore so is the volume. We can therefore write $V(t) = 0.15\pi\big[r(t)\big]^2$

    b Explain why $\frac{dV}{dt} = 12\,\text{m}^3\text{s}^{-1}$

    c **i** By using the chain rule, show that if we differentiate y^2 with respect to x, the result is $2y\frac{dy}{dx}$.

       ii Write down the result when you differentiate y^3 with respect to x.

d Hence, show that when $V = 0.15\pi r^2$ – and V, r are both functions of t:

$\frac{dV}{dt} = 0.3\pi r \frac{dr}{dt}$

e Deduce that when $r = 0.5$ km, the radius of the slick is changing at a rate of 91.7 metres per hour.

f Calculate, to the nearest minute, how long it will take for the tanker to completely empty. What can you say about the values of $\frac{dV}{dt}$ and $\frac{dr}{dt}$ at this point?

- I can find the greatest and least values of a function on a closed interval. ★ Exercise 16A Q1

- I can find the optimal solution to a problem. ★ Exercise 16B Q 1–4

- I can solve problems involving displacement, velocity and acceleration. ★ Exercise 16C Q1, 2

For further assessment opportunities, see the Preparation for Assessment for Unit 3 on pages 370–372.

17 Applying integral calculus

This chapter will show you how to:

- find the area between a curve and the x-axis
- find the area between a straight line and a curve or between two curves.

You should already know:

- how to integrate functions reducible to an expression in powers of x
- how to integrate expressions of the form $(x + q)^n$ and $(px + q)^n$
- how to integrate expressions of the form $p\sin(qx + r)$ and $p\cos(qx + r)$

Find the area between a curve and the x-axis

In the real world, areas and volumes need to be calculated where the shapes involved are not made up from simple polygons or parts of a circle. More specifically, the curved portions of such shapes can be described using mathematical functions.

What happens if we want to find the area enclosed by a curve and the x-axis?

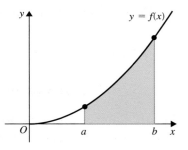

Consider a curve with equation $y = f(x)$ and define $F(x) = \int f(x)\,dx$. $F(x)$ is an **anti-derivative** of $f(x)$.

To find the area enclosed by the curve and the x-axis, from $x = a$ to $x = b$, an area function $A(x)$ is defined. We don't know what this function is, but it will be shown that $A(x) = F(x)$.

Since we are looking to find the area between $x = a$ and $x = b$ we define $A(x) = 0$ when $x = a$. The value of $A(b)$ represents the area under the curve.

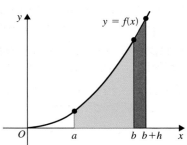

Now consider $A(b + h)$. The value of $A(b + h)$ represents the area under the curve $y = f(x)$ between $x = a$ and $x = b + h$.

The area of the grey strip shown is given by $A(b + h) - A(b)$.

When h is small, the area of the strip is also approximately equal to $h \times f(b)$.

It follows that $A(b + h) - A(b) \approx hf(b)$ and so $\dfrac{A(b + h) - A(b)}{h} \approx f(b)$.

As h gets smaller and smaller, i.e. as $h \to 0$, $f(b) = \lim\limits_{h \to 0} \dfrac{A(b+h) - A(b)}{h}$.

From Chapter 9 we have that $f(b) = \lim\limits_{h \to 0} \dfrac{A(b+h) - A(b)}{h} = A'(b)$.

The argument made above works for any value of x and so we can conclude that:

$A'(x) = f(x)$ and hence $A(x) = \int f(x)\,dx$.

We now have $A(x) = \int f(x)\,dx$ and so $A(x) = F(x) + c$.

When $x = a$, $A(x) = 0$ and so $0 = F(a) + c$ from which we conclude $c = -F(a)$.

When $x = b$, we have $A(b) = F(b) + c$, so:

$A(b) = F(b) - F(a)$

However, $\displaystyle\int_a^b f(x)\,dx = \left[F(x)\right]_a^b = F(b) - F(a) = A(b)$

It now follows that the area under the curve $y = f(x)$, between $x = a$ and $x = b$, is given by:

$$\int_a^b f(x)\,dx$$

The analysis given here gives an overview of the **fundamental theorem of calculus**, which states:

> **When $f(x)$ is** continuous **throughout a closed interval $a \le x \le b$:**
>
> $$\int_a^b f(x)\,dx = F(b) - F(a)$$
>
> **where $F(x)$ is any anti-derivative of $f(x)$**

Essentially, the fundamental theorem of calculus shows that differentiation and integration are inverse processes.

It should be noted that the theorem has not been proved. This needs much more complex and rigorous mathematics, but the previous analysis is a good start.

Chapter 9 showed how **differentiation** is used to find the **gradient** of curves. Similarly, **definite integration** is used to find the **area** under curves.

When trying to find the area enclosed by a curve and the x-axis, there are three scenarios to consider:

1 The curve lies entirely above the x-axis between $x = a$ and $x = b$.

2 The curve lies entirely below the x-axis between $x = a$ and $x = b$.

3 Part of the curve lies above the x-axis and part lies below.

These three options are explored in the following examples.

Example 17.1

The diagram shows part of the graph with equation $y = (x - 4)^2 + 2$. Find the shaded area.

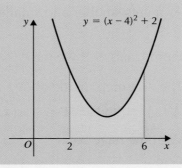

$$\int_{2}^{6} (x - 4)^2 + 2 \, dx$$ ● ──────── Write down the required integral. **S**

$$= \int_{2}^{6} x^2 - 8x + 18 \, dx$$ ● ──────── Express in integrable form. **P**

$$= \left[\tfrac{1}{3}x^3 - 4x^2 + 18x \right]_{2}^{6}$$ ● ──────── Integrate and evaluate in the usual way. **P**

$$= \left(\tfrac{1}{3}(6)^3 - 4(6)^2 + 18(6) \right) - \left(\tfrac{1}{3}(2)^3 - 4(2)^2 + 18(2) \right)$$

$$= \left(\tfrac{1}{3}(216) - 4(36) + 18(6) \right) - \left(\tfrac{1}{3}(8) - 4(4) + 18(2) \right)$$

$$= (72 - 144 + 108) - \left(\tfrac{8}{3} - 16 + 36 \right)$$

$$= (36) - \left(22\tfrac{2}{3} \right) = 13\tfrac{1}{3}$$

Area $= 13\tfrac{1}{3}$ square units.

Since the curve lies entirely **above** the x-axis between $x = 2$ and $x = 6$ we expect a **positive** answer when evaluating the definite integral. **P**

Make a clear statement about the area. **C**

Example 17.2

The diagram shows the graph with equation $y = x^2 - 6x + 5$. The graph intersects the x-axis at P and Q. Find the shaded area.

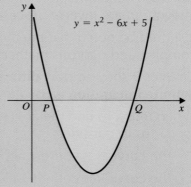

$x^2 - 6x + 5 = 0$ ● ──────── Set $y = 0$ to find the x-coordinates of P and Q. **S**

$(x - 1)(x - 5) = 0$

$x = 1$ or $x = 5$ ● ──────── We now have the limits for integration. **P**

$$\int_1^5 x^2 - 6x + 5 \; dx$$

Write down the required integral. **S**

$$= \left[\tfrac{1}{3}x^3 - 3x^2 + 5x \right]_1^5$$

$$= \left(\tfrac{1}{3}(5)^3 - 3(5)^2 + 5(5) \right) - \left(\tfrac{1}{3}(1)^3 - 3(1)^2 + 5(1) \right)$$

$$= \left(-\tfrac{25}{3} \right) - \left(\tfrac{7}{3} \right) = -\tfrac{32}{3} - 10\tfrac{2}{3}$$

We expect a **negative** answer because the area lies entirely **below the** x-axis. **P**

Area $= 10\tfrac{2}{3}$ square units.

Make a clear statement about the area. Do not be tempted to write down something like $-10\tfrac{2}{3} = 10\tfrac{2}{3}$. You will be penalised in the exam for doing this. **C**

Example 17.3

Evaluate the shaded area in the diagram.

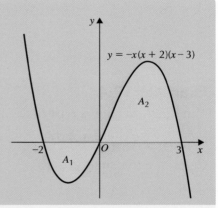

$y = -x(x + 2)(x - 3)$

In a situation like this we need to evaluate two separate integrals and then work out the area.

Let A_1 be the area between the curve and the x-axis from $x = -2$ to $x = 0$.

Let A_2 be the area between the curve and the x-axis from $x = 0$ to $x = 3$.

Note firstly that the answer is **not** $\int_{-2}^3 -x(x + 2)(x - 3) \; dx$.

If you evaluate $\int_{-2}^3 -x(x + 2)(x - 3) \; dx$ the answer obtained is equivalent to $-A_1 + A_2$. What we want to find is $A_1 + A_2$:

$$\int_{-2}^0 -x(x + 2)(x - 3) dx$$

The area between the curve and the x-axis from $x = -2$ to $x = 0$ lies entirely below the x-axis so we expect a negative value for this integral. **S**

$$= \int_{-2}^0 -x^3 + x^2 + 6x \; dx$$

Write $-x(x + 2)(x - 3)$ in integrable form. **P**

$$= \left[-\tfrac{1}{4}x^4 + \tfrac{1}{3}x^3 + 3x^2 \right]_{-2}^0$$

(continued)

$$= \left(-\tfrac{1}{4}(0)^4 + \tfrac{1}{3}(0)^3 + 3(0)^2\right) - \left(-\tfrac{1}{4}(-2)^4 + \tfrac{1}{3}(-2)^3 + 3(-2)^2\right)$$

$$= (0) - \left(\tfrac{16}{3}\right) = -\tfrac{16}{3}$$

Note the negative answer as expected. **P**

$$A_1 = \tfrac{16}{3} \text{ square units}$$

Make a clear statement about A_1. **C**

$$\int_0^3 -x^3 + x^2 + 6x \, dx$$

We now evaluate our second integral. **P**

$$= \left[-\tfrac{1}{4}x^4 + \tfrac{1}{3}x^3 + 3x^2\right]_0^3$$

$$= \left(-\tfrac{1}{4}(3)^4 + \tfrac{1}{3}(3)^3 + 3(3)^2\right) - \left(-\tfrac{1}{4}(0)^4 + \tfrac{1}{3}(0)^3 + 3(0)^2\right)$$

$$= \left(\tfrac{63}{4}\right) - (0) = \tfrac{63}{4}$$

$$A_2 = \tfrac{63}{4} \text{ square units}$$

Make a clear statement about A_2 **C**

$$A_1 + A_2 = \tfrac{16}{3} + \tfrac{63}{4} = 21\tfrac{1}{12} \text{ square units}$$

Add the two individual areas to obtain the total area. **C**

Example 17.4

The diagram shows part of the graph with equation $y = \dfrac{1}{(2x-1)^4}$ where $x \geq 1$.

Calculate the shaded area.

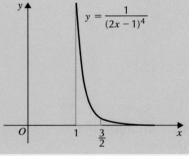

$$\int_1^{\frac{3}{2}} \frac{1}{(2x-1)^4} \, dx = \int_1^{\frac{3}{2}} (2x-1)^{-4} dx$$

Write $\frac{1}{(2x-1)^4}$ in integrable form. **P**

$$= \left[\frac{(2x-1)^{-3}}{2(-3)}\right]_1^{\frac{3}{2}} = -\frac{1}{6}\left[\frac{1}{(2x-1)^3}\right]_1^{\frac{3}{2}}$$

$$= -\frac{1}{6}\left[\frac{1}{\left(2\left(\frac{3}{2}\right)-1\right)^3} - \frac{1}{\left(2(1)-1\right)^3}\right]$$

$$= -\frac{1}{8}\left[\frac{1}{8} - 1\right] = -\frac{1}{6}\left(-\frac{7}{8}\right) = \frac{7}{48}$$

Integrate and evaluate, noting the expected positive answer. **P**

$$\text{Area} = \frac{7}{48} \text{ square units}$$

Example 17.5

The diagram shows part of the graph with equation
$y = 4\sin x$ where $0 \le x \le 2\pi$.
Calculate the shaded area.

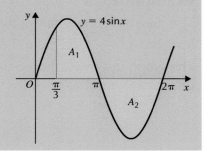

Let A_1 be the area between the curve and the x-axis from $x = \frac{\pi}{3}$ to $x = \pi$

Let A_2 be the area between the curve and the x-axis from $x = \pi$ to $x = 2\pi$

Note firstly that the answer is **not** $\int_{\frac{\pi}{3}}^{2\pi} 4\sin x \ dx$.

So for area A_1:

$$\int_{\frac{\pi}{3}}^{\pi} 4\sin x \ dx = \left[-4\cos x\right]_{\frac{\pi}{3}}^{\pi}$$

$$= (-4\cos\pi) - \left(-4\cos\frac{\pi}{3}\right)$$

$$= (-4(-1)) - \left(-4\left(\tfrac{1}{2}\right)\right) = 4 + 2 = 6$$

Note the positive answer **P**
as expected.

$$A_1 = 6 \text{ square units}$$

Make a clear statement about A_1. **C**

For area A_2:

$$\int_{\pi}^{2\pi} 4\sin x \ dx = \left[-4\cos x\right]_{\pi}^{2\pi}$$

$$= (-4\cos 2\pi) - (-4\cos\pi)$$

Note the expected negative answer. **P**

$$= (-4(1)) - (-4(-1)) = -4 - 4 = -8$$

Make a clear statement about A_2. **C**

$$A_2 = 8 \text{ square units}$$

$$A_1 + A_2 = 6 + 8 = 14 \text{ square units}$$

Add the two individual areas to **C**
obtain the total area.

Example 17.6

The diagram shows part of the graph with equation
$y = 2\cos\left(3x - \frac{2\pi}{3}\right)$ where $0 \le x \le \pi$.
The graph intersects the x-axis at P, Q and R. Calculate
the shaded area.

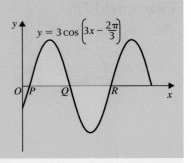

$2\cos\left(3x - \frac{2\pi}{3}\right) = 0$ ⟶ Set $y = 0$ to find the x-coordinates of P, Q and R. **S**

$3x - \frac{2\pi}{3} = \frac{\pi}{2}, \frac{3\pi}{2}, \frac{5\pi}{2}, \frac{7\pi}{2}$

Start to solve. **P**

$3x = \frac{7\pi}{6}, \frac{13\pi}{6}, \frac{19\pi}{6}, \frac{25\pi}{6}$

Since $0 \le x \le \pi$, we now bring all solutions for $3x$ into the **S**
range $0 \le 3x \le 3\pi$. Note that $\frac{19\pi}{6}$ is now rejected since it is

$3x = \frac{\pi}{6}, \frac{7\pi}{6}, \frac{13\pi}{6}$

out of range. Note also that an angle of $\frac{25\pi}{6}$ is equivalent to
an angle of $\frac{\pi}{6}$ because $\frac{25\pi}{6} = 2\pi + 2\pi + \frac{\pi}{6}$

$x = \frac{\pi}{18}, \frac{7\pi}{18}, \frac{13\pi}{18}$

Complete the solution to obtain the x-coordinates of P, Q and R. **P**

Let A_1 be the area between the curve and the x-axis from $x = \frac{\pi}{18}$ to $x = \frac{7\pi}{18}$.

Let A_2 be the area between the curve and the x-axis from $x = \frac{7\pi}{18}$ to $x = \frac{13\pi}{18}$.

$$\int_{\frac{\pi}{18}}^{\frac{7\pi}{18}} 2\cos\left(3x - \frac{2\pi}{3}\right) = \frac{2}{3}\left[\sin\left(3x - \frac{2\pi}{3}\right)\right]_{\frac{\pi}{18}}^{\frac{7\pi}{18}}$$

$$= \frac{2}{3}\left[\sin\left(3\left(\frac{7\pi}{18}\right) - \frac{2\pi}{3}\right) - \left(\sin\left(3\left(\frac{\pi}{18}\right) - \frac{2\pi}{3}\right)\right)\right]$$

$$= \frac{2}{3}\left[\sin\left(\frac{\pi}{2}\right) - \sin\left(-\frac{\pi}{2}\right)\right]$$

$$= \frac{2}{3}[1 - (-1)] = \frac{2}{3}[2] = \frac{4}{3}$$

Note the positive answer **P**
as expected.

$A_1 = \frac{4}{3}$ square units

Make a clear statement about A_1. **C**

$$\int_{\frac{7\pi}{18}}^{\frac{13\pi}{18}} 2\cos\left(3x - \frac{2\pi}{3}\right) = \frac{2}{3}\left[\sin\left(3x - \frac{2\pi}{3}\right)\right]_{\frac{7\pi}{18}}^{\frac{13\pi}{18}}$$

$$= \frac{2}{3}\left[\sin\left(3\left(\frac{13\pi}{18}\right) - \frac{2\pi}{3}\right) - \left(\sin\left(3\left(\frac{7\pi}{18}\right) - \frac{2\pi}{3}\right)\right)\right]$$

$$= \frac{2}{3}\left[\sin\left(\frac{3\pi}{2}\right) - \sin\left(\frac{\pi}{2}\right)\right]$$

Note the negative answer as expected. **P**

$$= \frac{2}{3}[-1 - 1] = \frac{2}{3}[-2] = -\frac{4}{3}$$

Make a clear statement about A_2. **C**

$A_2 = \frac{4}{3}$ square units

Add the two individual areas to obtain **C**
the total area.

$A_1 + A_2 = \frac{4}{3} + \frac{4}{3} = \frac{8}{3}$ square units

Exercise 17A

1 Evaluate the shaded areas.

a
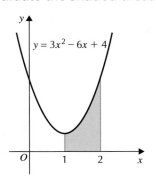
$y = 3x^2 - 6x + 4$

b
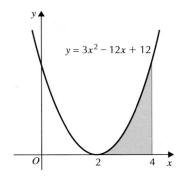
$y = 3x^2 - 12x + 12$

c
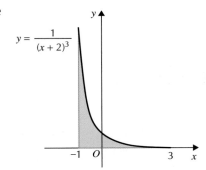
$y = x^3 - 6x^2 + 5x + 12$

d
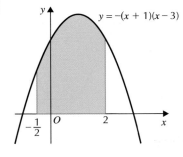
$y = -(x + 1)(x - 3)$

e
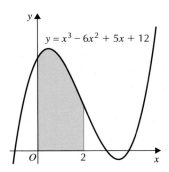
$y = \dfrac{1}{(x + 2)^3}$

f
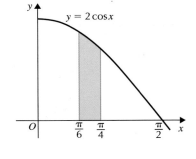
$y = 2\cos x$

2 Evaluate the shaded areas.

a
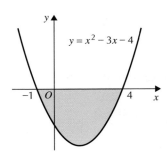
$y = x^2 - 3x - 4$

b
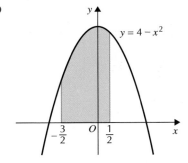
$y = 4 - x^2$

c

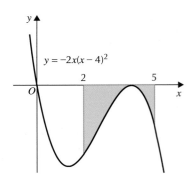

$y = -2x(x - 4)^2$

d

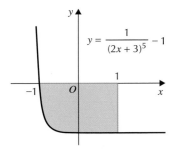

$y = \dfrac{1}{(2x + 3)^5} - 1$

e

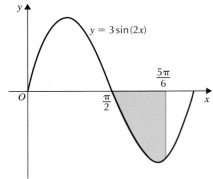

$y = 3\sin(2x)$

f

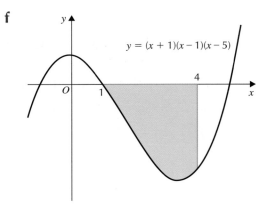

$y = (x + 1)(x - 1)(x - 5)$

3 Evaluate the shaded areas.

a

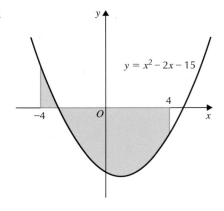

$y = x^2 - 2x - 15$

b

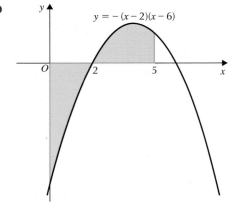

$y = -(x - 2)(x - 6)$

c

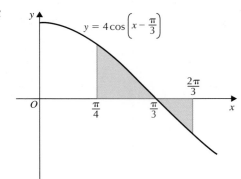

$y = 4\cos\left(x - \dfrac{\pi}{3}\right)$

d

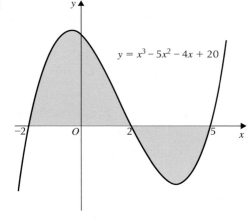

$y = x^3 - 5x^2 - 4x + 20$

e

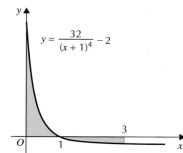

$y = \dfrac{32}{(x+1)^4} - 2$

f

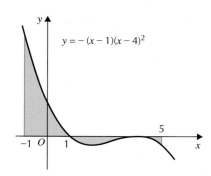

$y = -(x-1)(x-4)^2$

★ 4 The parabola shown has equation $y = 10 + 3x - x^2$.
It intersects the x-axis at P.

    a Determine algebraically the coordinates of P.

    b Hence calculate the shaded area.

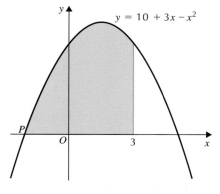

$y = 10 + 3x - x^2$

5 The parabola shown has equation
$y = \frac{1}{5}(x^2 - 6x + 5)$. It intersects the x-axis at
A and B.

    a Find algebraically the x-coordinates of A and B.

    b Calculate the shaded area.

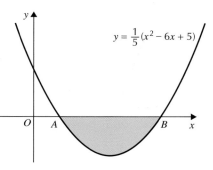

$y = \frac{1}{5}(x^2 - 6x + 5)$

6 a i Show that $(x + 2)$ is a factor of
$x^3 - 3x^2 - 6x + 8$.

    ii Hence fully factorise $x^3 - 3x^2 - 6x + 8$.

  b The diagram shows the cubic with
equation $y = x^3 - 3x^2 - 6x + 8$.
The graph intersects the x-axis at
K, L and M.

    i Determine the coordinates of
K, L and M.

    ii Hence evaluate the shaded area.

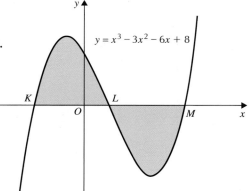

$y = x^3 - 3x^2 - 6x + 8$

7 **a** A curve has equation $y = x^3 - 2x^2 - 4x - 3$. P and Q are the stationary points of the curve.

    i Find the coordinates of the stationary points.

    ii Determine the nature of the stationary points.

    b Find the area enclosed by the curve, the x-axis and the lines $x = -1$ and $x = 2$.

8 The diagram shows the parabola with equation $y = -x^2 + 6x - 9$ and a straight line with equation $y = -5 + 2x$.

The line is a tangent to the parabola at P.

a Determine algebraically the coordinates of P.

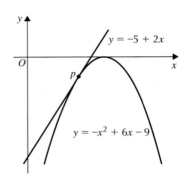

b The parabola touches the x-axis at Q. Calculate the shaded area.

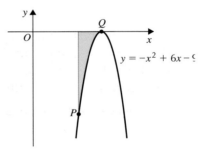

9 The graph shown has equation $y = a\sin(bx + c)$, where $-\frac{\pi}{2} < c < 0$. $\left(\frac{3\pi}{8}, 3\right)$ and $\left(\frac{7\pi}{8}, -3\right)$ are stationary points of the graph.

a Find the values of a, b and c.

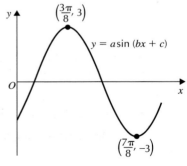

b The graph of $y = a\sin(bx + c)$ intersects the x-axis at P and Q. Calculate the shaded area.

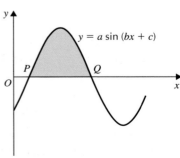

10 A parabola has equation $y = a(x + b)^2 + c$. The parabola intersects the y-axis at $\left(0, \frac{5}{2}\right)$ and has a minimum turning point at $(3, -2)$.

a Find the values of a, b and c.

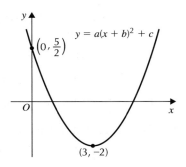

The parabola intersects the x-axis at R and S.

b **i** Determine algebraically the x-coordinates of R and S.

    ii Calculate the shaded area.

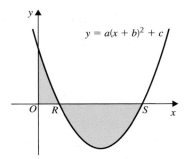

11 A parabola has equation $y = -x^2 + bx + c$. The point $(1, 10)$ lies on the parabola and at the point P, where $x = 3$, the tangent to the parabola makes an angle of $135°$ with the positive direction of the x-axis.

a Determine the values of b and c.

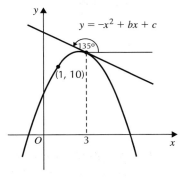

The area enclosed by the parabola, the x-axis and the lines $x = 0$ and $x = k$, where $0 < k < 6$ is $50\frac{5}{6}$ square units.

b **i** Show that $2k^3 - 15k^2 - 36k + 305 = 0$.

    ii Determine algebraically the value of k.

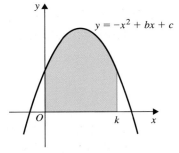

12 A geologist examining a section of coastline finds that it can be modelled by the equation $y = 0.0001x^3 - 0.02x^2 + 0.5x + 120$, where y represents the distance, in metres, of the coastline from a fixed line.

In the diagram, the fixed line is represented by the x-axis and is 90 m long.

a Calculate the shaded area.

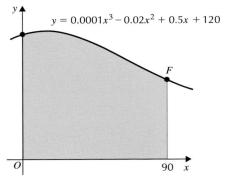

Data suggests that the coastline is eroding at a rate of 0.5 metres per year, so the coastline moves 0.5 m closer to the fixed line every year.

b **i** Write down an equation for the coastline after 20 years.

ii What assumptions did you make when writing down the equation?

c Determine the area of land lost because of coastal erosion after 20 years.

13 A local authority plans to repaint the brickwork surrounding an old railway arch. Relative to coordinate axes, the arch is represented by the equation $y = 12\left(1 - \frac{x^2}{25}\right)$. The parabola meets the x-axis at P and Q.

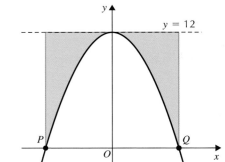

The line $y = 12$ is a tangent to the parabola.

On each axis, 1 unit represents 3 m.

The area of brickwork to be repainted is represented by the shaded area.

The local authority has the following information:

• labour for the job costs £4 per square metre
• a 5 litre tin of paint covers 20 square metres and costs £28.75
• other equipment total £650.

Based on this information, the local authority allocates £3000 to cover the entire cost of the job. Has enough money been set aside? Justify your answer.

Find the area between a straight line and a curve or between two curves

In many applications we will need to find the area between curves or between curves and lines. For example, the shaded area in the diagram represents the uniform cross-section of the storage space of a bulk carrier (Carriers store cargo in the hull of the ship).

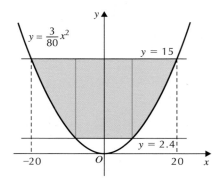

If we know the shape of the hull and the length, we can work out the volume of the cargo space available (Exercise 17B Question 15).

Consider now the curves with equations $y = f(x)$ and $y = g(x)$.

The diagrams show the area enclosed by each curve with the x-axis between $x = a$ and $x = b$. The curves also intersect when $x = a$ and $x = b$.

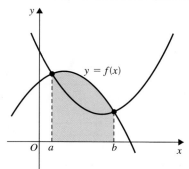 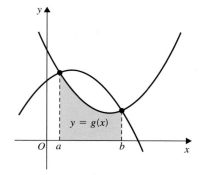

The area enclosed by the curve $y = f(x)$ and the x-axis, between $x = a$ and $x = b$, is

given by: $\displaystyle\int_a^b f(x)\,dx$

The area enclosed by the curve $y = g(x)$ and the x-axis, between $x = a$ and $x = b$, is

given by: $\displaystyle\int_a^b g(x)\,dx$

If the area under the curve $y = g(x)$ is subtracted from the area under the curve $y = f(x)$, we get the area between the curves $y = f(x)$ and $y = g(x)$, between $x = a$ and $x = b$, as shown in the diagram.

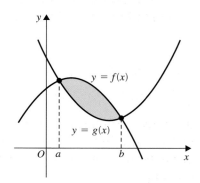

This area is given by $\left(\displaystyle\int_a^b f(x)\,dx\right) - \left(\displaystyle\int_a^b g(x)\,dx\right)$

which is simply $\displaystyle\int_a^b \big(f(x) - g(x)\big)\,dx$

Note that on the interval $a < x < b$, $f(x) > g(x)$ and for this reason we refer to $y = f(x)$ as the **upper curve** and $y = g(x)$ as the **lower curve**.

It doesn't matter if part of the area between the two curves lies below the x-axis. We can still find the area in exactly the same way.

When finding the area between two curves,

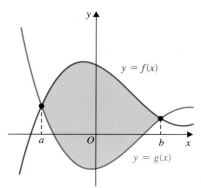

$\displaystyle\int_a^b \big(f(x) - g(x)\big)\,dx$ will always give a positive answer.

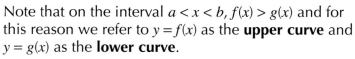

Example 17.7

The diagram shows the straight line with equation $y = x + 2$ and the parabola with equation $y = x^2 + 3x - 1$. The line and the curve intersect at $(-3, -1)$ and $(1, 3)$. Calculate the shaded area.

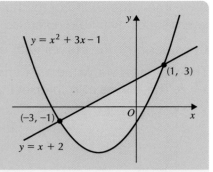

Upper curve: $y = x + 2$ Lower curve: $y = x^2 + 3x - 1$

$$\int_{-3}^{1} x + 2 - \left(x^2 + 3x - 1\right) dx$$

> **S** For $-3 < x < 1$, the line is above the parabola.

> **S** Write down the correct integral. Note the brackets around $(x^2 + 3x - 1)$.

$$= \int_{-3}^{1} x + 2 - x^2 - 3x + 1 \, dx$$

> **P** Remove the brackets. Remember that $-(x^2 - 3x - 1)$ is the same as $-1(x^2 - 3x - 1)$.

$$= \int_{-3}^{1} -x^2 - 2x + 3 \, dx$$

> **P** Simplify before integrating.

$$= \left[-\tfrac{1}{3}x^3 - x^2 + 3x \right]_{-3}^{1}$$

$$= \left(-\tfrac{1}{3}(1)^3 - (1)^2 + 3(1) \right) - \left(-\tfrac{1}{3}(-3)^3 - (-3)^2 + 3(-3) \right)$$

> **P** Integrate and substitute.

$$= \left(\tfrac{5}{3} \right) - (-9) = 10\tfrac{2}{3}$$

> **P** Note the positive answer. If you get a negative answer at this stage then you need to go back and check your work.

Area $= 10\tfrac{2}{3}$ square units.

> **C** Make a clear statement about the area.

Example 17.8

The diagram shows the straight line with equation $y = 2x + 1$ and the parabola with equation $y = x^2 - 3x + 5$.

The line and the parabola intersect at R and S.

a Find algebraically the coordinates of R and S.

b Calculate the shaded area.

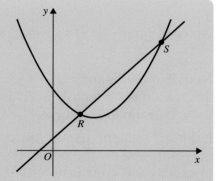

a $x^2 - 3x + 5 = 2x + 1$

> **S** Set the equations equal.

$x^2 - 5x + 4 = 0$

> **S** Write the quadratic equation in standard form.

$(x - 1)(x - 4) = 0$

$x = 1, x = 4$

> **P** Solve to find the x-coordinates of R and S.

When $x = 1$, $y = 2(1) + 1 = 3$

When $x = 4$, $y = 2(4) + 1 = 9$

Find the y-coordinates of R and S, noting that it is simpler to use the equation of the line. **P**

$R\,(1, 3)$, $S\,(4, 9)$

Clearly communicate the coordinates of R and S. **C**

b Upper curve: $y = 2x + 1$

Lower curve: $y = x^2 - 3x + 5$

For $1 < x < 4$, the line is above the parabola. **S**

$$\int_1^4 2x + 1 - \left(x^2 - 3x + 5\right)\,dx$$

Write down the correct integral. Note the brackets around $x^2 - 3x + 5$. **S**

$$= \int_1^4 2x + 1 - x^2 + 3x - 5\,dx$$

Remove the brackets. Remember that $-(x^2 + 3x - 5)$ is the same as $-1(x^2 + 3x - 5)$. **P**

$$= \int_1^4 -x^2 + 5x - 4\,dx$$

Simplify before integrating. **P**

$$= \left[-\tfrac{1}{3}x^3 + \tfrac{5}{2}x^2 - 4x\right]_1^4$$

$$= \left(-\tfrac{1}{3}(4)^3 + \tfrac{5}{2}(4)^2 - 4(4)\right) - \left(-\tfrac{1}{3}(1)^3 + \tfrac{5}{2}(1)^2 - 4(1)\right)$$

Integrate and substitute. **P**

$$= \left(\tfrac{8}{3}\right) - \left(-\tfrac{11}{6}\right) = 4\tfrac{1}{2}$$

Area $= 4\tfrac{1}{2}$ square units.

Make a clear statement about the area. **C**

Example 17.9

The diagram shows the parabolas with equations $y = \tfrac{3}{2}x^2 - 4x - 1$ and $y = -\tfrac{1}{2}x^2 + x + 2$.

The parabolas intersect at A and B.

a Find algebraically the x-coordinates of A and B.

b Calculate the shaded area.

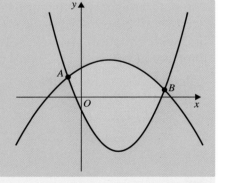

a $\tfrac{3}{2}x^2 - 4x - 1 = -\tfrac{1}{2}x^2 + x + 2$

Set the equations equal. **S**

$3x^2 - 8x - 2 = -x^2 + 2x + 4$

$4x^2 - 10x - 6 = 0$

Write the quadratic equation in standard form. **S**

$2(2x + 1)(x - 3) = 0$

$x = -\tfrac{1}{2}$, $x = 3$

Solve to find the x-coordinates of A and B. **P**

$x_A = -\tfrac{1}{2}$, $x_B = 3$

Communicate clearly. **C**

(continued)

b Upper curve: $y = -\frac{1}{2}x^2 + x + 2$

Make sure you are clear about the upper and lower curves. **S**

Lower curve: $y = \frac{3}{2}x^2 - 4x - 1$

$$\int_{-\frac{1}{2}}^{3} -\frac{1}{2}x^2 + x + 2 - \left(\frac{3}{2}x^2 - 4x - 1\right) dx$$

Write down the correct integral. Note the brackets around $\left(\frac{3}{2}x^2 - 4x - 1\right)$. **S**

$$= \int_{-\frac{1}{2}}^{3} -\frac{1}{2}x^2 + x + 2 - \frac{3}{2}x^2 + 4x + 1 \, dx$$

Remove the brackets. **P**

$$= \int_{-\frac{1}{2}}^{3} -2x^2 + 5x + 3 \, dx$$

Simplify before integrating. **P**

$$= \left[-\frac{2}{3}x^3 + \frac{5}{2}x^2 + 3x\right]_{-\frac{1}{2}}^{3}$$

Integrate and substitute. **P**

$$= \left(-\frac{2}{3}(3)^3 + \frac{5}{2}(3)^2 + 3(3)\right) - \left(-\frac{1}{3}\left(-\frac{1}{2}\right)^3 + \frac{5}{2}\left(-\frac{1}{2}\right)^2 + 3\left(-\frac{1}{2}\right)\right)$$

$$= \left(\frac{27}{2}\right) - \left(-\frac{5}{6}\right) = 14\frac{1}{3}$$

Area $= 14\frac{1}{3}$ square units.

Make a clear statement about the area. **C**

Example 17.10

The diagram shows the graphs of a cubic with equation $f(x) = x^3 - 4x^2 + 2x + 3$ and a straight line with equation $g(x) = 3x - 1$. The graphs intersect at $(-1, -4)$, $(1, 2)$ and $(4, 11)$.

Calculate the shaded area.

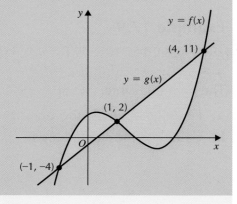

Be careful in situations like this. The answer is **not** $\int_{-1}^{4} f(x) - g(x) \, dx$.

For $-1 < x < 1$, $f(x)$ is the upper curve and $g(x)$ is the lower curve. However, for $1 < x < 4$ the opposite is true. We need to do two separate area calculations.

1 For $-1 < x < 1$:

$$\int_{-1}^{1} x^3 - 4x^2 + 2x + 3 - (3x - 1) \, dx = \int_{-1}^{1} x^3 - 4x^2 + 2x + 3 - 3x + 1 \, dx$$

$$= \int_{-1}^{1} x^3 - 4x^2 - x + 4 \ dx$$

$$= \left[\tfrac{1}{4}x^4 - \tfrac{4}{3}x^3 - \tfrac{1}{2}x^2 + 4x \right]_{-1}^{1}$$

$$= \left(\tfrac{1}{4}(1)^4 - \tfrac{4}{3}(1)^3 - \tfrac{1}{2}(1)^2 + 4(1) \right) - \left(\tfrac{1}{4}(-1)^4 - \tfrac{4}{3}(-1)^3 - \tfrac{1}{2}(-1)^2 + 4(-1) \right)$$

$$= \left(\tfrac{29}{12} \right) - \left(\tfrac{-35}{12} \right) = 5\tfrac{1}{3}$$

> This gives the area between the curve and the line from $x = -1$ to $x = 1$. **P**

2 For $1 < x < 4$:

$$\int_{1}^{4} 3x - 1 - \left(x^3 - 4x^2 + 2x + 3 \right) dx = \int_{1}^{4} -x^3 + 4x^2 + x - 4 \ dx$$

> Write down the integral which will give the area from $x = 1$ to $x = 4$ **S**

$$= \left[-\tfrac{1}{4}x^4 + \tfrac{4}{3}x^3 + \tfrac{1}{2}x^2 - 4x \right]_{1}^{4}$$

$$= \left(-\tfrac{1}{4}(4)^4 + \tfrac{4}{3}(4)^3 + \tfrac{1}{2}(4)^2 - 4(4) \right) - \left(-\tfrac{1}{4}(1)^4 + \tfrac{4}{3}(1)^3 + \tfrac{1}{2}(1)^2 - 4(1) \right)$$

$$= \left(\tfrac{40}{3} \right) - \left(-\tfrac{29}{12} \right) = 15\tfrac{3}{4}$$

> This gives the area between the curve and the line from $x = 1$ to $x = 4$. **P**

$$\text{Total area} = 5\tfrac{1}{3} + 15\tfrac{3}{4}$$

$$= 21\tfrac{1}{12} \text{ square units}$$

> Add the two separate areas to give the total area and communicate clearly. **P**

Exercise 17B

⊞ ★ **1** Evaluate the shaded areas.

a

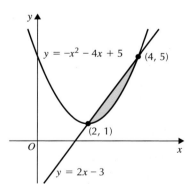

b

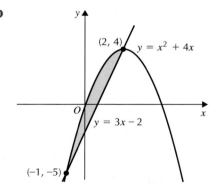

c

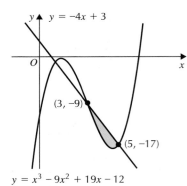

$y = -4x + 3$

$(3, -9)$

$(5, -17)$

$y = x^3 - 9x^2 + 19x - 12$

d

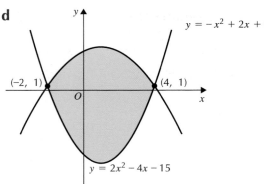

$y = -x^2 + 2x + 9$

$(-2, 1)$

$(4, 1)$

$y = 2x^2 - 4x - 15$

★ 2 Each of the following diagrams contains the graphs of two functions.

For each:

i find the *x*-coordinates of the points of intersection

ii calculate the shaded area.

a

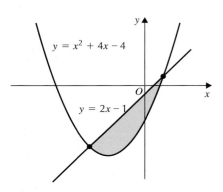

$y = x^2 + 4x - 4$

$y = 2x - 1$

b

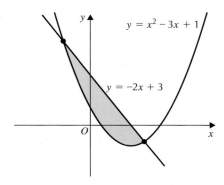

$y = x^2 - 3x + 1$

$y = -2x + 3$

c

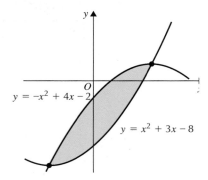

$y = -x^2 + 4x - 2$

$y = x^2 + 3x - 8$

d

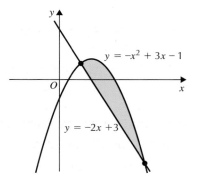

$y = -x^2 + 3x - 1$

$y = -2x + 3$

3 A cubic has equation $g(x) = x^3 - 2x^2 - 6x + 8$ and a parabola has equation $f(x) = 2x^2 + x - 2$.

The curves intersect at $(-2, 4)$, $(1, 1)$ and $(5, 53)$.

Calculate the shaded area.

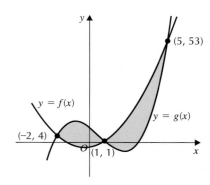

$(5, 53)$

$y = f(x)$

$y = g(x)$

$(-2, 4)$

$(1, 1)$

4 The diagram shows a parabola with equation $y = a(x - b)^2 + c$. The parabola has a minimum turning point at $P\,(4,\,1)$ and passes through $Q\,(1,\,7)$.

   a Find the values of a, b and c.

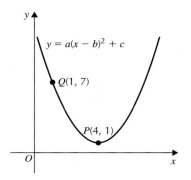

The straight line $y = 2x - 7$ passes through P and meets the parabola again at R.

   b Determine algebraically the coordinates of R.

   c Calculate the shaded area.

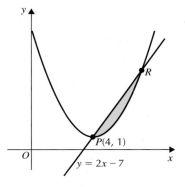

5 a Show that $(x + 3)$ is a factor of $x^3 - x^2 - 8x + 12$.

   b Hence, or otherwise, fully factorise $x^3 - x^2 - 8x + 12$.

   c The diagram shows the curve with equation $f(x) = x^3 - x^2 - 6x + 9$.

   i Show that the line $y = 2x - 3$ is a tangent to the curve at M.

   ii Determine the coordinates of M.

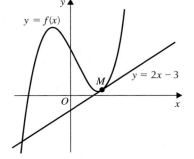

   d This tangent meets the curve again at N. Find the x-coordinate of N.

   e Calculate the shaded area.

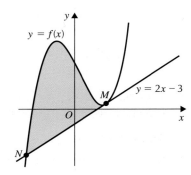

6 Functions f and g are defined, on suitable domains, by $f(x) = 2x^3 + kx^2 - 12x + 11$ and $g(x) = 2(x-1)^2 + 4$.

The graphs of $y = f(x)$ and $y = g(x)$ have a common stationary point at C.

a Find algebraically the value of k.

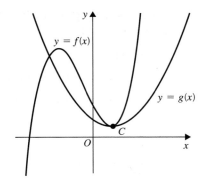

b The graphs also intersect at P.

   i Show that for intersection $2x^3 + x^2 - 8x + 5 = 0$.

   ii Hence, or otherwise, find the x-coordinate of P.

c Calculate the shaded area.

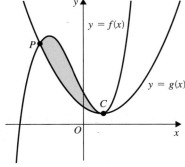

7 The graphs with equations $y = 2\cos 3x$ and $y = -\sqrt{3}$ intersect at Q and R as shown in the diagram.

a Determine the coordinates of Q and R.

b Calculate the shaded area.

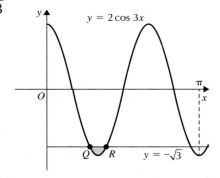

8 The diagram shows part of the graph of a quadratic function with equation $y = f(x)$.

The graph passes through the origin and has a maximum turning point at $(3, 9)$.

a Express $f(x)$ in terms of x.

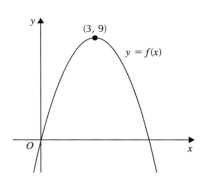

The line *l* is a tangent to the graph of $y = f(x)$ at *P*. The gradient of *l* is −4.

b Find:

　i the *x*-coordinate of *P*

　ii the equation of *l*.

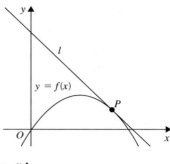

c Evaluate the shaded area.

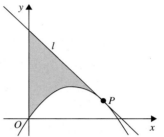

9 The diagram shows part of the curve having equation $f(x) = x^3 + x^2 - 2x + 1$. The tangent to the curve at *P* has gradient 3.

a Determine the *x*-coordinate of *P*.

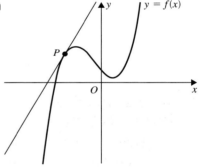

The tangent to the curve at *R* is parallel to the tangent at *P* and meets the curve again at *Q*.

b **i** Write down the *x*-coordinate of *R*.

　ii Determine the coordinates of *Q*.

c Calculate the shaded area.

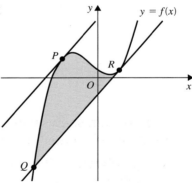

10 The diagram shows part of the graph of the quadratic function $y = x^2 - 4x - 12$.

The graph intersects the *x*-axis at *P* and *Q* and intersects the *y*-axis at *R*.

The line *l* passes through *R* and is parallel to the *x*-axis.

Determine the shaded area.

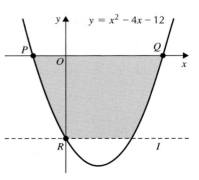

367

11 The diagram shows the graphs with equations $y = \sin 2x$ and $y = -\cos x$, $0 \le x \le \frac{3\pi}{2}$

The graphs intersect at A and B.

a Determine algebraically the coordinates of A and B.

b Calculate the shaded area.

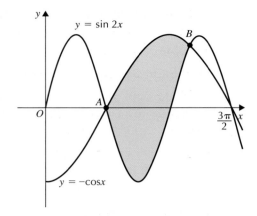

12 The diagram shows the graphs with equations $y = 3\cos 2x$ and $y = -7\cos x + 2$, $0 \le x \le 2\pi$.

The graphs intersect at A and B.

a Determine algebraically the coordinates of A and B.

b Calculate the shaded area.

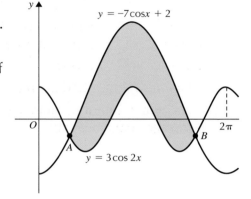

13 The diagram shows the parabola with equation $y = f(x)$.

The parabola has a minimum turning point at $(2, 1)$.

a Given that $f(x) = (x - a)^2 + b$, write down the values of a and b.

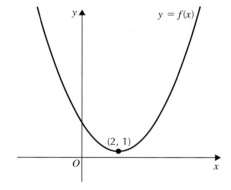

The diagram shows the curves with equations $y = f(x)$ and $y = g(x)$, where $g(x) = -f(2x) + 27$. These curves intersect at M and N.

b i Express $g(x)$ in terms of x.

ii Find algebraically the coordinates of M and N.

c Calculate the shaded area.

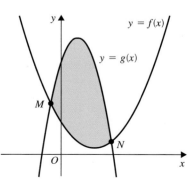

14 The diagram shows the graph with equation $y = \sqrt{2x - 3}$. The graph intersects the x-axis at P, where $x_P = a$.

A tangent to the curve is drawn at Q, where $x_q = b$. This tangent has gradient $\frac{1}{3}$.

The shaded area represents the area between the curve and the tangent from $x = a$ to $x = b$.

Calculate the shaded area.

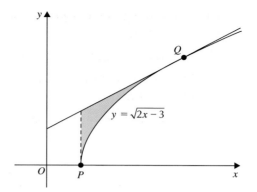

15 The shaded area represents the uniform cross-section of the storage space of a bulk carrier.

The hull is in the shape of a parabola having equation $y = \frac{3}{80}x^2$ and is 40 m wide at its highest. The carrier is 150 m long.

a Determine the shaded area.

b Calculate, in cubic metres, the total volume of storage space available.

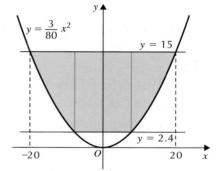

- I can find the area between a curve and the x-axis. ★ Exercise 17A Q1-4
- I can find the area between a straight line and two curves or between two curves. ★ Exercise 17B Q1, 2

For further assessment opportunities, see the Preparation for Assessment for Unit 3 on pages 370–372.

Unit 3 Applications

The questions in this section cover the minimum competence for the content of the course in Unit 3. They are a good preparation for your unit assessment. In an assessment you will get full credit only if your solution includes the appropriate steps of process and accuracy, so make sure you show your thinking when writing your answers.

Remember that reasoning questions marked with the symbol ⚙ expect you to interpret the situation, select an appropriate strategy to solve the problem and then clearly explain your solution. If a context is given you must relate your answer to this context.

The use of a calculator is allowed.

Applying algebraic skills to rectilinear shapes, to circles and to sequences (Chapters 13, 14 and 15)

1 Find the equation of the line that is perpendicular to the line $y = 3x + 6$, passing through the point $(0, 1)$.

2 Find the equation of the line that is parallel to the line $4x + 2y - 5 = 0$, passing through the point $(2, 1)$.

3 Find the equation of the line that is perpendicular to $y - 5 = 0$, passing through the point $(2, 2)$.

⚙ 4 The quadrilateral $ABCD$ has vertices $A(0, 5)$, $B(5, -5)$, $C(1, -7)$ and $D(-4, 3)$. Show that $ABCD$ is a rectangle.

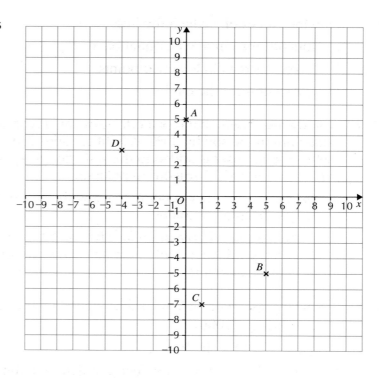

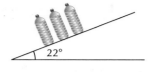

5 The design for a section of conveyer belt in a bottling plant is shown in the diagram. If the gradient of the belt to the horizontal is greater than 0.4, the bottles will fall over causing a jam. Is the design compliant with this limitation? You must justify your answer.

6 What acute angle does the line $3x + 4y - 8 = 0$ make with the x-axis and the y-axis?

7 Write down the equation of the circle with centre $(5, -3)$ and radius 6.

8 A circle has equation $(x - 5)^2 + (y + 7)^2 = 125$.

    a Write down the coordinates of the centre and radius of the circle.

    b Show that the point $(3, 4)$ lies on the circle.

9 Find the coordinates of the centre and the radius of the circle with equation $x^2 + 6x + y^2 - 2y + 6 = 0$.

10 The line AB is the diameter of a circle, where A and B are $(5, 4)$ and $(-1, 0)$ respectively.

    a Find the coordinates of the centre of the circle.

    b Find the equation of the circle.

    c Find the equation of the line that is perpendicular to AB.

11 Show that the line $x + 3y - 11 = 0$ is a tangent to the circle $(x + 1)^2 + (y + 6)^2 = 90$.

12 Find the values of c which the line $y = c - x$:

    a touches the circle with equation $x^2 + y^2 - 4x - 2y - 3 = 0$

    b intersects the circle at two points

    c does not meet the circle.

13 A study has shown that repeatedly growing a type of wheat reduces the available potassium in the soil by 30%. A fertiliser adds 25 mg of potassium per cubic metre. The fertiliser is added to the soil after every harvest. At the beginning of the study the soil had $300\,\text{mg/m}^3$ of potassium.

    a Set up a recurrence relation to show the amount of potassium in the soil after every fertiliser treatment.

    b If the amount of potassium falls below $40\,\text{mg/m}^3$ then the field will no longer become productive. Will the current fertiliser treatment maintain the required level of potassium?

14 The deer population of Scotland is defined by the recurrence relation $u_{n+1} = 0.95u_n + 3000$.

    a If $u_1 = 4000$, calculate u_3.

    b What will happen to the deer population in the long term?

Applying calculus skills to optimisation and area (Chapters 16 and 17)

15 A window has a semi-circular arch of radius x metres above a rectangular section as shown. The perimeter of the window must be 6m. If the area of the window is given by the equation $A = 6x - \frac{1}{2}x^2(4 + \pi)$, find the value of x that will maximise the area of the glass.

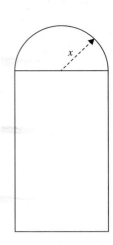

16 The curve with equation $x(x - 3)(x + 1)$ is shown below. Calculate the shaded area.

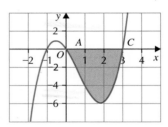

17 The graphs with equations $y = 4 - x$ and $y = -4 + 5x - x^2$ are shown on the diagram. The line and the curve intersect where $x = 2$ and 4. Calculate the shaded area.

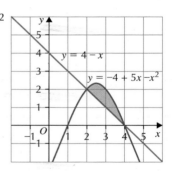

18 The curves $y = x^2 + x + 2$ and $y = 6 - x - x^2$ are shown on the diagram. They meet at the points where $x = -2$ and 1. Calculate the area between the curves.

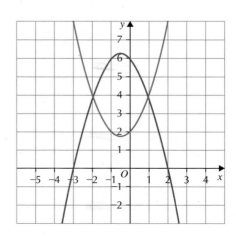